Critical Neuroscience

*A Handbook of the Social and Cultural
Contexts of Neuroscience*

Edited by

Suparna Choudhury and Jan Slaby

WILEY Blackwell

Library of Congress Cataloging-in-Publication Data

Paperback ISBN: 9781119237891

Critical neuroscience : a handbook of the social and cultural contexts of neuroscience / edited by Suparna Choudhury and Jan Slaby.
    p. ; cm.
  Includes bibliographical references and index.
  ISBN 978-1-4443-3328-2 (hardcover : alk. paper)
1. Neurosciences–Social aspects. 2. Mental illness. 3. Social adjustment. I. Choudhury, Suparna. II. Slaby, Jan.
  [DNLM: 1. Neurosciences. 2. Mental Disorders. 3. Mind-Body Relations, Metaphysical. 4. Social Adjustment. WL 100]
  RC343.3.C75 2012
  362.196′89–dc23
                                        2011013466
A catalogue record for this book is available from the British Library.

Cover image: [Production Editor to insert]
Cover design by [Production Editor to insert]

Set in 10/12.5pt Galliard by SPi Global, Pondicherry, India

1   2016

Critical Neuroscience

Contents

Credits

Figure 5.1   Sketch by Wilder Penfield to illustrate a hypothetical scanning mechanism in the hippocampi. Reproduced by permission of the Literary Executor of the estate of Wilder Penfield.

Chapter 9   This chapter is a slightly revised and shortened version of chapter 3 of the author's monograph *Picturing Personhood: Brain Scans and Biomedical Identity*. Princeton University Press, 2004. The material is re-produced with kind permission from Princeton University Press.

Chapter 11   This chapter is a revised translation of Nicolas Langlitz (2010), 'Kultivierte Neurochemie und unkontrollierte Kultur. Über den Umgang mit Gefühlen in der psychopharmakologischen Halluzinogenforschung.' Zeitschrift für Kulturwissenschaften, no. 2. The original German article is followed by a debate between the author and five natural and cultural scientists (Malek Bajbouj, Hartmut Böhme, Ludwig Jäger, Boris Quednow, & Sigrid Weigel).

Chapter 12   This chapter draws on a Target Article published by the author in *Neuropsychoanalysis*, with kind permission.

Figure 14.1   Cultural neuroscience: Parsing universality and diversity across levels of analysis. In S. Kitayama & D. Cohen (Eds.), Handbook of Cultural Psychology (pp. 237–254). Chiao, J. Y. & Ambady, N. (2007). Copyright Guilford Press. Adapted with permission of The Guilford Press.

List of Illustrations

Figures

About the Editors

Suparna Choudhury is Junior Professor at the Max Planck Institute for the History of Science and the Berlin School of Mind and Brain, Germany. Trained as a cognitive neuroscientist with interdisciplinary research experience, her current work examines the emergence of the 'neurological adolescent' and the relationship between neuroscience, culture and society.

Jan Slaby is Junior Professor in Philosophy of Mind and Emotion at Freie Universität Berlin, Germany. The author of a book exploring the world-disclosing nature of human emotions, he has also been involved in research and teaching on the philosophy of psychiatry, with a particular focus on affective disorders and background feelings.

Contributors

Cornelius Borck, M.D., Ph.D., Professor, Institute of History of Science and Medicine, University of Lübeck, Germany.

Bobby K. Cheon, MSc., Gaduate Student, Interdepartmental Neuroscience Program, Department of Psychology, Northwestern University, Evanston, IL., USA.

Joan Y. Chiao, Ph.D., Assistant Professor, Departments of Brain, Behavior and Cognition, and Social Psychology, Northwestern University, Evanston, IL., USA.

Suparna Choudhury, Ph.D., Junior Professor, Max Planck Institute for the History of Science and Berlin School of Mind & Brain, Humboldt University, Berlin, Germany.

Simon Cohn, Ph.D., Senior Lecturer, Institute of Public Health, University of Cambridge, Cambridge, UK.

Joseph Dumit, Ph.D., Associate Professor, Department of Anthropology and Director of Science and Technology Studies, University of California, Davis, USA.

Thomas Fuchs, M.D., Ph.D., Professor, Centre for Psychosocial Medicine, Universitätsklinikum Heidelberg, Heidelberg, Germany.

Shaun Gallagher, Ph.D., Professor, Department of Philosophy, University of Memphis, Tennessee, USA.

Ian Gold, Ph.D., Associate Professor, Departments of Philosophy and Psychiatry, McGill University, Montreal, Quebec, Canada.

Martin Hartmann, Ph.D., Professor, Department of Philosophy, University of Lucerne, Lucerne, Switzerland.

Laurence J. Kirmayer, M.D., Ph.D., Professor and Director, Division of Social & Transcultural Psychiatry, McGill University and Director, Culture and Mental Health Research Unit, Sir Mortimer B. Davis — Jewish General Hospital, Montreal, Quebec, Canada.

Nicolas Langlitz, M.D., Ph.D., Assistant Professor, Department of Anthropology, New School for Social Research, New York, USA.

Daniel S. Margulies, Ph.D., Max Planck Research Group Leader, Max Planck Institute for Human Cognitive and Brain Sciences, Leipzig, Germany.

Francisco Ortega, Ph.D., Associate Professor, Institute for Social Medicine, State University of Rio de Janeiro, Rio de Janeiro, Brazil.

Eugene Raikhel, Ph.D., Assistant Professor, Department of Comparative Human Development, University of Chicago, Chicago, USA.

Amir Raz, Ph.D., ABPH, Professor, Montreal Neurological Institute, McGill University and Institute of Community & Family Psychiatry at the Sir Mortimer B. Davis — Jewish General Hospital, Montreal, Quebec, Canada.

Steven Rose, Ph.D., Professor, Department of Life Sciences, The Open University and Department of Anatomy & Developmental Biology, University College London, London, UK.

Jan Slaby, Ph.D., Junior Professor, Department of Philosophy, Freie Universität, Berlin, Germany.

Max Stadler, Ph.D., Postdoctoral Fellow, Department of Science Studies, ETH Swiss Federal Institute of Technology, Zurich, Switzerland.

Fernando Vidal, Ph.D., Research Scholar, Max Planck Institute for the History of Science, Berlin, Germany.

Allan Young, Ph.D., Professor, Departments of Social Studies of Medicine and Psychiatry, McGill University, Montreal, Quebec, Canada.

Preface

The story of critical neuroscience began on a bus in the outskirts of Berlin, where the editors first met. The spirit of excitement of the first discussion would soon be followed with frustration—not simply in response to the growing neuromania in the natural and human sciences, but also about the seemingly intractable differences between our disciplines and the difficulties in articulating how, and to what ends, to be "critical."

These tensions gave rise to the growth of an energetic group of young scholars with backgrounds in neuroscience, philosophy, history of science, anthropology, sociology, and psychology, who began to meet weekly in seminar rooms, cafes, bars, and apartments in Berlin. What first emerged was a shared sense of irritation about the hubris of neuroscience and the reverberations of "brain overclaim" in areas of everyday life far beyond the lab. What eventually followed, after months of wrestling with diverse concepts, vocabularies, and standpoints, was a consensus that what is needed is an understanding of how these neurophenomena are worked out, circulated, and applied; and to figure out how analyzing the social and cultural context of the neurosciences might help to push experimental work in alternative directions. Taking seriously the relevance, but rejecting the primacy, of the brain in understanding behavior, we asked ourselves whether such analysis might contribute to more complex, theory-rich, nuanced explanations of behavior.

Four years later, we are still asking questions, and certainly have no firm answers. The outcome of the debates has, however, been fruitful in numerous ways, for example in leading us to call for a "reality check" on the neurosciences. In what ways are we witnessing insights that are entirely novel, potentials that are revolutionary, applications that are empowering or threatening to human beings? To begin to approach these questions in such a way that was from the outset neither besotted with neuroscience nor suspicious of its practitioners, it became clear that close engagement with neuroscience and neuroscientists was central to our task.

This volume collects the preliminary results of these reflections since the project's inception. Its chapters serve to open up a discursive space for critical analysis and, we hope, subsequent practical engagement with neuroscientific approaches. Our aim is to address neuroscientists, sociologists, anthropologists, and philosophers at various levels of research, practitioners in fields such as medicine, education, law, and social policy, as well as representatives of funding agencies and the public at large. The volume marks the first step towards articulating an empirically informed theoretical and strategic alternative to the widespread over-confidence in the transformative power of the new neurobiologism.

We are enormously grateful to our colleagues and friends who came together during a conference at UCLA, organized by the Foundation for Psychocultural Research and McGill University in January of 2009. We thank Rob Lemelson for providing the opportunity, with much enthusiasm, for us all to meet in Los Angeles to debate these issues. The chapters in this volume are a result of the conference papers and speak directly to the questions critical neuroscience raises in thoughtful, creative, and at times challenging essays. The authors of the chapters have helped to develop our ideas and questions, and we express sincere thanks for their encouragement and their generosity in helping to create a space of openness and reflexivity (beginning at the Division of Social & Transcultural Psychiatry, McGill University in July 2008) in which this project could take shape. In particular we benefited from prolonged conversations with, and feedback from, Laurence Kirmayer, Ian Gold, Martin Hartmann, Allan Young, and Shaun Gallagher.

We express our gratitude to the Volkswagen Foundation in Hannover, Germany, for funding our early work in critical neuroscience within their European Platform for junior scholars in the Life Sciences, Mind Sciences, and Humanities. This grant, which funded the project originally called "Neuroscience in Context," enabled us to carry out workshops and conferences and gather a network of scholars that led to the ideas laid out in this volume. In particular, we thank Henrike Hartmann and Thomas Brunotte of the Foundation for helping to facilitate the administration of our activities.

Most of all we are grateful to the original collective of researchers in Berlin who have, with imagination, good humor, mutual support, and hard work, sustained the project. We have spent many lively hours talking cerebral with the group, and are indebted to them for refining the ideas expressed in our proposal for a critical neuroscience in Chapter 1. We owe particular thanks to Max Stadler, who has kept us on our toes with his rigorous critique of our own critique, contributing considerably to the very character of our approach and its content. We thank Saskia K. Nagel with whom we collaborated closely in the early stages of the project, and who continues to provide us with insights about the social implications of neuroscience. We are also immensely grateful to group members Lukas Ebensperger, Lutz Fricke, Jan-Christoph Heilinger, Daniel Margulies, and Moritz Merten, whose contributions, both intellectually and in spirit, were fundamental to the development of the project.

We also thank Beate Eibisch at the Institute of Cognitive Sciences, Osnabrück University, for the administration of our activities, and for making it possible for us to teach two graduate courses in Critical Neuroscience. The students of the lively courses continue to push us to think in different directions and to clarify our thinking. We have profited from the support of the following individuals along the way: Isabelle Bareither, Cornelius Borck, Felicity Callard, Simon Cohn, Christoph Demmerling, Nicole Golembo, Philipp Haueis, Kelly McKinney, Alessandra Miklavcic, Laura Moisi, John Protevi, Steven Rose, Fabian Stelzer, Achim Stephan, Ulas Türkmen, Fernando Vidal, Philipp Wüschner, Matthew Young and the original

VW *Neuroscience in Context* Group including Thorsten Galert, Ahmed Karim, Felicitas Krämer, Lambros Malafouris, and Stephan Schleim.

Katrin Maclean's patience and attention to detail have been invaluable throughout the process of preparing this volume. We are very grateful for her good humor and hard work in copyediting the chapters. We also thank Karen Shield for her assistance during the production process at Wiley-Blackwell.

Finally, we invite readers to continue conversations about the topics raised in this volume through our website at www.critical-neuroscience.org.

S. Choudhury & J. Slaby
Berlin, December 2010

Introduction

Critical Neuroscience—Between Lifeworld and Laboratory

Suparna Choudhury and Jan Slaby

Critical neuroscience arose in response to the tremendous pace of developments in neuroscience[1] during the last two decades, in particular the increasing emphasis of its findings in the social and cultural life of human beings. Indeed, the developments in neuroscience research have elicited a surge of interest from medicine, policy, and business. Furthermore, the last two years have seen a number of well-documented methodological controversies within the field, along with the emergence of ethical, historical, and social scientific projects *on* neuroscience. Many social scientists have claimed that notions of personhood among people in medicalized contexts are being radically transformed, replaced with the idea that "we are our brains" (Vidal, 2009) or that we are "neurochemical selves" (Rose, 2003, 2007). Neuroscience is therefore not only expanding as a field, and arguably as a culture, but is also increasingly discussed and contested within and beyond the academic sphere. There are, as a result, a number of different voices—some claiming the societal threats, others the revolutionary potential, and others still the banality of insights from research in neuroscience. How then should we make sense of the many growing discourses about neuroscience in society? How should we evaluate its effects?

While there is no doubt that we are better off in our knowledge about processes in the brain in health and disease since the explosion of the neurosciences, we are—in spite of the resounding optimism—still far from reaching an understanding of the

---

[1] We use the term "neuroscience" in a broad sense to denote neuroscientific approaches and subfields that deal with higher-level mental and behavioral phenomena *in humans*. This volume is especially concerned with "social cognitive neuroscience"—a field that is now increasingly seen to encompass subfields such as social, affective, and the newly emerging cultural neuroscience. If not explicitly stated otherwise, uses of the term "neuroscience" in this volume refer to social, cognitive, and affective neurosciences in this broad sense. We are well aware, however, that construing neuroscience as merely *cognitive* neuroscience is problematic (this issue is discussed by Max Stadler in Chapter 6). However, it is exactly this conflation and the focus on cognitive phenomena that is assumed by most contemporary literature that celebrates or problematizes the implications of neuroscience research-literature that constitutes the focus of this volume.

Critical Neuroscience: A Handbook of the Social and Cultural Contexts of Neuroscience, First Edition.
Edited by Suparna Choudhury and Jan Slaby.
© 2012 John Wiley & Sons, Ltd. Published 2016 by John Wiley & Sons, Ltd.

brain that would reliably enable changes to our lives that are noteworthy—in terms of practices, technologies, and institutions. Moreover, it is not clear how neural processes manage to realize subjective experience (Chalmers, 1996; Levine, 1983, 1993) nor is there consensus about the relationship between neural processes and cognitive, social, and emotional capacities captured in their full complexity (Gold & Stoljar, 1999). More importantly, we ought to ask whether these philosophical conundrums are at all sensible questions to ask of a science of the nervous system. Perhaps the conviction that the "big riddle of humanity"—the relationship between brain processes and subjective experience—at long last awaits its scientific solution is part of the problem surrounding today's neurosciences. It would be a misrepresentation of neuroscience to claim that its chief goal is the solution of the (philosophical) mind–body problem. There are even voices claiming that neuroscience, for the most part, is not about "the mind" at all (see Stadler, this volume).

Regardless of these unresolved issues, "neurotalk" (Illes et al., 2010)[2] pervades several domains of our everyday lives, beginning to exert various impacts on us through evolving "neuropolicies"[3] and in some cases, by starting to transform our understanding of ourselves—as patients, consumers, students, teachers, and decision makers (Cohn, 2010; Dumit, 2004; Martin, 2009; Ortega, 2009; Rose, 2007; Singh & Rose, 2009; Vrecko, 2006). A field that is garnering so much attention, accumulating resources, and pledging to revise our understanding of the very features of our life we take to define us, warrants special analysis. The goal of critical neuroscience is to create a space within and around the field of neuroscience to analyze how the brain has come to be cast as increasingly relevant in explaining and intervening in individual and collective behaviors, to what ends, and at what costs (Choudhury, Nagel, & Slaby, 2009). It encourages an empirical approach that seeks to go beyond the rhetoric of uncritical embrace or rejection of neuroscience, testing the commonly cited claims that our lifeworlds, language, and habits are already being subtly transformed by findings from neuroscience.[4]

---

[2] Several neuroscientists and neuroethicists have urged for greater public outreach between experts and lay audiences to make research more transparent, disseminate findings, and improve public literacy about the brain. Increasingly, funding agencies encourage "public engagement" as part of the scientific research process, and university press offices are firmly in place to deliver findings to the media. See for example Illes et al. (2010) and Herculano-Houzel (2002). Analysis of press content reporting functional neuroimaging (fMRI) results by Racine and colleagues demonstrated an overriding sense of neurorealism and neuroessentialism in the reportage; that is, images and interpretations conveyed uncritical reality and a sense of objectivity, at the same time as readily equating fMRI results with personal identity (Racine, Bar-Ilan, & Illes, 2005; Racine, Waldman, Rosenberg, & Illes, 2010).

[3] fMRI data are increasingly used to promote political agendas. From interest groups in the United States that have used experimental findings as an evidence base to argue against the use of pornography (see www. lightedcandle.org) to exceptionally well funded university departments dedicated to brain research aimed to inform the law (for example, at Stanford Law School) and the recent opening of a government backed research unit in France that use concepts from neuromarketing to develop public policy, findings, including preliminary data, have enormous appeal to contribute to evidence-based policy-making processes.

[4] For example, we suggest that the "impact" of neuroscience on subjectivity is, and will likely continue to be, manifold. The appeal of (or resistance to) neuroscience in constructing identities depends on what is gained or lost politically, economically, and in the meaning of the category, and it is certainly not a given that people exposed to neuroscience in popular or clinical contexts will come to understand themselves as brains (see Ortega and Choudhury, in press).

The aim is to achieve an understanding of the situatedness, leading assumptions, conceptual and explanatory resources, historical developments, and social implications of the emerging neuroindustry and of the new culture they are—or are not—in the process of establishing. Our claim is that a sustained engagement with neuroscience is necessary to provide a more accurately informed picture of what is *actually* happening in and around the neurosciences. It is this kind of engagement we want to cultivate: on the one hand tracing the journeys of "brain facts" between neuroscience laboratories and their various sites of appropriation and application in the institutions, discourses, and practices that constitute our human lifeworld;[5] and on the other hand probing whether contextual knowledge gained in this way can be reflexively applied to the practice of neuroscience to complement existing approaches, by inspiring enriched paradigms and broadening interpretive possibilities. Preserving and integrating the forms of expertise and the discourses about human nature and the human lifeworld that philosophy, anthropology, sociology, history, and other humanities disciplines provide, is necessary in the face of neuroscience's expansion and unquestioned cultural and institutional capital. This will ultimately benefit neuroscience itself as it may be productively aligned with—instead of opposed to—those more traditional canons of knowledge that still, and rightly we believe, form the foundation of our scientific, cultural, and political self-understanding (see Nussbaum, 2010).

To analyze a "hybrid of hybrids" field such as the neurosciences (Abi-Rached & Rose, 2010) requires critical neuroscience to be necessarily heterogeneous in its conceptual languages and methodological tools. What holds this assemblage of tools and concepts together, however, is on the one hand a rejection of the individualistic, reductionistic scientism that differentiates itself from the culture of knowledge and society and permeates much of the literature and its surrounding "neuro-hype". On the other hand, the goal is to work towards an integrated approach to behavior that situates the brain and cognition in the body, the social milieu, and the political world. As such, the notion of critique employed in critical neuroscience is constructive and *engaged with* neuroscience research, instead of merely assessing the field from disengaged standpoints. With these aims, critical neuroscience is crucially different both from "neuroethics" and Science and Technology Studies (STS). From neuroethics it differs chiefly through its skepticism towards the projection of futuristic scenarios and assumptions of an impending "neurorevolution"—a revolution that will inevitably create "ethical issues" calling for a new neuroethical expertise. In addition,

---

[5] We use the term "lifeworld" as it has been used by phenomenologist Edmund Husserl (1970) to describe the broad sphere of lived reality that is, in the first instance, the pretheoretic reality in which we all live and from which we draw our prescientific understanding of ourselves and of the world. On the one hand, the lifeworld provides the (often unnoticed) foundation of scientific rationality, scientific concepts, and practices; but on the other hand, it is itself not only shot through with, but even (at least nowadays) in large parts *constituted by* the results, tools, practices, and understandings that emanate from the sciences and their technological applications (with this latter idea we surely depart from Husserl's thinking on the matter). The term lifeworld is also used in a normatively laden sense by Habermas, when he speaks of the pervasive "colonizing" of the lifeworld by "system imperatives," among them chiefly those of science and technology (Habermas, 1971). As will become clear in the course of this introduction and also in Chapter 1, critical neuroscience is in an indirect sense taking up the intuition that today's neuroscience is in part threatening to be a colonizing force with regard to certain domains of today's lifeworld, especially to aspects of our self-understanding as free and responsible actors (see Hartmann, Chapter 3 in this volume).

critical neuroscience differs through its conscious distancing from institutional entanglements with neuroscience foundations and associations (in which neuroethics has not been overly shy to engage; see de Vries, 2007). Critical neuroscience aims to go beyond localist modes of inquiry in STS that are too often detached and apolitical. Instead, critical neuroscience strives to establish a hands-on approach that does not stop short of direct involvement in empirical research. In addition, while STS generally takes an agnostic stance in its analysis of scientific research, critical neuroscience makes explicit its commitments to views of the brain and cognition as situated and contingent (see Chapter 1).

This volume is a collective effort among a group of multi-disciplinary scholars around the globe to contribute diverse strands of inquiry that help to understand how particular intellectual, economic, and political conditions hold in place current views of the brain, and how these models of the brain and neurocentric practices may in turn produce ontological impacts in society. What kinds of ideas, hopes, methods, and institutions come together to produce what will count as facts about the brain? And what sorts of ideas, people and institutions do these facts go on to produce? Some of the chapters attempt to flesh out how alternative ontologies of the ecological brain can take shape, and how these analyses open up possibilities of experimenting with, and interpreting, the nervous system in ways that avoid reifying either the biological or the social realm; other contributions chart less known historical developments in neuroscience with the aim of questioning aspects of today's self-understanding of the field; furthermore, there are chapters that analyze the trends and tendencies in the field that can be shown to be immediately problematic from (variously articulated) political or social standpoints.

There are (at least) two risks involved in any such critical endeavor: first, being too confrontational as observers or commentators and engaging in what may be understood as unproductive polemic; and second, not being "critical" enough, especially in light of institutional dependencies (as pertains today to most scholars in newly *neuro* prefixed disciplines); or, in light of it being fashionable again among certain factions in the humanities, to enthusiastically buy into a certain biologism or scientism in the name of "interdisciplinarity," the rarely questioned watchword of the neoliberal university.

Aware of these tensions, this volume is less about providing ready-made answers, than an attempt to provoke more (and more critical, more empirical) investigations into the conditions that enable and sustain the current expansion of the "neurosciences," whether discursive or in practice. It is synthetic in bringing together a number of existing historiographical, sociological, philosophical, and ethnographic research programs pertaining to the neurosciences, and explicit about its driving force: a challenge to narrow neurobiological programs that privilege the molecular, cellular, synaptic, or functional realm of the brain in explaining human behaviors and disorders. This narrowness establishes essentialized differences between "kinds of people" on the assumption of distinct types of brains and constitutes the basis for behavioral and institutional reforms, thus participating in masking the life experiences and social structures that equally may account for them.

Our aim for this introduction, then, is to draw out the starting premises of this project, to gesture at our approach to critique (which is further elaborated in Chapter 1), and to summarize some of the ways in which the contributors have attempted to tackle these goals.

Imagined Futures (or, What Revolution?)

Talk of a "neurorevolution" has been in the air for a while.[6] When George H. W. Bush proclaimed the start of the "Decade of the Brain" in 1990, grand scale initiatives were set in motion to shed new light on the workings of the human brain ultimately in order to "conquer brain disease."[7] The neurosciences have promised much more than the alleviation of brain disease since then; cognitive neuroscientists now offer novel biological approaches to explain the core human capacities to reason, interact, and emote, as well as our cultural habits and beliefs.

While pharmaceutical drugs are being developed to eliminate unhappiness by way of neurotransmission, or intelligence agencies promise to root out terrorism by imaging malevolent intentions, neuroscience is not only making waves at the level of social institutions. Under the attentive gaze of the media, cultural critics, and ethicists, the neurosciences have brought to the horizon new technologies that are being mobilized to make us healthier, smarter, and happier. Within the reach of many of our everyday lives in medicalized societies, a new kind of neuroscientific wisdom has in this way become pervasive: whether or not we take seriously education initiatives that aim to enhance creativity through the stimulation of "brain buttons"[8] or explanations of the appeal of love poetry in terms of neurons (Byatt, 2006), manuals that urge social workers to use neuroscience to deal with family predicaments (Farmer, 2008) or advertizing campaigns that persuade us to choose one drink over another based on what our "brains prefer" (McClure et al., 2004), it is not an overstatement to point out the widespread invocation of the brain to lend credence to explanations of the way we are[9] and prescriptions of the way we should live.[10] Where there are questions

---

[6] Lynch (2009) is perhaps the most enthusiastic author in this area; popular neuroscience and the blogosphere are rife with references to the "neuroscientific revolution." Accounts that likewise are not hesitant in using revolutionary rhetoric in connection with the new brain sciences include those by: Crick (1994), Churchland (2002), Edelman (1992), Metzinger (2009), and Zeki (2008).

[7] See Presidential Proclamation 6158 http://www.loc.gov/loc/brain/proclaim.html. Retrieved on June 25, 2010.

[8] See http://www.braingym.org/. Retrieved on June 25, 2010.

[9] For a historical account of "brainhood"—the conception of the self that identifies brain with self, consonant with modern Western forms of naturalized and singularized individuals, see Vidal, 2009.

[10] The fact that neuroscience is increasingly called upon to build prescriptions about how to live and how to organise various sectors of society is impressively evidenced by two recent government initiatives in Britain and France respectively. The British "Foresight Project on Mental Capital and Wellbeing" and the French government initiative to improve public health, focus on broad risk assessments, monitoring and prevention policies and on how to "optimize" mental resources in the population at large. Both initiatives can serve as paradigm examples for the recent trend towards neuroscience-inspired policy initiatives in neoliberal settings. For the British case, see Beddington et al. (2008) and for the French case see Oullier and Sauneron (2010–http://www.strategie.gouv.fr/IMG/pdf/NeuroPrevention_English_Book.pdf).

unanswered, or applications as yet unrealized, the academic and popular literatures carry the breathless conviction that within a few years technological advancements will ensure their fruition, and knowledge from the brain sciences will subsequently begin to supersede social, cultural, philosophical, political, literary, or other "folk" explanations of behavioral phenomena.

It would be fair to say that most contributors to this collection share a certain *ennui* about this revolutionary rhetoric. At the same time, the expansion of the brain sciences is occurring within the context of some tangible change: changes in the ways lives are lived, changes in the ways science is practiced, how it is embedded and applied in society and financially endowed seem to be happening in concert with trends that implicate the neurosciences or other biological approaches to "human nature;" shifts in prestige and cultural capital in the academic sphere, with the humanities globally declining, revived discourses about human nature, evolution, resilience, new emphasis on emotional intelligence, human resources, and "mental capital," all of which have flourished around the idea that new evidence from neuroscience is transforming notions of human nature.

Where we depart from many of the current problematizations of neuroscience is that we do not believe that existing ideas of human behavior and social life are really called into question by neuroscience *per se*. We believe that it is not only *what* is being claimed by the neuroindustry that deserves analysis, but the fact *that* these claims are being floated in the first place, further still that they are heard—within academia and beyond. An analysis of these conditions leads us beyond the question of whether or not the outcomes of neuroscience can really fulfill their promises, and towards a critical engagement with the assumptions and visions of neuroscience on which such scenarios are built; and, hence, to explore the reasons as to how and why findings from an inchoate science manage to portend radical reinventions of notions of human nature and structures of social institutions (Choudhury, Gold, & Kirmayer, 2010).

In short, given the discrepancies between theories of the brain and theories of mental life, it is not at all clear why existing knowledge of the brain should lead us to shape social life according to it. This project, therefore, aims to perform a "reality check," problematizing the discourses and the phantom debates—both alarmist and enthusiastic—that thrive within and around it (Quednow, 2010). Our insistence on empirical engagement with neuroscience will, we believe, avoid the futurism which frequently serves to obscure rather than illuminate processes that drive current developments, such as political reforms in the academic system and in science funding. Based on the assumption that most of our conceptions of our selves, our societies, and our ways of life happen in spite of the momentum and promissory character of the neurosciences, the project is alert to the fact that neuroscience is a historically situated enterprise, always already enmeshed in a broader realm of the social and cultural.

To avoid reifying either the neuroscientific "threat" or the conception of human nature allegedly "under siege," it is important to enter the gaps between hypothesis and discovery, discovery and application, and to attend to the "back stories" that give them life and appeal (Young, this volume). It is especially important to see that neuroscientific knowledge and expertise, in order to smoothly operate as applied knowledge, requires a naturalistic construal of a biological substrate that is supposedly substructuring a realm that is "cultural" and "social," making it amenable to technological intervention. This assumption of a stable, accessible, and manipulable cerebro-substrate of personal, social, and cultural processes, often not explicit, is a maneuver of simplification and purification that obscures the complicated conceptual and ontological entanglement between things natural and things social and cultural.[11] In effect, this initial move assumes contested philosophical issues to be settled from the outset, without acknowledging theoretical alternatives. Instead of opening up discursive spaces to belabor these entanglements and possibilities for creative engagement, the leading naturalistic assumption forecloses meaningful debate and moves right on to programs of technocratic intervention (see Mitchell, 2002).[12]

Neuroscience, Society, and Personhood

At the core of critical neuroscience is the goal to examine the reciprocal interactions between neuroscience and social life, and the diversity of factors that come together not only to breathe life into neurobiological theories and fuel their journeys beyond the lab, but which create and sustain such divisions in the first place—those between "social life" and "neuroscience," or more broadly "science" and "society", and those which shape how and where "interactions" are located, defined, or framed (Choudhury, Nagel, & Slaby, 2009; Slaby, 2010). Since these journeys increasingly include hospitals, schools, law courts, and our vocabularies about who we are, and since the stakes are much greater than the knowledge itself, our analyses must pay careful attention to the ways in which neuroscience increasingly functions as a screen upon which to project everyday values about mental life, personhood, and kinds of people. How do certain metaphors begin to frame, and even shape, our understanding of the brain? How do these metaphors become tenable in the first place? These questions are taken up for example by Martin Hartmann (this volume), in his discussion of the correspondence between the discourse of management and human resources in late-modern institutions centered on non-hierarchical organization of companies, social networks, soft skills, flexibility, and lifelong learning (Boltanski & Chiapello, 2007; Hartmann & Honneth, 2006), and the recent (popularly simplified) neuroscientific discourse about the brain's organization as—precisely—a non-hierarchical network without center, a malleable, plastic structure capable of adaptation, constantly rewiring to fit new conditions and demands, and increasingly seen as an emotional brain instead of

[11] Arguably, mainstream naturalistic philosophy of mind has done some damage by fostering, uncritically, the very idea of it being uncontroversial to assume a material substrate with these properties. The whole field has for a long time worked under the assumption that the ontology, accessibility, and manipulabilty of the physical side of human reality, despite its complete ontological detachment from subjectivity, discourse, and "mind," poses little or no problem, while the alleged "other side"—mental states, personal traits, the cultural, and the social—is deemed scientifically mysterious (as it is unclear how exactly it relates to its alleged physical base). We believe, on the contrary, that the idea of even conceptually separating both alleged "realms" is highly troublesome, but cannot argue this here. For similar views, see Rouse (2002) and Barad (2007), and S. D. Mitchell (2009).

[12] Although not dealing with the brain/mind relationship, Timothy Mitchell's reflections about entanglements of natural and human realities are quite instructive—especially his well-argued contention that the construction of expert knowledge often involves purifications and simplifications that are not merely artifacts of theoretical abstraction but are performed consciously to serve political purposes (see Mitchell, 2002, ch. 1).

a classically rational one. *Honi soit, qui mal y pense* (Hartmann, this volume; Karafyllis & Ulshöfer, 2008; Malabou, 2008).

Our analytical perspective rests partly on a historical ontology of subjectivity and personhood. This view understands the make-up of human beings to be, in an important sense, historically constituted—through processes of situated self-interpretation of human subjects in material settings and in relation to social structures and practices (Foucault, 1973; Taylor, 1985, 1989; see also Brinkmann 2005, 2008). Properly spelled out, such a perspective need not break with a naturalistic understanding of the human world (see Rouse, 1996, 2002).[13] In particular, we agree with Ian Hacking in the assumption that science, medicine, education, and other institutions and powerful areas of social practice and policy are key contributors in "creating" kinds of people through processes of "classificatory looping" (Hacking, 1995, 1999). Classificatory terms come bundled with certain norms and expectations about the objects collected under their scope, and objectifying an identity, stage of life, culture, or behavior in those terms can interact with the experience of that which is classified. In other words, classifications can be taken up into the self-understanding of those classified. These processes can lead, in turn, to the emergence of new practices, new alliances, new institutions that interact with the persons in question—in establishing and sustaining habits, thought patterns, forms of conduct, and schemes of judgment. "Classificatory looping" is a circular interaction between the categories used to classify groups of people, these people's behavior, attitudes, and understanding of themselves in response to these classifications, and the modification of the original categories as a result of the classified subjects' altered behaviors and ways of being. These processes are obviously complex and involve much more than an idea being voiced or a concept applied— successful classifications are richly situated both materially and institutionally. What results can be a "new" type of person in a new "social niche" in which this way of being a person finds a stable habitat (Hacking, 1998, p. 13).[14]

The idea that kinds of people are historically "made" through powerful classifications gains additional relevance when placed in the context of what sociologist Anthony Giddens has called *institutional reflexivity*: the routine incorporation of new knowledge into environments of organized action that are in this way constantly transformed and reorganized (see Giddens, 1991, p. 243)—a central working principle of institutions in late-modern societies. Expert knowledge, variously mediated, interacts in multiple settings and through complex feedback loops with the practices and self-understanding of subjects, to the extent that these interactions are no longer

---

[13]   Joseph Rouse's pragmatist naturalism is in many ways inspirational to our approach, as it breaks with dominant assumptions that contrast nature in a dualistic manner either with mind, the social, or the normative. Instead, he articulates a position that steers completely clear of these oppositions. In this way, Rouse manages to incorporate important anti-naturalistic insights by authors such as Charles Taylor, John McDowell, and Robert Brandom into a naturalistic outlook. Obviously, Foucault is an important inspiration to this line of thought, as can be seen in Rouse's early work (1987).

[14]   Hacking adopts the notion of an "ecological niche" for the purposes of a historical ontology of mental illness: "I argue that one fruitful idea for understanding transient mental illness is the ecological niche, not just social, not just medical, not just coming from the patient, not just from the doctors, but from the concatenation of an extraordinarily large number of diverse types of elements which for a moment provide a stable home for certain types of manifestation of illness" (Hacking, 1998, p.13).

recognized for what they are and are taken as natural givens (Ward, 2002).[15] It is well documented that the modern life sciences, especially psychology, have been a crucial element in processes of this kind (Danziger, 1990; N. Rose, 1996; Richards, 1996; Ward, 2002).[16]

Increasingly, today, the neurosciences are entering into the loop as the "new image of man" discourse becomes increasingly widespread, and a wealth of brain-based approaches exerts its influence upon medicine, education, advertising, and recreation. In addition they influence other domains of knowledge production, such as the burgeoning *neuro* disciplines—from neuroeconomics to neurotheology or neuroliterary criticism. Not only are powerful new styles of scientific thought emerging, but also new forms of thinking about life itself—about subjectivity, ethics, and politics—that pertain to many areas of today's social life.[17] Increasingly noticeable, for example, is the enthusiasm with which neuroscience is received within many of the humanities and the social sciences, revealing the scientistic reformatting of discourses on human nature that is currently underway.

While neuroscience officially promises to penetrate to the ultimate level of human functioning—the "first nature" of the central nervous system—in fact, importantly and probably unwittingly, it participates in the construction of a powerful "second nature:" an institutional, informational, and "ideational" environment that breeds practices and institutions of subjectification. These practices in these settings "make up people" (Hacking, 2002; Hartmann, this volume). A central task for critical neuroscience is to make these construction processes explicit, with the goal of scrutinizing their formative assumptions and underlying commitments.[18]

That being said, it is important to see that we are not advocating unconstrained social constructionism or historicism with regard to human nature or human forms of life. Indeed, it is because we believe that significant changes are underway, that we take the phenomenon "neuroscience" seriously; what we reject, however, is the notion that neuroscience, entangled as it is in much wider processes of transformation, is the *sole* cause, driver, or solution to a set of relevant social, cultural, and political changes

---

[15] Again, Mitchell's (2002) reflections upon the construction of expert knowledge are relevant here as they forcefully call to mind the frequent uses of gross abstractions, purifications, and outright misconstruals that are involved in them. Making these processes explicit and keeping a "balance sheet" weighing epistemic losses and practical frictions against practical benefits and epistemic advances is a crucial ingredient in any endeavor of social critique under present-day circumstances.

[16] For a well-worked out reconstruction of the various approaches to the history of psychology in relation to a historical ontology of the psychological subject, see Brinkmann (2008).

[17] Nobody has charted these developments more rigorously than Nikolas Rose (2007), who uses the powerful term "neurochemical selfhood" to describe the medicalized and molecularized ways of thinking about oneself evident in today's techno-medical assemblages. The depth of Rose's socio-historical studies is reflected in the range of social domains covered: various areas of business, medicine, advertizing, education, criminal law, and social policy (see also Abi-Rached and Rose, 2010; Rose, 2010), and not least the very formation of a new conception of ethico-spiritual thinking, an ideal of "somatic ethics" that increasingly takes the place of classical religious or otherwise transcendent systems of meaning. While we are indebted to Rose for providing rich analytical perspectives in these areas, we will take issue with aspects of his approach in Chapter 1.

[18] We will say more on this projected re-politicization of classificatory looping processes below. In fact, as will become evident shortly, spelling this out coincides in important parts with what is meant by "critique."

and problems. Moreover, what is made and molded in processes of classificatory looping and in situated self-interpretation is a natural entity. Hacking's approach helps clarify how situated processes of classification interact with the biological substrate underlying personal traits and ways of being, hereby rendering stark oppositions between the "social/historical" and the "natural" obsolete (see also Langlitz, this volume). He considers the possibility that medical diagnoses—such as one of depression—interact not only with the self-understanding of the patient, but also with the *biological processes* related to the condition diagnosed. Upon being diagnosed, a depressed person might adopt a specific behavioral regime, abandon hazardous routines, avoid stress, and so forth and as a consequence the neurological condition underwriting his or her depressive symptoms might change, so that another categorical modification is called for. Classificatory looping is in this way revealed as an instance of *biolooping* (Hacking, 1999, p. 123). As an inherently social and culturally mediated process, biolooping is, in turn, disparate with problematizations that would myopically center on the (alleged) "impacts" or "implications" of neuroscience, on worrisome advances in what is known about the brain—and on what is possible for future applications. Instead, biolooping is a key part of the complex process of interaction between individual persons, social systems, and institutions, mediated self-understandings and the results produced in the human sciences—it therefore points to some of the processes that become chief objects of critical neuroscience.

Countering the Cerebral Subject: Embodied Experience and the Politics of Situated Subjectivity

Our focus on the social and historical ontology of personhood connects our reflections to a broader trend in the philosophy of the cognitive sciences: the increased tendency to leave behind narrowly mentalistic, Cartesian approaches to behavioral experience such as computational/representational theories of mental activity. The emerging alternative picture has been labeled the "4EA approach:" the mind as *embodied, embedded, enacted, extended, and affective* (Protevi, 2009, p. 4). This view—of which we can only provide a very rough outline here—breaks with the mentalist legacy of assuming strict dichotomies between mind and body, body and world, and one person's mind and the minds of others (Clark, 1997, 2008; Gallagher, 2005; Haugeland, 1998; Rowlands, 2010; Thompson, 2007).[19] This perspective stands in sharp contrast to conceptions of "cerebral subjectivity," that is, approaches that combine traditional Cartesian mentalism with the assumption of a strict explanatory dependency of mental processes on neuronal processes alone, culminating in Crick's (among others) famous exclamation that "you are your brain"[20] (Crick, 1994; Metzinger, 2003, 2009; Revonsuo, 2005).

---

[19] See also the texts collected in the two recent anthologies by Robbins and Aydede (2009) and Menary (2010).

[20] This is not the original wording, but it comes close. Here is what Crick *did* write: "You, your joys and your sorrows, your memories and your ambitions, your sense of personal identity and free will, are in fact no more than the behavior of a vast assembly of nerve cells and their associated molecules." (Crick, 1994, p. 3) For a useful contextualization and critique of the "cerebral subjectivity" paradigm see Vidal (2009), Ortega and Vidal (2011), and Vidal and Ortega (Chapter 17 in this volume).

The counter-ontology of critical neuroscience resonates with the "4EA view," which assumes that mental processes are understood as constitutively embodied and environmentally embedded such that they cannot be properly characterized without reference to their bodily dimensions and relations to the physical and social environment (Gallagher, 2005; Haugeland, 1998). In addition, the assumption of a strict separation between experience (perception, emotion, sensation) and action is abandoned in favor of an action-oriented understanding of embodied experience (Brooks, 1991; Clark, 1997; Hurley, 1998; Noë, 2005; O'Regan & Noë, 2001). *Enaction* refers to the dynamic integration of perception, cognition, and knowledge with action, so that there is no non-arbitrary distinction between perception and action—"enaction" denotes the unified sensorimotor activity that takes the place of what formerly was conceptualized as distinct capacities. The resulting image is an integrative, holistic understanding of how an embodied cognitive agent is constitutively embedded in its environment. Enactive approaches are anti-representationalist in their conception of an agent's relation to its world not as a spectatorial view of an "outside" reality, but as an interactive process in which an intimate organism-environment mutuality is established (or, in other words, "enacted").

An enactive understanding of the mind sidelines the classical "sandwich model" that long dominated cognitive science[21]—the obsolete strict distinction between perceptual input, central cognition (often conceived as computationally manipulated mental representations), and behavioral output—thereby abandoning the assumption of clearly identifiable interfaces between mind, body, and world (Noë, 2005, 2009; Thompson, 2007; Varela, Thompson, & Rosch, 1991).

A further focus of the emerging picture is on *intersubjectivity*: human experience consists of modes of relating to the world that are socially shared, while the experienced world itself is, in this way, revealed as a social lifeworld from the outset (de Jaegher & di Paolo, 2007; Gallagher, 2008, 2009, this volume). To be sure, the 4EA discourse is itself potentially at risk of becoming the sally port of some rather uncritical reception within the humanities and, elsewhere, of (popular) neuroscience. For example, the sudden, widespread focus on emotions and affective capacities (see Damasio, 1995; LeDoux, 1996)—sometimes strikingly simplistic—has been eagerly taken up by a popular self-management literature (see, for example, Goleman, 1995; for helpful critique, see the essays in Karafyllis & Ulshöfer, 2008; see also Malabou, 2008). Thin conceptualizations of "social intelligence" (again, Goleman, 2006) have proliferated in the domains of education, popular psychology, and business management; similarly, the recent resurgence of discourses on embodiment and bodily capacities bear traces of problematic biologism. It is important to maintain the complexity of these themes and to examine how they are appropriated. Critical neuroscience is thus committed to putting those theories, discourses, and trends that it draws on itself under constant scrutiny.[22]

Historically, many of the approaches sketched above continue the legacy of the phenomenological tradition, drawing on Husserl, Heidegger, and Merleau-Ponty, and

[21] The expression "sandwich model" in this regard was coined by Susan Hurley (Hurley, 1998, p. 401).
[22] Crucially, critique is also in an important sense always *self*-critique. We say more on the necessarily self-reflexive nature of critique in Chapter 1.

their early sociological followers such as Schütz (1974), Gurwitch (1931/1978), and Berger and Luckmann (1966). Much of the new work that links the phenomenological tradition with recent research in the cognitive sciences has been focused on the nature of experience, especially on the ways of embodiment, the integration of motor skills with perceptual capacities, and externalist approaches to mental content. Only recently have some scholars started to address the broader consequences of the situatedness and social embeddedness of cognitive capacities (Gallagher, this volume; Gallagher & Crisafi 2009; Protevi, 2009). If it is true that experience, cognition, action, and personhood are intelligible only as constitutively situated, as emerging from and co-varying with our natural and social environments, then it becomes a task of great importance to understand and analyze how all those "cognitive extensions" are organized, how they develop and by whom they are managed. Reflexive knowledge of this kind is a precondition in a project of active engagement and conscious participation in the construction, critique, and re-construction of the social and institutional environments that create our modern lifeworld. The broad ensemble of social institutions, of shared practices, symbol systems, predominant habits, the public spaces of possibilities as established and regulated by the economy, the media, the educational and medical systems are crucial scaffolds of subjectivity with immediate relevance for all of our lives.

Critical neuroscience is explicit about the political dimension that emerges from this theoretical perspective. Just like the social world, the human mind is partly of our own, historical making—critical reflexivity about the situatedness of subjectivity, and equally of the role of novel technoscientific developments, allows us to be aware of (and ready to intervene in) the various processes that shape it. While strands of cognitive science and philosophy of mind have been re-focused towards insights from the phenomeno-logical tradition, the social and political dimensions of our mental constitution have not yet garnered enough attention, scholarly effort, and reformist initiative. These are among the key dimensions of our notion of critique in the project of critical neuroscience. We come back to this in much more detail in Chapter 1.

Outline of Chapters

This volume serves preparatory purposes. The collective chapters focus on developments in and around the neurosciences from diverse disciplinary perspectives, with some authors honing in on potentially problematic aspects of research and its applications, while others explore initial ideas as to how a constructive engagement between the human sciences and neuroscientific theory and practice could take shape. Some of the chapters actively interrogate possible approaches to critique and to constructive enrichment of neuroscience, demonstrating the necessary self-reflexivity of critical perspectives. Overall, the texts collected here serve to open up a discursive and—subsequently—practical space for a critical analysis and constructive engagement with neuroscientific approaches. They address neuroscience researchers who develop paradigms and interpret data, historians studying the development of the brain sciences and the metaphors of mind–brain, sociologists tracing the economic and cultural contexts of contemporary "brain facts" and their application, anthropologists observing the practices of scientists who operationalize and disseminate neuroscientific phenomena, philosophers engaged in drawing larger consequences from current work in the human sciences, practitioners in fields such as medicine, education, or the law, policy makers and representatives of funding agencies, and—not least—the public at large. Such a broad, inclusive, discursive space has so far been absent from institutionalized neuroscience research and training.

Specifically, this collection assembles contributions from the areas of philosophy, history of science, anthropology, psychiatry, and of course neuroscience itself to provide an informed picture of the current situation at the intersection between cognitive, affective, and social neuroscience, the humanities and various areas of social practice and policy.

In Part I, entitled Motivations and Foundations, the basic assumptions and premises behind the idea of a critical neuroscience are explored. Not surprisingly, most chapters in this first part of the volume are predominantly philosophical in nature as they outline what it could mean to integrate "critique" into neuroscience research, and analyze the conceptions of nature and naturalness that are put forward by neuroscientists. What we hope to bring into focus here is the potential for mutual enrichment of critical theorizing with empirical approaches in the neuro and cognitive sciences.

In Chapter 1, we extend the ideas of this introduction and offer a programmatic proposal for a critical neuroscience. In particular, we focus on the concept of critique and on the possible ways it could be implemented in the vicinity of actual research processes. Obviously, things have changed a lot since the heyday of social critique in the 1960s and 1970s: the geopolitical changes in the past 20 years alone have altered the political climate, while the university system and research have undergone clear structural changes, in line with changes in the capitalist economy[23]. Openly political forms of critique within academia or science have largely fallen into disrepute, and many of the catchwords of social critique such as positivism, objectivism, instrumental rationality, or interest dependence have lost their currency. However, it would be wrong to suggest that the problems to which these initial critical movements responded have disappeared, let alone been resolved.

In the opening chapter, we propose a dual strategy for critical neuroscience: on the one hand a constructive approach to enrich research perspectives by assembling construals of phenomena captured in the full fabric of meaningful relations that contribute to their significance as focal "matters of concern"—in effect a call to adopt a hands-on approach that embeds and involves the critic within interdisciplinary research. On the other hand, we formulate a proposal for a multi-dimensional critical investigation of neuroscience-in-context that reckons with various biases, ideological influences, interest-driven "overclaim," skewed representations of research findings by practitioners and the media, tacit schemes and frames of judgment that distort rather than illuminate relevant phenomena, institutional "pathologies" such as colonizing tendencies of research agendas and the construction and politically problematic deployment of expert knowledge to serve specific—for example corporate—interests.

Clearly, this dual strategy is not without intrinsic tensions, but, as we will argue, it is the only viable response to the highly ambivalent and immensely complex institutional

---

[23] In Britain, these changes are highlighted by recent controversy over pressures on the academy to produce research pertaining to a political brand: the Conservative government's "Big Society" agenda has been set as a priority for the Arts and Humanities Research Council (AHRC).

and cultural context of today's neuroscience, which in itself obviously comprise a heterogeneous multitude of approaches, techniques, and institutions, embedded within multiple disciplinary and corporate affiliations. The opening chapter concludes with the outline of several contributing activities that, when implemented together, could fuel the idea of a self-reflective and socially responsive scientific practice in the neuroscience lab.

In Chapter 2, neurobiologist and public commentator on neuroscience in society, Steven Rose, provides an assessment of some of the most problematic tendencies he has observed in his discipline, in particular the problem of turning methodological necessities into philosophical, even metaphysical, commitments. He describes how the sensible research strategy of isolating single components out of the vast complexity of the overall nervous system in its natural context (methodological reductionism) too often degenerates into crude ideology when its experimental data are later taken as accurate descriptions of the original phenomena under study. The concept of consciousness is a case in point: as an object of neuroscientific study, consciousness is often conceptually reduced to mere "awareness," while all the richer connotations that link it to history, culture, group, class, or deeper aspects of personality are lost from view. Rose emphasizes that reductionist ideologies become particularly disturbing in combination with novel neurotechnological developments such as smart drugs or brain-based monitoring devices and the increasing political push for their application in society. Rose urges neuroscientists to develop a biosocial understanding of the person as embodied and culturally embedded to counter the neurocentrism of exclusively focusing on isolated brains.

Martin Hartmann's contribution (Chapter 3) relates some of the goals of critical neuroscience to the tradition of Frankfurt School critical theory. Hartmann poses the question of whether there can be a "critical theory of the neurosciences" and whether "neuroscience is positivistic." To answer such questions, Hartmann revisits several stages in the development of critical theory, starting with Max Horkheimer's founding documents written in the 1930s, and spanning both early and later periods of Jürgen Habermas' writings. Hartmann concludes that the traditional forms of critique cannot be applied in a straightforward manner to the current methods and theories in the neurosciences, primarily because these have moved well beyond the detached, theoretical, and value-neutral inquiry characteristic of older "positivistic" science. Importantly, today, many neuroscientists readily engage in intervention-oriented, or applied, research, proposing social reforms on the grounds of alleged insights into the natural workings of human beings. In response to these novel "normative first-nature arguments," Hartmann calls for a modified approach to critique—an approach that places neuroscientific construals of nature or naturalness under scrutiny. As an example, Hartmann points to the striking parallels between descriptions of brain organization and prescriptions for the ideal employee in today's corporate capitalism. Is the focus in both on flexibility, non-hierarchical networks, self-organization, and adaptability merely accidental? Or is it a symptom of a tendency of a larger-scale naturalization of social categories in which neuroscience unwittingly takes a leading role?

Continuing the discussion of thought originating in Frankfurt School critical theory, phenomenologist Shaun Gallagher, in Chapter 4, reverses the direction of questioning, suggesting that the relation between critical theory and cognitive neuroscience could be a two-way street. Agreeing that critical theory can aid in the assessment of current neuroscientific work, he suggests in addition that it might itself benefit from being more closely aligned with current empirical work in neuroscience—on the condition that these latter approaches avoid reification of human capacities and crude reductionism. Specifically, Gallagher explores approaches to intersubjectivity and social cognition that can help provide an empirical footing for approaches in critical theory. At the same time, he makes use of phenomenological considerations to critique certain problematic empirical and conceptual approaches to understanding others; for instance the exaggerated mentalism and universalism of both "theory theory" and "simulation theory" in the understanding of other minds. Defending his own enactivist interaction theory of social cognition, Gallagher puts research on "mirror neurons" in perspective, divorcing it from problematic conceptual baggage and notorious over-interpretation. With this well-informed theoretical and empirical perspective in hand, he returns to the writings of critical theorists, notably Habermas and Honneth, to suggest improvements with regard to the conceptualizations of intersubjectivity these employ.

Part II, Histories of the Brain, collects chapters from three historians who provide evidence that neuroscience, as it is commonly understood today—the discipline which investigates mind–brain problems and which will provide biological solutions to human nature—has not always been so. They chart historical developments in metaphors, models, narratives, and disciplines to offer a sobering antidote to the tone of self-confidence and conviction that permeate contemporary neuroscience and drive its expansion and applications. These contributions elegantly demonstrate how the relation of mind and body and notions of human nature, and their relationship with the brain, have relied—and continue to rely—on our available cultural metaphors at any moment in time, guiding our theories and investigations of brain function in particular directions. Such insights push us to step back a little, reminding us that neuroscientific questions, models, and results are not simply driven by scientific advances but are always historically and culturally contingent, challenging us not to take today's solutions as the final answers.

Cornelius Borck demonstrates, for example, in Chapter 5, how the brain, the organ for understanding the *condition humaine* over the last two hundred years, has been analogized by neuroscientists with an array of different tools, each one serving to explain the brain and accentuate specific functions, be it in terms of a psychic tape recorder, a telephone, radio, or an inscription device. Charting the changing metaphors of the brain from the late eighteenth century to contemporary neuroscience, Borck shows how machines, communication technology, or the computer have functioned as metaphorical linkages, mediating between the world of biological functions and the realm of everyday-life experiences, and structuring neuroscientists' view of the make-up and function of the brain in terms of their technical functionality as well as by their cultural significance. The instability of a metaphor for the brain in neuroscience, compared to other organs such as the heart—likened for a long time to a pump—reflects, according to Borck, the cultural status given to the brain, as an organ holding answers to mysteries about human nature, so complex that it escapes stable analogies. As the computer metaphor wanes and we enter the realm of the plastic brain viewed "at work" through neuroimaging, Borck concludes with challenging questions about the next top model for the brain, and about our relationships to the models and metaphors in the age of "brainhood," when "we *are* our brains."

In Chapter 6, Max Stadler argues that many such historical narratives, which focus for example on cybernetics, take a myopic view of neuroscience, conflating neuroscience with *the brain*. Such a view, Stadler claims, serves to conceal rather than reveal the more "mundane determinants" of the field of neuroscience during the last half century, which belong in the realms of molecular biology, physics, and engineering. Stadler's insistence on a more empirically informed history aims to set straight existing narratives about the history of neuroscience which tend not only to view the field brain-centrically (rather than attending to the decidedly less exciting parts of the nervous system such as reflexes) but also to represent its trajectory as a revolutionary one, culminating, thanks to new technology, in solutions to societal problems through a newly-arrived exposure of the "true" human nature. Historians, he cautions, need to remain wary of reinforcing rather than deflating the novelty rhetoric and the sense of exigency—and in order to do so, it is necessary to contextualize neuroscience within the broader scheme of intellectual and socio-political sea-changes including, for example, transformations in the academic research sector. In shifting our gaze away from the wildest visions of neuroscience's future to the subtleties of its more prosaic past, Stadler makes explicit the dilemmas for critical neuroscience to maintain its critical impetus, raising challenging questions about the meaning and goals of critique within the current academic climate.

In Chapter 7, Allan Young proposes that contemporary fMRI research in social neuroscience is giving rise to a new conception of human nature based on a neurally-based, natural, pro-social benevolence. Young provides a historical perspective on social neuroscience's discovery of empathy, arguably the most important concept in cognitive neuroscience, as it purportedly distinguishes humans from other animals and, Young argues, marks a shift from the Enlightenment notion of human nature ("Human Nature 1.0") to the new version, still emerging through evidence from fMRI studies ("Human Nature 2.0"). While the former version 1.0 characterizes the mind as self-contained and in its normal state, rational, the new version 2.0 is characterized by a capacity to directly communicate, or resonate, with other mind–brains. This new inter-penetration between minds occurs in the form of mirroring or empathy, via the recently discovered mirror neuron system, a capacity of normal humans, which when absent, manifests as disorders such as autism, schizophrenia, and psychopathy. Young returns to nineteenth- and twentieth-century neurology to set up the problematic for the future, the set of puzzles about human nature and the brain's evolution that are no longer questioned in social neuroscience. Describing three narratives that are integral to the "social brain," Young deftly demonstrates how modern neuroscience attempts to answer recurring questions about the mind–brain relationship, the cognitive arms race, and the formation of stable societies in terms of the social brain. However, Young's analysis of empathy and its construals in neuroscience experiments bring him to the problem of empathic cruelty, a form of empathy selectively excluded from the "social glue" theory of empathy and human nature in neuroscience. Once again, a historical perspective is invaluable in showing how brain function and structure, and conceptions of human nature, cannot be dissociated from the norms and values of discourses structuring societies at particular moments in time.

Ethnographic research is a crucial ingredient in the methodological portfolio of critical neuroscience. It focuses on the practices, behaviors, and attitudes of various parties involved in current research: the practitioners of neuroscience, study participants, and psychiatric patients and members of the public who interact with (embrace or resist) the messages neuroscience delivers thanks to coverage in the popular media and increasing exposure to language and applications of neuroscience in medical settings. In Part III of this volume, called Neuroscience in Context: From Laboratory to Lifeworld, all these dimensions are touched upon, with a focus on researchers and the complicated technical procedures they operate, and on psychiatric patients who are often at the center of scientific as well as public attention. Overall, these chapters sketch a picture of the self-understanding of neuroscientists and biological psychiatrists, of the intricate technical details of their day-to-day work, of tacit assumptions built into the research process at various stages, and not least of the immense appeal that novel and technologically developed neuroscience exerts on parts of the public. Anthropological research thus allows us to glance beyond official declarations to the minute realities of regular practice and in this way forms a central aspect of the "reality check" that is to be performed on today's neuroscience. Crucially, some of the work collected in this part moves beyond a mere description of the status quo to formulate proposals for enriching neuroscientific research on the basis of ethnographic data.

In Chapter 8, medical anthropologist Simon Cohn presents material from interviews conducted with both neuropsychiatrists and their patients who participated as research subjects in non-clinical MRI studies. These patients were handed copies of their brain scans to take home, while being informed that the scans served purposes of basic research with no direct relevance to diagnosis or treatment. However, in spite of this information, Cohn relates his observations of surprisingly strong emotional reactions among the participants—reactions that charged the images with personal significance and turned them into focal points of narratives of hope. Furthermore, these brain scans were also taken as definite signifiers of illness identity, providing "objective" legitimacy for what prior to the scan were unstable self-images, both in relation to participants' sometimes shaky diagnoses and the reactions they faced from peers and family. Cohn interprets his findings as evidence for a potentially radical alteration in the understanding of mental illness and of psychiatric practice: a turn away from the messy realities of social encounters towards robust and objective categories—both in self-understanding of practitioners and in the imagination of patients. Through these vignettes, Cohn illustrates that maintaining the notion that neuroscience is on the verge of uncovering the biological bases of mental illness, neuroimaging research might effect a shift in the emotional climate that surrounds psychiatry. He suggests that MRI-generated brain images play an important role here as pictorial emblems of technological capabilities, objectivity, and progress, despite their artificiality and indirectness and regardless of their (at least to date) limited practical value.

In 2004, Joseph Dumit published a landmark ethnographic account of PET brain imaging research entitled *Picturing Personhood. Brain Scans and Biomedical Identity*. In Chapter 9, an adaptation of a chapter from his book, Dumit hones in on the indirectness of neuroimaging data that Cohn alludes to. To exemplify, he covers in minute detail the multitude of processes involved in PET research, starting with the design of an experiment and the selection of appropriate subjects and spanning the technical details of data acquisition and the complicated processes of data selection, normalization, and analysis. In addition he examines the processes of interpretation, image production, and selection for purposes of publication, and ends with the "looping journeys" of the published PET images beyond the scholarly sphere into the wider public arena—both through media representations and through contexts of practice such as medicine, education, or the law.

Dumit's strategy consists in putting detailed technical explanations alongside ethnographic interviews with practitioners reflecting upon the technical procedures, their range and limitations, as well as the sources of confusion that might be encountered along the way. In this manner, the intrinsic complexity of PET research, the variations between different research sites, and the significant degree of critical reflexivity within the community of researchers comes to the fore, providing valuable glimpses inside the "black box" of experimental neuroscience. At the same time, it becomes evident that despite a high level of critical awareness among practitioners and despite well-established systems of disciplinary rigor, exaggerations and misinterpretation loom large, often initially occasioned by representational styles focusing on extreme instead of average images. Those misconstruals are often amplified once the colorful images embark upon their journeys beyond the labs, where they are used to stabilize stereotypes or specific, often interest-driven, ways of classifying people into kinds.

In Chapter 10, Eugene Raikhel takes us on a journey to Russia in order to relate the history and development of addiction medicine, as established under the communist regime of the Soviet Union and still making its presence felt today in specific forms of biological psychiatric practice, and in some of its theoretical and conceptual underpinnings. Raikhel focuses on the complex field of addiction—a ripe example to illustrate the ways in which natural science, political ideology, societal developments, medical practice, and individual self-understandings of both practitioners and patients intersect in manifold ways. Raikhel's perspective helpfully complements the other contributions' focus on developments in psychiatry and medical policy in Western societies. From this uncommon angle, he illustrates a specific historical trajectory and cultural appropriation of a specific materialist ideology that is, in the end, not so radically different from some of the ideas currently brought forth within Western approaches to biological psychiatry.

The nature/culture dichotomy has often been criticized on conceptual grounds, but few scholars have so far provided concrete suggestions as to how empirical research could in fact move beyond this divide. In Chapter 11, Nicolas Langlitz attempts to do just that by discussing observations from his ethnographic fieldwork in two laboratories concerned with research on the effects of hallucinogenic drugs on humans and rodents. A key insight of this research, according to Langlitz, is that different reactions to psychotropic drugs show that drug-induced physiological processes vary depending on cultural and environmental context, suggesting a constitutive role of non-physiological factors in the enabling conditions of the drug-induced experiences. Since the effects on conscious experience of substances such as psilocybin seem to depend crucially on the subject's cultural background, this research seems suited to explore ways of integrating culture, controlled environmental conditions and carefully recorded subjective experience into experimental designs. In pointing towards a forgotten proposal to this end developed by Anthony Wallace in the 1950s, Langlitz asks whether, and how, anthropological second-order observation of scientific practice might be fed back into first-order research—in this way addressing one of the central ambitions of critical neuroscience. Langlitz's chapter and reappraisal of Wallace's work in the context of contemporary neuroscience establishes an important step towards surpassing the conceptual and practical divide that still separates scientific from humanities approaches to human reality. The question about feeding ethnographic observations into first-order observations in the lab is an invitation to neuroscientists and anthropologists alike to consider how such integrative experimental work might function.

Part IV addresses how cognitive neuroscience can have a powerful role to play in the critical project in more than one way: to enrich but also to subvert. This section, entitled Situating the Brain: From Lifeworld back to Laboratory?, brings together voices from within brain imaging research and can be seen as an attempt to stimulate a debate about just how neuroscience itself can hold a pivotal position in realizing the reflexivity at the core of critical neuroscience—in linking lifeworld and laboratory through engaged analysis and critical practice. The idea to feed insights from the human sciences back into neuroscience sets this project apart from existing research agendas focused on neuroscience and society, and this particular aim is crucial in embodying our position with regard to neuroscience as well as our notion of critique. The authors of these chapters remind us of the necessity to remain empirically engaged in order to make informed judgments about the degree of threat or promise posed by neuroscientific findings in society. Moreover, this section attempts to push further on the issue raised in Part III about the potential function of second-order observations from anthropology in first-order observations of brain activity, using the case example of cultural neuroscience. So, what can *being critical* mean for cognitive neuroscientists?

In Chapter 12, cognitive neuroscientist Amir Raz proposes that neuroscientists have an active role to play in cautious and accurate readings and representations of functional MRI results. As a methodology whose "seductive allure" has ensured its status as a mainstay in neuroscience research and propelled neuroscience into the public eye, with colorful brain scans becoming iconic representations of disease, difference, personhood, and arguably neuroscience more broadly, fMRI requires particular rigor in performing and interpreting. Raz uses the example of a recent *New York Times* op-ed column by neuroscientist Marco Iacoboni and his colleagues to elaborate his argument about the need for scientific scrutiny and careful representation. The published study which sought to read swing voters' political leanings using fMRI in the run-up to the last US presidential election in 2007, captures the dangers of oversimplification of data and perhaps more importantly, of capitalizing, as scientists, on the cachet of neuroscience and the apparent readability of brain scans among wider audiences. It also leads Raz to discuss the "indirect and crude" nature of fMRI, which like tea leaves in a cup can give rise to coincidental patterns that lead to compelling narratives. To illustrate, he provides a discussion about the conceptual slippages that result from using the reverse-inference approach to make predictions about mental states. However, Raz holds on to the promise of neuroimaging for providing useful insights into brain-behavior relationships, and offers readers a glimpse of how machine-learning approaches might overcome these limitations. Still, he cautions readers that fMRI can only be useful as one of many tools to study behavior, given that it is limited by its correlational logic.

In Chapter 13, neuroscientist Daniel Margulies tackles this question head-on, in the context of a lively evaluation of the eventful months of methodological controversy that swept through the neuroscience community between 2008 and 2009. The events now commonly dubbed "the voodoo correlations" saga and the "dead salmon" affair revolved around the legitimacy of fMRI as a tool for investigating brain-behavior relationships in humans. Margulies' incisive treatment of the dissemination (rather than the content) of the statistical scandals that surrounded social neuroscience examines the visceral responses of the scientists who were "named and shamed" by Vul and his critical colleagues (Vul, Harris, Winkielman, & Pashler, 2009) and analyzes the dramatic style of the critics' campaign. In doing so, he considers how such debates should take shape, touching on important changes occurring in neuroscience such as the increasing media-driven sensationalism and the expansion of the blogosphere as a forum for presentation and discussion of findings. In particular, by contrasting Vul et al.'s critique of statistical practice with Bennett et al.'s scanning study of the dead fish, Margulies raises questions about how best to "do" critique in such a way that is not only heard but also constructive. The most effective strategy, he suggests, is one of creative playfulness. Bennett et al.'s approach did not involve naming names, or distinguishing between camps, but simply replacing the normal human subject with a dead salmon in a standard fMRI experiment. Margulies argues that some simple experimental irony by neuroscientists and a limp Atlantic salmon suffice as powerful tools to caution fellow neuroscientists about the risk of red herrings.

Joan Chiao, a pioneer of cultural neuroscience, suggests in Chapter 14 how some of the goals of critical neuroscience concerning reflexivity in experimental research might be put into practice. With cultural neuroscience, a burgeoning area of research that uses neuroimaging techniques to investigate how mental and neural events vary as a function of culture, culture and ethnicity have resurfaced as objects of biological investigation. However, these study designs are rendered potentially problematic because of categorization of experimental subjects and the assumption that cognitive processes and culture can be captured by patterns of neural activity. Certainly, the findings of these studies are marked by tensions in their varied ways of operationalizing culture—a highly complex and fluid category. In this chapter, Chiao asks how researchers can do justice to the complexity of culture, and the meanings and values associated with emotions, distress, or perceptual processes, while at the same time maintaining sensitivity towards the rigor of neuroscientific empiricism and the constraints of its tools. Finally, Chiao explores the potentials of cultural neuroscience for understanding cultural differences in somatization of mental illness. While her field offers the possibility to explore how cultural traits may be embodied, and how the brain may mediate certain culturally specific experience relevant for psychiatry, Chiao emphasizes the need, as this young field unfolds, for researchers to remain open to methodological and theoretical tools from anthropology, and alert to the social implications of biological approaches to culture through constant self-reflexivity.

Part V, Beyond Neural Correlates: Ecological Approaches to Psychiatry, brings together contributions that analyze and contest the recent neurobiological push in psychiatry. The chapters are united in posing challenges to the logic of various neuroreductionist programs and the implicit conceptions of personhood, experience, illness, and health that these programs are built upon. All three chapters in this section go beyond mere criticism by articulating and defending theoretical and practical alternatives to reductionism, borrowing from systems biology, complexity theory, transcultural psychiatry, ecological, and enactivist approaches as well as approaches that make use of narrative and the systematic study of first-person experience.

In Chapter 15, Laurence Kirmayer and Ian Gold raise challenges to the logic of biological psychiatry, starting from the premise that mental distress or illness reflects the interaction of biological and socio-cultural systems from the outset. They argue against Insel and Quirion's claim that psychiatry is in the end not much more than "clinically applied neuroscience" (Insel & Quirion, 2005). Their chief objection is that biological psychiatry reduces the social suffering and the phenomenology of mental illness to a list of symptoms and signs, while it conceptualizes the social world that the patients inhabit as a set of mere learned behaviors, attitudes, and contingencies. To prevent psychiatry from turning into a discipline that is "both mindless and uncultured," Gold and Kirmayer call for more nuanced conceptualizations of pathological processes and their phenomenology. The authors oppose reductionist approaches with an argument to the effect that the social environment plays a fundamental role in human mental life and that it is frequently implicated among the causes of mental illness. After a critical review of the reductionist tendencies in contemporary psychiatry, they supply illustrations of the importance of the social world in psychopathology, especially by pointing to evidence from studies of cultural difference. Moreover, Kirmayer and Gold's argument is informed by an in-depth discussion of different varieties of reductionism and of levels of explanatory complexity. This helps readers to distinguish sensible methodological strategies from spurious metaphysical assumptions and to imagine a future in cognitive neuroscience that allows for a higher degree of theoretical complexity.

Fernando Vidal and Francisco Ortega, in Chapter 16, take issue with research aimed at identifying neural correlates of mental illness. Their strategy is to combine three strands of critical inquiry. First, they illustrate the broader context in which psychiatry is becoming increasingly neurologized—an increasingly hyped "neuroculture" in academia and public life. Second, they reconstruct and critically assess the widespread attempt to identify so-called *neural correlates of consciousness* (NCC), ultimately arguing against the viability of this approach by appealing to insights from enactivist approaches to conscious experience. And third, they trace how the idea of NCC underwrites a reductionist research program in biological psychiatry that sets out to identify the neural bases of depression. In particular, the authors point to a wealth of expensive but methodologically problematic neuroimaging studies and to the strategies and rhetoric used by researchers in the field in spite of the highly contingent and much debated status of the results obtained so far. Vidal and Ortega build up a case against the neurocorrelational research program: by continuously presenting preliminary and highly contestable results as a basis for "future research" which carries "prospects for novel forms of effective treatment," against a background of increasing social currency and legitimacy of neuroscientific approaches and not least a profit-hungry pharmaceutical industry, leading practitioners manage to stabilize and even expand neurocorrelational research whose prospects for explanatory success is, according to Vidal and Ortega, highly doubtful.

To close Part V, psychiatrist and philosopher Thomas Fuchs, in Chapter 17, elaborates a related anti-reductionist line of thought, first in theoretical terms and subsequently through a discussion of therapeutic practice. Fuchs argues against the notion that "mental illnesses are diseases of the brain," by demonstrating that both altered subjective experience and dysfunctional interpersonal relations, while constitutive for mental illnesses, are irreducible to brain processes. To support this systemic-ecological view of psychopathology, Fuchs introduces the notion of circular causality, operative both vertically in intra-organismic causal loops spanning brain, body and subjective experience and horizontally, linking the individual organisms to its social environment in circles of causal influence. In both kinds of causal loops, the brain, according to Fuchs, functions as an "organ of transformation and mediation" that is continually shaped and modified by psychosocial interactions and multiple bodily processes. Importantly, Fuchs claims that subjective experience exerts a structuring influence on the neural substrate. If correct, this claim would amount to a radical departure from the deep-seated dogma of a one-way explanatory dependency of experience upon underlying brain processes. This view has significant implications for therapeutic practice: Fuchs concludes by discussing the effectiveness of talk- and interaction-based therapy that his position predicts by citing empirical studies of the measurable effects of different kinds of therapeutic intervention.

In the volume's closing chapter, Laurence Kirmayer starts off with a comprehensive vision of critical neuroscience, spinning together some key threads of theory and practice motivating and shaping the endeavor outlined in this volume. He focuses on the multiple looping processes that span neuroscience laboratories, the media, various contexts of practice, and the self-understanding and imagination of laypeople. Psychiatry surfaces as a central problem area with its inherent tendency to classify people into scientifically stabilized kinds, its hidden forms of social control, its increasing focus on neurobiological approaches, and its Big Pharma-sponsored push towards expanding drug treatments for common disorders around the globe. Kirmayer reminds readers of research in genetics and its appropriation in simplified models and fantasies of biotechnological intervention. He once again highlights how mainstream media simplifies and severely distorts even those scientific results that had originally been presented by their scientist authors in careful, balanced, and variously qualified ways. Looking ahead to developments likely to be of relevance in the foreseeable future, Kirmayer urges practitioners of critical neuroscience to pay specific attention to novel neurotechnological intervention possibilities. Echoing many of the authors in this volume, he concludes by pointing again to theories of the extended mind and by stressing the extent to which the human mind is constitutively dependent upon factors different from, and external to, the individual's brain. This final push for a distributed and situated understanding of human mindedness, along with a call to practitioners to apply these insights in experimental designs, interpretation of data, and disciplinary theorizing is a fitting finale for a volume that strives to broaden the theoretical and conceptual horizon of current neuroscientific research.

We hope that this collection of chapters is the beginning of an intensified debate about the potentials and problems associated with human-level neuroscience research, and of rigorous attempts to enrich the disciplinary perspectives of the neurosciences such that they can be informed by sociological, historical, anthropological, and philosophical approaches. Conversely, researchers in the social sciences and humanities at large are themselves urged to enter into constructive collaboration with the open-minded majority of neuroscientists.

References

Abi-Rached, J. M., & Rose, N. (2010). The birth of the neuromolecular gaze. *History of the Human Sciences, 23*(1), 1–26.

Barad, K. (2007). *Meeting the universe halfway: Quantum physics and the entanglement of matter and meaning.* Durham, NC/London: Duke University Press.

Beddington, J., Cooper, C. L., Field, J., Goswami, U., Huppert, F. A., Jenkins, R., & Thomas S. M. (2008). The mental wealth of nations. *Nature, 455,* 1057–1060.

Berger, P., & Luckmann, T. (1966). *The social construction of reality.* Garden City/New York: Doubleday.

Boltanski, L., & Chiapello, E. (2007). *The new spirit of capitalism.* (G. Elliot, Trans.) London: Verso.

Borck, C. (this volume). Toys are us: Models and metaphors in brain research.

Brinkmann, S. (2008). Changing psychologies in the transition from industrial society to consumer society. *History of the Human Sciences, 21*(2), 85–110.

Brinkmann, S. (2005). Human kinds and looping effects in psychology: Foucauldian and hermeneutic perspectives. *Theory & Psychology, 15*(6), 769–791.

Brooks, R. (1991). Intelligence without representations. *Artificial Intelligence, 47,* 139–159.

Byatt, A. S. (2006, September 22). Observe the neurones. *Times Literary Supplement, 104.*

Chalmers, D. (1996). *The conscious mind.* Oxford: Oxford University Press.

Choudhury, S., Gold, I., & Kirmayer, L. (2010). From brain image to the Bush doctrine: Critical neuroscience and the political uses of neurotechnology. *American Journal of Bioethics: Neuroscience, 1*(2), 17–19.

Choudhury, S., Nagel S. K., & Slaby, J. (2009). Critical neuroscience: Linking neuroscience and society through critical practice. *BioSocieties, 4*(1), 61–77.

Churchland, P. (2002). *Brain-wise: Studies in neurophilosophy.* Cambridge, MA: MIT Press.

Clark, A. (2008). *Supersizing the mind: Embodiment, action, and cognitive extension.* New York/Oxford: Oxford University Press.

Clark, A. (1997). *Being there: Putting brain, body and world together again.* Cambridge, MA: MIT Press.

Cohn, S. (this volume). Disrupting images: Neuroscientific representations in the lives of psychiatric patients.

Cohn, S. (2010). Picturing the brain inside, revealing the illness outside: A comparison of the different meanings attributed to brain scans by scientists and patients. In J. Edwards, P. Harvey & P. Wade (Eds.), *Technologized images, technologized bodies: Anthropological approaches to a new politics of vision.* Oxford: Berghahn Press.

Crick, F. (1994). *The astonishing hypothesis: The scientific search for the soul.* New York: Touchstone Press.

Damasio, A. R. (1995). *Descartes' error: Emotion, reason, and the human brain.* New York: HarperCollins.

Danziger, K. (1990). *Constructing the subject: Historical origins of psychological research.* Cambridge: Cambridge University Press.

De Vries, R. (2007). Who guards the guardians of neuroscience? Firing the neuroethical imagination. *EMBO Reports, 8,* 1–5.

De Jaegher, H., & Di Paolo, E. (2007). Participatory sense-making: An enactive approach to social cognition. *Phenomenology and the Cognitive Sciences, 6*(4), 485–507.

Dumit, J. (this volume). Critically producing brain images of mind.

Dumit, J. (2004). *Picturing personhood: Brain scans and biomedical identity.* Princeton: Princeton University Press.

Edelman, G. M. (1992). *Bright air, brilliant fire: On the matter of the mind*. New York: Basic Books.

Farmer, R. L. (2008). *Neuroscience and social work practice: The missing link*. London: Sage.

Foresight Mental Capital and Wellbeing Project (2008). Final project report. The Government Office for Science, London.

Foucault, M. (1973). *Madness and civilization: A history of insanity in the age of reason*. New York: Vintage.

Gallagher, S. (this volume). Scanning the lifeworld. Toward a critical neuroscience of action and interaction.

Gallagher, S., & Crisafi, A. (2009). Mental institutions. *Topoi, 28* (1), 45–51.

Gallagher, S. (2009). Two problems of intersubjectivity. *Journal of Consciousness Studies, 16* (6–8), 289–308.

Gallagher, S. (2008). Direct perception in the social context. *Consciousness and Cognition, 17,* 535–543.

Gallagher, S. (2005). *How the body shapes the mind*. Oxford: Oxford University Press.

Giddens, A. (1991). *Modernity and self-identity: Self and society in the late modern age*. Stanford: Stanford University Press.

Goleman, D. (2006). *Social intelligence: The new science of human relationships*. New York: Bantam.

Goleman, D. (1995). *Emotional intelligence*. New York: Bantam.

Gold, I. & Stoljar, D. (1999). A neuron doctrine in the philosophy of neuroscience. *Behavioral and Brain Sciences, 22*(5), 809–830.

Gurwitch, A. (1978). *Human encounters in the social world*. Pittsburgh: Duquesne University Press. (Original work published 1931)

Habermas, J. (1971). *Knowledge and human interests*. London: Heinemann.

Hacking, I. (2002). *Historical ontology*. Cambridge, MA: Harvard University Press.

Hacking, I. (1999). *The social construction of what?* Cambridge, MA: Harvard University Press.

Hacking, I. (1998). *Mad travelers: Reflections on the reality of transient mental illnesses*. Cambridge, MA: Harvard University Press.

Hacking, I. (1995). *Rewriting the Soul: Multiple personality and the sciences of memory*. Princeton: Princeton University Press.

Hartmann, M. (this volume). Against first nature: Critical theory and neuroscience.

Hartmann, M., & Honneth, A. (2006). Paradoxes of capitalism. *Constellations, 13*(1), 41–58.

Haugeland, J. (1998). Mind embodied and embedded. In *Having thought: Essays in the metaphysics of mind* (pp. 207–237). Cambridge, MA: Harvard University Press.

Herculano-Houzel, S. (2002). Do you know your brain? A survey on public neuroscience literacy at the closing of the Decade of the Brain. *Neuroscientist, 8*(2), 98–110.

Hurley, S. (1998). *Consciousness in action*. Cambridge, MA: Harvard University Press.

Husserl, E. (1970). *Crisis of European sciences and transcendental phenomenology*. (D. Carr, Trans.). Evanston: Northwestern University Press.

Illes, J., Moser, M. A., McCormick, J. B., Racine, E., Blakeslee, S., Caplan, A., & Weiss, S. (2010). Neurotalk: Improving the communication of neuroscience research. *Nature Reviews Neuroscience, 11*(1), 61–69.

Insel, T. R., & Quirion, R. (2005). Psychiatry as a clinical neuroscience discipline. *Journal of the American Medical Association, 294,* 2221–2224.

Karafyllis, N. C., & Ulshöfer, G. (Eds.). (2008). *Sexualized brains: Scientific modeling of emotional intelligence from a cultural perspective*. Cambridge, MA: MIT Press.

Kirmayer, L. J. (this volume). The future of critical neuroscience.

Kirmayer, L. J., & Gold, I. (this volume). Re-socialising psychiatry: Critical neuroscience and the limits of reductionism.

Langlitz, N. (this volume). Delirious brain chemistry and controlled culture: Exploring the contextual mediation of drug effects.

LeDoux, J. (1996). *The emotional brain: The mysterious underpinnings of emotional life.* New York: Simon & Schuster.

Levine, J. (1993). On leaving out what it's like. In M.Davies & G.Humphreys (Eds.), *Consciousness: Psychological and philosophical essays.* Oxford: Blackwell.

Levine, J. (1983). Materialism and qualia: The explanatory gap. *Pacific Philosophical Quarterly, 64*, 354–361.

Lynch, Z. (2009). *The neuro revolution: How brain science is changing our world.* New York: St. Martin's Press.

Malabou, C. (2008). *What should we do with our brain?* New York: Fordham University Press.

Martin, E. (2009). Identity, identification, and the brain. Presented at the Workshop 'Neurocultures', Max Planck Institute for the History of Science, Berlin.

McClure, S. M., Li, J., Tomlin, D., Cypert, K. S., Montague, L. M., & Montague, P. R. (2004) Neural correlates of behavioral preference for culturally familiar drinks. *Neuron, 44*, 379–387.

Menary, R. (Ed.). (2010). *The extended mind.* Cambridge, MA: MIT Press.

Metzinger, T. (2009). *The ego-tunnel: The science of the mind and the myth of the self.* New York: Basic Books.

Metzinger, T. (2003). *Being no-one: The self-model theory of subjectivity.* Cambridge, MA: MIT Press.

Mitchell, S. D. (2009). *Unsimple truths: Science, complexity, and policy.* Chicago: University of Chicago Press.

Mitchell, T. (2002). *Rule of experts: Egypt, techno-politics, modernity.* Berkeley and Los Angeles, CA: University of California Press.

Noë, A. (2009). *Out of our heads: Why you are not your brain and other lessons from the biology of consciousness.* New York: Hill and Wang.

Noë, A. (2005). *Action in perception.* Cambridge, MA: MIT Press.

Nussbaum, M. C. (2010). *Not for profit: Why democracy needs the humanities.* Princeton: Princeton University Press.

O'Regan, J. K., & Noë, A. (2001). A sensorimotor account of vision and visual consciousness. *Behavioural and Brain Sciences, 24*(5), 883–917.

Ortega, F. (2009). The cerebral subject and the challenge of neurodiversity. *BioSocieties, 4*(4), 425–445.

Ortega, F., & Choudhury, S. (in press). Wired up differently: Autism, adolescence and the politics of neurological identities. *Subjectivity*

Ortega, F., & Vidal, F. (Eds.). (2010). *Neurocultures: Glimpses into an expanding universe.* Berlin, New York: Peter Lang.

Oullier, O., & Sauneron, S. (2010). *Improving public health prevention with behavioural, cognitive and neuroscience.* Centré d'Analyse Stratégique, French Government. See http://www.strategie.gouv.fr/IMG/pdf/NeuroPrevention_English_Book.pdf

Protevi, J. (2009). *Political affect: Connecting the social and the somatic.* Minneapolis: University of Minnesota Press.

Quednow, B. B. (2010). Ethics of neuroenhancement: A phantom debate. *BioSocieties, 5*, 153–156.

Racine, E., Bar-Ilan, O., & Illes, J. (2005). fMRI in the public eye. *Nature Reviews Neuroscience, 6*(2), 159–64.

Racine, E., Waldman, S., Rosenberg, J., & Illes, J. (2010). Contemporary neuroscience in the media. *Social Science & Medicine, 71*(4), 725–733.

Raikhel, E. (this volume). Radical reductions: Neurophysiology, politics and personhood in Russian addiction medicine.

Revonsuo, A. (2005). *Inner presence: Consciousness as a biological phenomenon.* Cambridge, MA: MIT Press.

Richards, G. (1996). *Putting psychology in its place: An introduction from a critical historical perspective.* London: Routledge.

Robbins, P., & Aydede, M. (Eds.). (2009). *The Cambridge handbook of situated cognition.* Cambridge: Cambridge University Press.

Rose, N. (2010). 'Screen and intervene': Governing risky brains. *History of the Human Sciences, 23*(1), 79–105.

Rose, N. (2007). *The politics of life itself: Biomedicine, power, and subjectivity in the twenty-first century.* Princeton: Princeton University Press.

Rose, N. (2003). Neurochemical selves. *Society, 41*(1), 46–59.

Rose, N. (1996). *Inventing our selves: Psychology, power, and personhood.* Cambridge: Cambridge University Press.

Rouse, J. (2002). *How scientific practices matter.* Chicago: Chicago University Press.

Rouse, J. (1996). *Engaging science: How to understand its practices philosophically.* Ithaka, NY: Cornell University Press.

Rouse, J. (1987). *Knowledge and power: Toward a political philosophy of science.* Ithaka, NY: Cornell University Press.

Rowlands, M. (2010). *The new science of the mind: From extended mind to embodied phenomenology.* Cambridge, MA: MIT Press.

Schütz, A. (1974). *Collected papers, The problem of social reality* (Vol. 1). Dordrecht: Kluwer.

Singh, I., & Rose, N. (2009). Biomarkers in Psychiatry. *Nature, 460*(9), 202–207.

Slaby, J. (2010). Steps towards a critical neuroscience. *Phenomenology and the Cognitive Sciences, 9,* 397–416.

Stadler, M. (this volume). The neuromance of cerebral history.

Taylor, C. (1989). *Sources of the self: The making of the modern identity.* Cambridge, MA: Harvard University Press.

Taylor, C. (1985). Self-interpreting animals. In *Philosophical Papers,* (Vol. 1), (pp. 45–76). Cambridge: Cambridge University Press.

Thompson, E. (2007). *Mind in life: Biology, phenomenology, and the sciences of the mind.* Cambridge, MA: Harvard University Press.

Varela, F., Thompson, E., & Rosch, E. (1991). *The embodied mind: Cognitive science and human experience.* Cambridge, MA: MIT Press.

Vidal, F. (2009). Brainhood: Anthropological figure of modernity. *History of the Human Sciences, 22*(1), 6–35.

Vidal, F., & Ortega, F. (this volume). Are there neural correlates of depression?

Vrecko, S. (2006). Folk neurology and remaking of identity. *Molecular Interventions, 6*(6), 300–303.

Vul, E., Harris, C., Winkielman, P., & Pashler, H. (2009). Puzzlingly high correlations in fMRI studies of emotion, personality, and social cognition. *Perspectives on Psychological Science, 4,* 274–290.

Ward, S. C. (2002). *Modernizing the mind: Psychological knowledge and the remaking of society.* Westport: Praeger.

Young, A. (this volume). Empathic cruelty and the origins of the social brain.

Zeki, Z. (2008). *Splendors and miseries of the brain: Love, creativity, and the quest for human happiness.* Oxford: John Wiley & Sons.

Part I
Motivations and Foundations

1

Proposal for a Critical Neuroscience

Jan Slaby and Suparna Choudhury

The label "critical neuroscience" captures an important—and, we believe, productive—tension. This tension represents the need to respond to the impressive and at times troublesome surge of the neurosciences, without either celebrating it uncritically or condemning it wholesale. "Critical" alludes, on the one hand, to the notion of "crisis," understood—in the classical Greek, predominantly medical sense of the term—as an important juncture and point of intervention, and, relatedly, to a task similar to that proposed by Kant (1992) in *The Conflict of the Faculties* (rather than in his more famous "Critiques"), where he defends a space of unconstrained inquiry into the continual pressures put on scientific knowing by the vagaries of the political sphere. This opens up a space for inquiry that is itself inherently and self-consciously political. On the other hand, the concept of "critique" raises important associations with Frankfurt School critical theory. While critical neuroscience does not directly follow a Frankfurt School program, nor the reduction of science to positivism espoused by early critical theory, it does share with it a spirit of historico-political mission; that is, the persuasion that scientific inquiry into human reality tends to mobilize specific values and often works in the service of interests that can easily shape construals of nature or naturalness. These notions of nature or of what counts as natural, whether referring to constructs of gender, mental disorder, or normal brain development, require unpacking. Without critical reflection, they appear as inevitable givens, universal and below history, and are often seen as a form of "normative facticity," making specific claims upon us in everyday life (see Hartmann, this volume).

In this chapter, we will spell out how our proposal for a critical neuroscience is not motivated by the aim to undermine the epistemological validity of neuroscience or debunk its motives, nor is it simply an opportunity to establish yet another neuro-prefixed discipline. Situated between neuroscience and the human sciences, our notion of critical neuroscience uses a historical sensibility to analyze the claim that we

Critical Neuroscience: A Handbook of the Social and Cultural Contexts of Neuroscience, First Edition.
Edited by Suparna Choudhury and Jan Slaby.
© 2012 John Wiley & Sons, Ltd. Published 2016 by John Wiley & Sons, Ltd.

are in the throes of a "neurorevolution" since the beginning of the Decade of the Brain in 1990. It investigates sociologically the motivations and the implications of the turn to the *neuro* in disciplines and practices ranging from psychiatry and anthropology to educational policy, and it examines ethnographically the operationalization of various categories in the laboratory. Investigating the historical and cultural contingencies of these neuroscientific categories, critical neuroscience analyzes the ways in which, and conditions through which, behaviors and categories of people are naturalized. It also traces how these "brain facts" are appropriated in various domains in society, starting with medicalized contexts of the West, but also using cross-national comparative methodology to understand the production and circulation of neuroscientific knowledge globally. Maintaining close engagement with neuroscience is, on the one hand, crucial for building accurately informed analyses of the societal implications of neuroscience, whilst, on the other hand, providing a connection, a reflexive interface, through which historical, anthropological, philosophical, and sociological analysis can feed back and provide creative potential for experimental research in the laboratory.

In attempting to build up a picture of what critique might look like for this project, we avail ourselves of a number of disciplines and sensibilities that can contribute as resources for critique. Our goal is to render critique amenable to a number of diverse disciplines—we propose that this versatile set of tools can contribute to reviving a critical spirit while also broadening the neuroscientist's gaze. That being said, we certainly do not intend to outline a fully-fledged, scholarly program or recipe for critique. Instead, we will try to sketch some building blocks for a mode of engagement, an ethos, that aims to raise awareness of the factors that come together to stabilize scientific worldviews that create the impression of their inevitability. Furthermore, critical engagement in neuroscience can increase the complexity of behavioral phenomena (for example, emotions, interaction, decision making, mental disorders), and motivate scholars to enrich conceptual vocabularies of behavior and mental illness, keeping debates from being foreclosed by the belief that the ontologically most fundamental level of explanation is by default the most appropriate one (see Mitchell, 2009).[1]

To bear relevance outside the narrow scholarly sphere, such an endeavor requires a self-reflexive hermeneutics that is necessarily multi-dimensional (or "undisciplined"). The result, we envisage, will not so much be an unpacking of the black boxes of the neurosciences as an assemblage of resources that ultimately widens the ontological landscape of a behavioral phenomenon under study. It is the plurality—reflecting the complexity of behavior as well as the many contingencies of neuroscience—of elements of this landscape that gives rise to the solidity of a claim, the "realness" of a fact. Contextualizing neuroscientific objects of inquiry—whether the "neural basis" of addiction, depression, sociality, lying, or adolescent behaviors—can, in this way,

---

[1] Recent debates about levels of explanation, reduction, and complexity in the philosophy of science demonstrate that the field is increasingly departing from the classical hierarchical models in which a fundamental physical level is deemed the only truly explanatory level, such that all higher levels of a complex system's organization have to be reduced to it. Sandra Mitchell's complexity theory-inspired argument for "integrative pluralism" is a helpful case in point (Mitchell, 2009). See also the useful charting of relevant debates in Brigandt & Love (2008).

demonstrate how such findings, whilst capturing an aspect of behavior in the world, are also held in place by a number of factors, co-produced by a collection of circumstances, social interests, and institutions (Hacking, 1999; Young, 1995). These circumstances and interests are often quite systematically ignored in neurodiscourse (see, for example, Heinemann & Heinemann, 2010).

However, we propose that critical neuroscience should not stop at description and complexification. Indeed, we share a sense of uneasiness, recently voiced within the field of Science and Technology Studies (STS) in particular (Anderson, 2009; Cooter, 2007; Cooter & Stein, 2010; Forman, 2010) about depoliticalization of scholarship in the face of the increasing commercialization of academia. In line with a broader cultural tendency favoring voluntarist conceptions of the "entrepreneurial self," centered around ideas of "resources" and personal "capital" (social, emotional, "mental"), we sense an implicit correspondence between scholarly discourse and economic imperatives and normative schemas.[2] Certainly, these are preliminary intuitions, and we will not impose ready-made answers. However, we share the conviction that a more radical and openly political positioning is needed in face of these trends. In the first instance, it is important to reinvigorate a sense of the impact that larger social, political, and economic dynamics have on the very shape of academic and scientific culture. We return to this below.

Assemblage: The Thickening of Brain-Based Phenomena

Bruno Latour, in his animated essay about critique and its effect of *weakening* scientific facts, appeals to his critically-oriented readers to "suspend the blow of the [critical] hammer" and calls for a renewal of a realist attitude oriented to matters of concern, rather than matters of fact (Latour, 2004). Matters of concern are those around which the human world revolves: they enthrall us, involve us, and challenge us to embrace or oppose them—they will be the focal point in practices, discourses, disputes. Critical neuroscience shares this constructive spirit, the "stubbornly realist attitude" and the focus on what matters in relation to scientific practices (Rouse, 2002). Importantly, critical neuroscience embraces the added dimension that enters the scene with the focus on matters of concern: values, conflicting moral outlooks and evaluative perspectives, changes in the attribution of relevance pertaining to a given phenomenon or scientific result, often contested among affected parties. Critical neuroscience thus emphasizes the politics implicit in scientific practices (see Rouse, 1987, 1996).

However, while Latour is helpfully non-dogmatic and quasi-democratic in giving a voice to participants in practices—both human and non-human—in the process of assembling their collectives (instead of silencing the actors behind grand-scale theoretical assumptions), in the end, he relinquishes too much—by sidelining entirely any non-local invocation of the social, the economic, or the political. By contrast, our proposal for critical neuroscience calls for a less detached attitude on the part of the critical investigator, a more active engagement, and, at times, a more confrontational

---

[2] How these postmodernist tendencies might have rendered explanations that invoke "social influences" less common and less valued in STS is helpfully discussed by Forman (2010).

response in cases of violation of scientific standards (Fine, 2010), strategies of ignorance (McGoey, 2009), imperialistic export of Western assumptions to Non-Western contexts (Watters, 2010), or the political use of preliminary data (Choudhury, Gold, & Kirmayer, 2010; Raz, this volume). Such responses need to be supported by attempts to identify and render explicit more subtle biases and frames of evaluation: the specific organization of public attention, patterns of distribution of affective energies, collectively sustained valuations and schemes of judgment that are instituted in subtle but pervasive ways in both scientific and popular discourses, in representations of scientific results, but also in spheres of public understanding at some distance from the practice of research. Notions such as the neural basis of adolescent risk taking, hard-wired sex differences, molecularized understandings of mental illnesses, or narratives about behavioral and emotive tendencies universally present in humans and set in stone by evolution are cases in point. Some of these narrative patterns solidify to form what Judith Butler has called "frames"—powerful but often unnoticed ways in which perception, knowledge, and normative judgment are preorganized so that some conceptualizations and evaluations are made likely while others are ruled out a priori (Butler, 2009). Critique here has the task of working against engrained habits of perception, thought, and judgment in order to enable alternative framings of matters of concern.

What we envisage as the practice of critique, therefore, starts with the activity of assembling (Latour, 2004, p. 246; Slaby, 2010). "Assembling" refers to the collection of material from multiple sources and perspectives to enrich scientific conceptualization as well as the broader intellectual horizon in which problems and issues are framed for empirical investigation and interpretation. Objects of neuroscientific investigation can, as a result, be situated in the full fabric of meaningful relations—while this very fabric is itself placed under scrutiny and has to be kept open for contestation. The social situatedness, cultural meanings, and various interests of affected groups all package the ontological landscape of neurocognitive phenomena. This view holds that what we see in the brain is at any time held in place by a rich web of factors within the epistemic culture (Knorr-Cetina, 1999; Young, 1995), and in the ambient society, which in turn mobilizes these findings beyond the laboratory. Insights from multiple disciplines can bring to light the internalized scientific ideals, or "epistemic virtues" (Daston & Galison, 2007) that direct the formulation of neuroscientific findings— the filtering of information, the criteria for, and goal of, objectivity, and the operationalization of chosen aspects of the lifeworld (Cooter, 2010).

To illustrate this, let us take the example of addiction. Addiction is increasingly understood as a disease of the brain, in which addictive substances cause malfunction of the frontal regulation of the limbic system, thus "hijack[ing] the brain's reward system" (Leshner, 2001) and potentially even altering gene expression (Kuhar, 2010). The goal of these brain-centered approaches to addiction is to locate candidate molecular mechanisms that can lead to effective new treatments (Hyman & Malenka, 2001). While these studies have yielded some notable findings, addiction is far more than (and different from) a mere change in brain chemistry. "Addiction" denotes a family of conditions that are inextricably tied up with social environments, drug markets, and cultural triggers (Campbell, 2010), and depend on collectively developed and sustained habits (Garner & Hardcastle, 2004) and also upon institutional practices that emerge in response, as a feedback, to the original phenomenon—through classificatory looping as described by Ian Hacking (Hacking, 1995, 1999, 2007; see also Raikhel, this volume).

Approaching addiction using an ecological systems view, through multiple epistemic cultures, would mean to re-inscribe and integrate these multiple causal factors. Such an approach would examine the linkages across levels of description using various methodologies and would include recording the cultural phenomenology of addictive behaviors. It would additionally attend to the political economy of addiction and the effects of industry on concepts of addiction (Rasmussen, 2010). Taken together, this integrative approach will yield an explanandum much richer than any of the single construals developed exclusively from a single scientific or medical perspective.[3] Clearly both registers—social and biological—are necessary to assemble a richer understanding of addiction. The more relevant questions for a critical neuroscience to work out will be how to overcome the gap between social and neural, how to develop conceptual vocabularies and frameworks that overcome this stark distinction, and how to empirically study phenomena like addiction with a view of the *situated* brain and nervous system. This goal would take as a premise that the brain and nervous system are nested in the body and environment from the outset and that their functions can only be understood in terms of the social and cultural environment (Choudhury & Gold, in press).[4]

How Does the Social Get Under the Skin?

Ethnographic work by Margaret Lock has provided powerful evidence for the need to collapse conventional dichotomies between the "inside" and "outside" of the human body. Her seminal study of the experience and physiological characteristics of menopause among Japanese and American women led her to the concept of "local biologies," a useful way to denote her finding that social context and culture can refashion human biology (Lock, 1993; Lock & Kaufert, 2001; Lock & Nguyen, 2010, ch. 4). Lock found that the cultural differences in menopause/*konenki* ran deep, manifesting on biological, psychological, and social levels. She argued that the different experiences of hot flushes were not simply due to differences in cultural expectations in relation to the body, but down to the biological effects of culturally determined behaviors such as diet. This finding challenges the tendency in biological science to draw boundaries at the skin, and demonstrates instead the ongoing dialectic between biology and culture. Laurence Kirmayer has extended these ideas to the brain and behavior through his concept of "cultural biology," which understands

---

[3] Phenomenological analysis can play an important role in these enriching constructions of behavioral phenomena—in the case of addiction and certainly with regard to many other objects of neuroscientific inquiry. See, for example, Gallagher (this volume), Ratcliffe (2008, 2009), and Zahavi (2004). On the other hand, it would be wrong to assume that phenomenological approaches alone could be the answer in amending the limitations and reductive tendencies of empirical investigation. Phenomenological construals themselves have to be reflexively questioned and balanced with social contestations to prevent the erection of the myth of a universal, ahistorical, and authoritative sphere of pure experience.

[4] For the more general background to this perspective, see Noë (2009), Protevi (2009), and Wexler (2006).

culture as a biological category in the sense that human beings have evolved a "biological preparedness to acquire culture ... through various forms of learning and ... neural machinery" (Kirmayer, 2006, p. 130). Lock and Kirmayer's concepts of local biologies and cultural biologies, respectively, capture a notion of central importance to critical neuroscience: biology and culture are mutually constraining and co-constitutive, such that they are each conditions of the other's determination and development.

Explanations that situate brain and cognitive function within the social and cultural environment of the person are, in fact, increasingly encouraged within psychiatry and neuroscience. Calls for interdisciplinary research that lead to integrative explanations are certainly heard within psychiatry as a route to developing multi-level theories of disease and their etiologies (Kendler, 2008). Advances in epigenetics have been especially influential in fueling major shifts in scientific thinking about the linkages between the body and its environment, between soma and society (Pickersgill, 2009). Research on epigenetics has begun to reveal how interactions between the genome and the environment over the course of development lead to structural changes in the methylation patterns of DNA that regulate cellular function. There is compelling evidence, for example, that early parenting experiences and social adversity alter the regulation of stress response systems for the life of the organism (Fish et al., 2004; McGowan et al., 2009; Meaney & Szyf, 2005; Weaver et al., 2004). Such studies provide biological evidence that lived experience, developmental histories, dynamic interactions, and cultural contexts are all fundamentally bound up with biological processes as "low level" as gene expression.

In parallel to these developments in genomics, social and cultural neuroscience have become the most rapidly-developing areas of cognitive neuroscience. While social neuroscience explores linkages between social interaction processes and the brain, cultural neuroscience investigates cultural variation in a range of psychological processes with respect to brain function. These research fields posit that the human brain is fundamentally a social brain, adapted for social learning, interaction, and the transmission of culture (Emery, Clayton, & Frith, 2010; Frith & Frith, 2010; Rizzolatti & Craighero, 2004). Moreover, its structural malleability is understood to be experience-dependent and long-lasting. Evidence of genomic and neural plasticity thus forces scientists to rethink the primacy given to biophysical levels of explanations, and challenges us to destabilize the dichotomy of nature/culture and instead address the fundamental interaction of mind, body, and society.

This concept of the situated brain brings up a number of possibilities and challenges for critical neuroscience. First of all, it requires the critic (or critics in collaboration) to act as a *bricoleur*, collecting data at a number of different levels, layering phenomena, such as menopause or addiction, with these different strands of inquiry that ultimately serve to enrich one another in their explanatory value. Secondly, the emerging discourses of "interaction" require critical analysis by sociologists and anthropologists of science. How exactly are aspects of social life, culture, and individual difference incorporated into scientific observations and methodologies? Furthermore, when the environment and biology are each assigned roles in the development of pathologies, such as schizophrenia or antisocial behavior, how are the social and cultural realms made relevant or visible in medical explanations? How might the more complex ontologies of mental disorders that result from these integrative explanations bring about new ethical and political challenges by opening up new spaces of intervention or creating new "at risk" populations (Pickersgill, 2009; Rose, 2010; Singh & Rose, 2009)?

Situating the brain and behavior in social and cultural contexts also underscores the importance of examining recursive loops between neurobiological and social/cultural processes such as the way in which explanatory theories of illness and behavior themselves interact with the physiological processes involved. This biolooping, as discussed in the introduction to this volume, refers to the ways that both culture and local biologies can transform one another, exerting their influence on the way we understand ourselves, the way we experience mental and bodily phenomena, and the way that this in turn shapes the corresponding biological processes. We return to these issues later in a discussion of what critical neuroscience can do for neuroscience itself.

Critical neuroscience research is thus understood as a broad, interpretative, and qualitative mode of inquiry. One important—though surely not the only—way to "operationalize" critique lies in the attempt to enrich the often necessarily limited, lab-based empirical perspective by providing science with themes of significance captured within a fabric of meaningful relations in cultural and social settings. The practice of critical neuroscience could in this way serve as a natural complement to the selective attitude and methodological reductionism of experimental approaches.

Re-invoking the Social in Studies of Neuroscience

Openly politicized forms of critique are no longer much in evidence, and may not currently seem very workable (Cooter, 2007; Latour, 2004). Prevalent, for example, in science studies and cultural studies are approaches that appear to trade in critical engagement for an aestheticization of scientific practices, stopping short of penetrating into manifestly pathological developments. One reason for this may be the increasing professionalization and differentiation of various metascientific approaches over the past 40 or so years. Are practitioners no longer "allowed" to operate on a broader, holistic level of social understanding that transcends clearly circumscribed local expertise?[5] It is likely that certain intellectual as well as political and economic developments support some of this academic quietism (Forman, 2010).

In opposition to these tendencies, critical neuroscience strives to regain room for scrutiny, in reckoning with perspective-bound and interest-specific constraints that belie, in some contexts at least, objectivist aspirations of neuroscience and of those enthusiastic about its applicability in everyday life. Certainly, the gathering of context in many cases may end up laying bare the economic and political imperatives that sustain particular styles of thought from "screening and intervening" to "essential

---

[5] This might be one reason why critique of scientific and medical malpractice and corporate influence has recently been more a business of journalists, popular writers, and non-academic intellectuals than of professional STS practitioners (recent examples: Fine, 2010; Greenberg, 2010; Watters, 2010).

differences" (Abi-Rached & Rose, 2010; Fine, 2010). It may also end up shedding light on the ways in which the very concepts and categories that produce new kinds of responsibility towards the "natural" make-up of our minds are—knowingly or unknowingly—themselves shot through with our projections, and give rise to "facts," worldviews and policies that may collude with social and political orders (Hartmann, this volume; Malabou, 2008). This is well illustrated by Cordelia Fine's recent book, *Delusions of Gender.* Fine, trained both as a cognitive neuroscientist and a science journalist, rigorously analyzes neuroscience experiments, their results, and their interpretations among media exegetes, that purport to show hard-wired differences in behavior between men and women. She demonstrates how biases creep into the assumptions involved in experimental paradigms, and how cultural stereotypes are reified by "brain facts," placing these trends in the context of the social conditions that maintain this prejudice in the form of a new neurosexism (Fine, 2010).

As variously indicated above, critical neuroscience puts particular emphasis on the social. Of course, it is important not to take "the social" as a static, homogenous thing, but rather to work with this notion as a proxy for the associations between scientists, laboratories, media, agencies, governments, and other constituencies. Non-modern approaches such as actor-network theory are in this context very helpful. They do not construe "the social" as the kind of stuff out of which phenomena are literally *made,* and equally steer clear of the opposite extreme of a scientistic naturalization of the social (Latour, 2005, pp. 87–120; see also Latour, 1993). Instead, phenomena, as matters of concern, are reconstructed by being placed in networks of actors and actants forming theme-related alliances and vastly distributed webs of relations. Scientific knowledge as such can be viewed as embodied in material alliances or what Rouse, alluding to Wartenburg's conception of socially distributed power, has called "epistemic alignments" (Rouse, 1996; Wartenburg, 1990). In an important sense knowledge only "exists" in the material–practical interactions between people, things, instruments, agencies, and policies; and thus cannot be understood in abstraction from "the various kinds of resistance posed by anomalies, inconsistencies, disagreements and inadequacies of skill, technique, and resources" (Rouse, 1996, p. 194).[6]

While no grand-scale invocations of "social factors" can substitute for precise analyses of particular interactions and alignments between social actors and material actants, it is important, we believe, to keep the bigger picture in view. It is here that we diverge from the localism of actor-network theory and the STS mainstream: epistemic and political alliances, as well as cognitive and affective frames and interpretive

---

[6] It is helpful to emphasize again that material objects are themselves integral ingredients in both social power relations and in those material—practical alignments that constitute scientific knowledge: "Things can break down, are unavailable when needed, convey confusing signals, and sometimes even get in the way. Things can also open new possibilities for resistance to the power relations they mediate. And when things do fall out of alignment in these ways, the effects on power relationships are quite comparable to those which follow the breakdown of social alignments. We avoid fetishism not by strictly separating the natural and the social or by reducing the natural to the social but by recognizing the artificiality of the distinction." (Rouse, 1996, p. 190–191). It is hard to not think here of the heavy and complicated machinery that the neuroscience inevitably have to mobilize in order to establish epistemic contact with their object of inquiry (see Dumit, 2004 and this volume).

schemes instituted by them, often operative through media representations or discursive practices that begin in local settings and are subsequently broadened, all contribute to a structure of secondary objectivity or second nature. These processes of solidification can easily escape the purview of science and its commentators because of the incremental nature and slow timescales of change, and because of the authoritative nature of the finished product: established, official, institutional knowledge. The "social" needs to be viewed not as an assumed explanatory factor but as the result of various micro- and meso-level operations and alignments between a wealth of actors, tools, quasi-objects, and agencies. In turn, the social re-emerges as a potential explanatory resource; for example in the mobilization and distribution of attention, of concern and relevance, and in the workings of tacit schemes of interpretation and normative judgment (Butler, 2009). In light of this it is not enough to merely point to ontological hybridization or celebrate one's having superseded modernist dualisms (Latour, 1993, 2005). Neither does it suffice, for our purpose, to neutrally chart the cartography of "emergent forms of life"—such as biological citizenship and neurochemical selfhood—nor simply to leave it upon others to "judge" these developments (Rose, 2007, p. 259).[7]

While such descriptive endeavors provide important staging for subsequent analysis, it is crucial to penetrate beneath the surface of emerging practices, relations, and styles into the dynamics of power that may shape or stabilize surface phenomena, facilitate or hinder certain alliances or actions. It is important to reckon with pathological developments, render explicit interest-driven biases, hegemonic schemes of judging, templates of knowing and classifying, dangerous blind-spots in interpretations, unquestioned narrative patterns, and various unholy material alliances.[8] For example, the neoliberal mobilization of "human resources" in the name of employability, flexibility, and soft skills has found a new space to take shape among neuroscientists performing the naturalization of social/economic categories, and increasingly biologized notions of personhood, human experience, and the good life. Subjectivity is parsed from the outset into economic categories and becomes a type of bio-economic capital that is in turn used to sort people into

---

[7]   We refer here to the puzzlingly moderate final remarks in Nikolas Rose's *The Politics of Life Itself.* Rose's proclamation of neutrality at the end of that work is surprising in face of the many blatantly critique-worthy developments he had charted so rigorously throughout the book. As Cooter and Stein put it, "It is a vagueness that is popular in today's academic world run as it is by the changing fashions and fortunes of grant-giving bodies, for it permits study of almost everything but commitment to nothing—hence, a valuable strategy for the retention of patronage. This is not to say that Rose is openly opportunistic, but he does seem to suggest that one can separate the empirical analysis of contemporary life from larger questions of collective human direction and purpose. He keeps his hands clean" (Cooter & Stein, 2010, p. 115).

[8]   Here critical neuroscience preserves what could be called historical solidarity with the project of critical theory: the similarity lies in the attempt to move beyond sporadic interventions towards a theoretically integrated account of a system of normative assumptions, interpretive patterns, and material conditions that jointly stabilize, on the scale of society or significant parts of it, a tacitly pathological status quo. The term "theory" in critical theory is no accident (Geuss, 1981; Honneth, 2009). Almost needless to say, we are currently far from advocating anything in the direction of a worked-out theoretical account of this kind. There is as of yet no critical theory of the neurosciences (on this, see also Hartmann, this volume).

kinds, construct risk profiles, and suggest enrolment in enhancement programs (Fricke & Choudhury, 2011).[9]

Needless to say, within this discourse characteristic of neoliberal think tanks, social experience is thoroughly individualized and cultural and behavioral phenomena are declared "natural" (Brinkmann, 2008). Is this something that we, as academic observers and affected individuals, should merely register in a neutral way?[10] In light of this, we argue that critical neuroscience must ask hard questions about conceptual and normative assumptions and strategic alliances, and work towards re-opening contestations and restaging alternative interpretations and evaluations.[11]

Structural Pathologies in Science and Society

The activity of assemblage, in our sense of the term, is thus an inherently political one. It allows the critic to identify something close to what Axel Honneth has called "social pathologies of reason" (Honneth, 2009, ch. 2):[12] such pathologies are defects or malfunctions in social systems, practices, and institutions—malfunctions that come into view against the background of some normative understanding of society and properly functioning institutions. In the case example of addiction, described earlier, one might come to reckon with diverging perspectives from medical professionals, pharmaceutical companies, health administrators, social workers, governments and political parties, the education sector, newly constituted "risk populations," and certainly "the addicts" themselves. However, "addict"—and similarly, other kind terms in use in neuroscientific research—must be seen as a category that is co-produced through dominant classifications, styles of thought, and cultural practices. Incisive analysis of the interactions which make possible these neurological categories give ground for active assertions about what is at stake in the case of "brain overclaim" or tangible corporate influences on scientific practice.

For example, as Laurence Kirmayer and Ian Gold (this volume) argue, there is a trend in mainstream Western psychiatry to employ increasingly narrow construals of mental suffering that neglect the situatedness of patients in distorted social environments and direct the focus away from cultural embeddedness towards assumed

---

[9] Take for example the UK Foresight Project's definition of "well-being:" "Mental well-being, [...], is a dynamic state that refers to an individual's ability to develop their potential, work productively and creatively, build strong and positive relationships with others and contribute to their community" (Beddington et al., 2008, p.1057; see also Foresight Report, 2008). A related, large-scale government sponsored project is currently being conducted in France, employing a strikingly similar rhetoric (see Oullier & Sauneron, 2010).

[10] What the word "we" refers to here is of course a non-trivial issue. Provisionally, what we mean is the broad group of potential "recipients" of the conceptual transformations alluded to here—in other words those affected by structural changes in the conceptions of subjectivity and well-being brought forth by the current alliance between some practitioners in the human sciences and the spin doctors of corporate culture. To clarify further the exact standpoint of critique is of course important—but on the other hand not as important as to be able to postpone the beginning of critical reflection indefinitely.

[11] The important theme of norms is taken up again below.

[12] We take up Honneth's notion in a rather loose manner, divorcing it from the specific context of a theory of rationality implicit in approaches to "critique" from a Frankfurt School perspective.

"neurological underpinnings" of illness, agency, and personhood. Ignoring the social and cultural contexts of phenomena under investigation can render neuroscientific research (unknowingly) complicit with problematic developments in the medical sector, despite the best intentions of individual practitioners. Scientists are not usually trained to be very sensitive to the subtleties of, and social conflicts within, political and institutional environments—as science prizes epistemic virtues of other kinds (Daston & Galison, 2007). This can lead to distorted interpretations of experimental results—with very real consequences in the lives and treatment choices of patients, for example. Continuing the above example of addiction research, a narrowly neuroscientific understanding of substance addiction might lead to the neglect of the conditions that stabilize addictive behavior, and thus encourage forms of practice and treatment less conducive to the well-being of those affected than those that become available through a more complex understanding of the condition. Moreover, such narrow explanations fail to acknowledge the role of politics in addiction and other forms of human suffering.

Likewise, the widespread fascination with brain-based approaches in parts of the wider public calls for more critical responses, since circulation of simplistic accounts systematically serves to obscure these wider and often inconvenient entanglements (Heinemann & Heinemann, 2010; Weisberg, Keil, Goodstein, Rawson, & Gray, 2008). Intensified media representation coupled with audiences increasingly trained, through continuous exposure, to be receptive to easy-to-digest narratives of self-objectification ("your brain made you do it") contribute to the distorted images of the person—as lacking in free will, possessing skewed decision-making powers, being driven instead by automatized emotions, and thus as not genuinely responsible for their acts (while simultaneously making them responsible for "managing" their brains). Pervasive media messages in this manner lead to a climate of opinion that singles out sensationalistic themes, often ideologically laden, and pushes towards simplified, technocratic solutions to social problems (Greenberg, 2010). Critical neuroscience aims to function as an informed voice opposing those distorted images. Importantly, Fine's critique of neurosexism mentioned earlier is made particularly strong by her close engagement with the experimental design and statistics as well as her skill to write compellingly for a broader audience. Given that the flawed findings she critiques have traveled into the popular cultural script of male/female differences, critical writing for a public audience is a vital move that can benefit the repertoire of critical neuroscience activities.

Whose Norms? Expertise, Participation, and Contestation

The goal to scrutinize and lay bare scientific conventions that are taken for granted, tacit knowledge, vested interests at work in neuroscience research or their impacts on people, opens up complex questions about norms. In order to identify social pathologies or "system malfunctions," any critical endeavor will inevitably operate in a normative space, reflecting particular assumptions about the conditions for both social organization and individual wellbeing. What we deem "pathological" depends on a contrast with non-trivial ideas of a non-pathological alternative—such as a well-functioning institution or, where individual subjects are concerned, an orientation towards an image of the "good life." However, no version of a critical neuroscience should simply impose a set of normative standards or values. Norms are ubiquitous, operative at any time, on various levels, in all forms of social organization, social practices, and individual ways of life. The critic's task in the first instance is to render these norms explicit, point to possible tensions between different normative outlooks, and, where necessary, measure institutional realities against the normative assumptions that legitimate them. This will raise questions of power, the constructions of expertise, the social distribution of knowledge, and the possibilities for participation in decision-making processes. Critical neuroscience thus needs to engage with the current debates about the transparency, accountability, and inclusivity of the new "science in society" communicators, and, not least, to examine their role (Strathern, 2004).

The last few years have seen a steep increase in numerous forms of popularization of neuroscience. Driven by various parties, including neuroscientists, funding agencies, and the media, public engagement in neuroscience has emerged in the form of outreach projects, popular science writing, and—not least—as interactive neuroscience exhibitions geared towards a range of audiences, with the aim of informing (and to varying degrees engaging) the lay citizen. If critical neuroscience advocates informed participation in the scientific process, then it will need to confront questions about representation, expertise, and agency of lay citizens, particularly in information societies characterized by a more demanding and active citizenry (Beck, 1997; Giddens, 1991). There is no doubt that efforts to "democratize" scientific processes this way pose difficulties. With hindsight, earlier optimism about the potential of a renewed politicization of society around issues of science and technology seems to have been premature (see Kerr & Cunningham-Burley, 2000).[13] Rather than an emerging "sub politics" (Beck, 1997)—grass root political engagement that responds to hazards of scientific and technological development—we increasingly witness restricted expert circles monopolizing the negotiation and regulation of relevant issues.

One way for critical neuroscience to attempt to establish (or challenge) normative conceptions—themselves always necessarily under reflexive scrutiny—is by creating a discursive space for debate both in professional and practical domains about the categories and applications of neuroscience, and about related social issues such as the organization of labor, conception of health and disease, goals and practices in parenting and education, issues about law and punishment, technological self-optimization, and much more. In order to make this move however, it needs to probe critically at ways in which the choices and views of the public are regulated, particularly amidst the growing clamor for "neurotalk" in public spheres (Illes et al., 2010). Expert counseling and state-run programs of screening and risk assessment (Rose, 2010), and the instant professional take-up of ethical concerns into an institutionalized "neuroethics" (de Vries, 2007), increasingly occupy the space for public engagement. In what ways might the space for "science in society" or neuroethics experts, as well as the domains of psychiatrists, doctors, and educators (connected to government, funders, or companies) act as intermediaries in aligning public opinions with scientific

---

[13] Probably the most optimistic voice in this area has been German sociologist Ulrich Beck, see Beck (1995, 1997) and Beck's opening essay in Beck, Giddens, & Lash (1994). See also Giddens (1991, ch. 7).

agendas, ratifying or legitimating neuroscientific research programs (Rose & Miller, 1992)? Who can legitimately make knowable what the public wants or thinks about neuroscience and its applications? How can participatory approaches avoid opening up new forms of stratification?

With such problems in mind, critical neuroscience aspires to open up discursive spaces that facilitate debate among practitioners, "stakeholders," and lay citizens about the goals, concerns, and normative standards that society wants its science to pursue or live up to: where the work of the critic involves not merely encouraging the accessible promotion of new ideas from neuroscience, but invites plural viewpoints and promulgates a degree of critical rigor through provocation—that is, by illuminating blind spots or limitations and by questioning assumptions and applications. It is vital that public neuroscientists conceive of audiences not as listeners or viewers but as potential speakers. It is at these sites of contestation that specific normative issues surrounding scientific matters of concern can emerge and take shape. This process pushes science beyond reliable knowledge—subject only to validation within its own disciplinary context—to the production of "socially robust knowledge;" that is, knowledge tested for validity both outside and inside the lab, developed through the involvement of socially distributed experts including those from different disciplinary and experiential backgrounds within and outside of academia, and knowledge produced through repeated testing, expansion, and modification (Nowotny, 2003). While the embeddedness in society and the iterative process of open contestation may render this knowledge more robust, the means of such forms of polycentric knowledge production in neuroscience must be carefully worked out (Jasanoff, 2003).

A model of "public" neuroscience such as this faces challenges within the changing structure of the university and changes in the organization and funding of professional research. Both are increasingly oriented towards a corporate, neoliberal management model (Giroux, 2007; Mirowski & Sent, 2005). How can critical neuroscience reach its goals in a system that places its values on outcomes and efficiency, increasingly fosters commercializable or applied research, and encourages corporate influences in the form of sponsorship, company spin-offs, profitable patents, and institutional joint ventures?

There are trends pulling neuroscience in different directions, certainly not all negative—a push towards applications and intensified collaboration can also bring synergies and create new perspectives. The ambivalence of the situation can be illustrated by reference to interdisciplinarity (a term that has become a powerful buzzword in academia, including neuroscience). Successful integration of distinct perspectives and methodological approaches can lead to unforeseen benefits and novel insights. However, genuine inter-, trans- and postdisciplinary research is constantly forced to acknowledge, and to work with, tensions between ontological and epistemological frameworks, and is thus necessarily slow, compared to conventional single-discipline research processes. The sustained and, as it were, "organic" integration of different disciplinary approaches and conceptual frameworks will be difficult in outcome-oriented environments dominated by short time frames and institutional structures of commercialized or translational research. In order to enable a reflexive ethos, and to keep open a space for critical inquiry in a context that favors "outcomes" in terms of revenues and commodities, and entrepreneurial over critical skills, critical neuroscience will need to continue discussing and analyzing structural transformations, and challenging the increasing dominance of market orientation in the wider academic arena.

What Difference Can Critique Make to Neuroscience?

The metaphor of the *looping journey*—of that which is taken to be a "brain fact"—can help to operationalize critique, opening up the many possibilities for thickening, or assembling, a given brain-based phenomenon. Whether we focus on the neural basis of addiction, depression, adolescence, culture, gender, morality, or violence, the journey can be traced using multiple methodologies, from the point of a theme's entry into—and treatment in—the lab, through various technical and knowledge practices, to the interaction with the media and policy, to its reception by the public. What we mean by a "brain fact" is not an absolute thing-in-itself, but a specifically conceptualized phenomenon or "local resistance" that emerges from the collective practices and directed cognition of neuroscientists working in a community at a given time and in a given context (see Choudhury, Nagel, & Slaby, 2009).[14]

With this in mind, it is important to ask what difference second-order observations of laboratory conditions, communities of scientists, and historical and cultural contingencies make to neuroscientists themselves, whose goal is to develop and test paradigms that ultimately contribute to mapping social, cultural, or perceptual processes on particular brain regions. Critical neuroscience renews the possibility for critical commentators to be engaged with, rather than estranged from, laboratory science. Functioning through the collaboration of work from multiple methodologies, it aims to find entry points for social theory, ethnography, philosophy, and history of science, in the laboratory. In the following, we put forward ways in which the latter fields can play a contributory role in both the *practice* of neuroscience in the lab and in the *representation* of neuroscience beyond the lab.

From educational initiatives for junior level researchers to the development of collaborative working groups[15] investigating behavioral phenomena from different disciplinary perspectives, critical neuroscience explores whether a kind of reflexivity can, through interdisciplinary training, be inscribed into experimental practice. The aim here is not to conduct a purer or "better" neuroscience. Instead, reflective practice includes social and historical contextualization and cross-cultural comparison of behavioral phenomena, within neuroscience. Examining these contingencies will

---

[14] We use the notion of a "brain fact" analogously to Ludwik Fleck's conceptualization of a scientific fact in his seminal study *Genesis and Development of a Scientific Fact* (see Fleck, 1935/1979). On the looping journeys of scientific facts in the context of neuroscience see also Dumit (2004).

[15] Since the emergence of critical neuroscience a handful of joint education opportunities have been started for junior researchers from diverse scientific and academic fields. Pursuing this through graduate courses, themed summer/winter schools, and collaborative workshops will sustain mutual learning and joint work on a number of themed topics related to neuroscience. See www.critical-neuroscience.org for upcoming activities.

generate alternative possibilities for findings in neuroscience, which on the one hand open up interesting empirical questions for neuroscientists, and on the other hand, function as a form of critique from within.

How should we conceive of the relationship between first-order (descriptions of brain and behavior) and second-order (descriptions of neuroscientists observing behavior) observations (Choudhury, Nagel, & Slaby, 2009; Langlitz, this volume; Roepstorff, 2004)? We believe engagement between these socio-cultural and historical studies and experimental neuroscience can be constructive in a number of ways:

(i)   demonstration of alternative possibilities of results of neuroscience experiments by modifying technical parameters or comparing and re(de)fining categories;[16]
(ii) exploring routes to empirically investigate social and cultural phenomena without assuming universal neural mechanisms from the outset;
(iii) enriching behavioral theories by allowing for pluralistic viewpoints and methodologies to result in layered explanations of complex phenomena; and
(iv) examining the subtle relationship and feedback loops between popular opinion or ideologies about the brain and findings in neuroscience.

Such goals can only be realistically achieved through collaborative work. Working groups, as initiated since the emergence of critical neuroscience, consist of the following.

Sociologists of science who observe communities of scientists and capture the thought styles that govern their cognition in studying the particular phenomenon in the lab (Fleck, 1935/1979). Fleck described the "tenacity" of systems of thought that govern scientific practices and explanatory styles, and that ultimately give rise to what from then on will count as fact. What solidifies a local resistance into a recognized "fact"? By studying the journey of a phenomenon in and around the neuroscience lab, we can study how the methods, concepts and theories involved in the development of a fact of neuroscience may be culturally conditioned; in addition we can identify the refractory effects of the thought collective that sustain it and the wider culture in which it functions (see, for instance, Dumit, 2004, this volume; Joyce, 2008). Neuroscientists are working at a time of unprecedented politicization through the commercialization of research (Wise, 2006), and sociological analysis can highlight the pressures that commercial, pharmaceutical, and military interests place on neuroscience (Greenslit, 2002; Healy, 2004; Moreno, 2006). Moreover, sociologists can begin to draw cross-national comparisons of the social structures of neuroscience. Comparing the international contexts of trends in neuroscience research and its representation will help to spell out the logic of the neuroindustry, that is, the institutional, historical, political, and ideological planes in which the rapid developments, the allure, and the influence on cultural formulations and other academic disciplines take place, over and above the events within neuroscience per se.

---

[16]   This is an example of how neuroscience itself can be used to subvert its own assumptions and demonstrate the contingencies of categories and methodologies it employs, a move we have called experimental irony. In Chapter 13, Daniel Margulies illustrates the power of this strategy of critique "from inside" through a review of the recent study by Bennett, Miller, & Wolford (2009) that used a dead Atlantic salmon in an fMRI scanner to highlight the high possibility of red herrings in brain imaging research.

Philosophy contributes the analysis of central phenomena under investigation (and their different, often competing, conceptualizations); for example, emotions, moral decisions, and responsibility. It also serves to clarify the content and status of notions such as determinism, reductionism, specificity, and consilience—concepts that have been floated in neuroscience and its critiques for a while, and require sharpening. Often, these and other concepts play key roles in what Hartmann (this volume) calls the hidden hermeneutics of the neurosciences: structural narratives that practitioners routinely employ as they describe their objects of investigation and construct interpretations of data, but that are rarely reflected upon explicitly. Ideas about "cerebral subjectivity" (Vidal, 2009) or the ubiquitous but often vague appeals to evolutionary theory are good examples (Richardson, 2007); similarly the new hype around the notion of cerebral plasticity (Malabou, 2008).

The task here is to elucidate a specific meta level: ascending from the manifest contents of theories, explanatory frameworks, and core concepts in current neuroscience to the analysis of latent assumptions and formative backgrounds, such as the implicit construal of the brain as the stable ontological foundation of both personal traits and social and cultural phenomena (to name just one, albeit crucial example). Philosophy also contributes to enriching the description of phenomena under study through phenomenological investigations, performing what has been called "front-loaded phenomenology" (Gallagher, 2003, this volume; Gallagher & Zahavi, 2008; Ratcliffe, 2008, 2009; Zahavi, 2004).

Cognitive neuroscientists contribute to technical and conceptual analysis of research processes, including methodological assessments. What are the potentials and limits of specific methodologies or tools such as fMRI and the associated statistical methods, and to what extent are these clear or made clear in different venues (Logothetis, 2008; Vul, Harris, Winkielman, & Pashler, 2009)? How are cultural, psychological, functional, and genetic models of cognitive phenomena mapped onto each other? Once a phenomenon enters the neuroscience lab, how do scientists break down the phenomenon into constituents that they are able to study within the constraints of their methodology? What efforts are involved in setting up experimental apparatuses and stabilizing the phenomena under study? Do researchers employ concepts that are sufficiently precise and that fully encompass the relevant dimensions of the phenomenon under study? How are the results analyzed and evaluated in comparison to other data from different experiments? How can data—quantitative and qualitative—from social science and humanities disciplines be brought to bear on the neurobiological results?

Cultural or medical anthropologists will draw on ethnographic data to develop cross-cultural comparisons of behavioral phenomena or symptoms and experimental paradigms (tasks, questionnaires) that have largely been studied on—or standardized using—particular groups of subjects deemed to represent the "norm" (Henrich, Heine, & Norenzayan, 2010).[17] Critical neuroscience draws on medical anthropology

---

[17]   In their recent comparative article, Heinrich, Heine, & Norenzayan (2010) use the acronym WEIRD to denote the White, Educated, Industrialized, Rich, and Democratic societies that behavioral science researchers take to be "standard subjects," in spite of the considerable heterogeneity across populations taken to be groups, and in spite of the fact that so called WEIRD populations are frequently *unusual* or *outliers.*

to supplement findings of neural correlates with phenomenological insights, biographical accounts of the person, and the *meaning*—that is, the social, cultural, moral, or spiritual significances—of behavioral phenomena, including mental illness and interventions (Cohn, this volume). Critical neuroscience resonates with cultural psychiatry, in emphasizing that the most fundamental level, using neuroscience in its current form, is not necessarily the most appropriate either for explaining or intervening in psychopathology. While neuroscientists and medical practitioners increasingly invoke the use of neuroscience in psychiatric nosology and clinical practice (Hyman, 2007; Insel & Quirion, 2005), critical neuroscience must find ways to consider how "meaning and mechanism" intersect via the brain (Choudhury & Kirmayer, 2009; Seligman & Kirmayer, 2008; Wexler, 2006).

The new subfields of social and cultural neuroscience have indeed just begun to investigate how aspects of cultural background may influence cognition, such as the expression and regulation of emotions and understanding of others (Chiao et al., 2008; Han & Northoff, 2008; Zhu, Zhang, Fan, & Han, 2007). As this area of neuroscience burgeons, critical neuroscience looks to anthropology to contribute to the conceptualization of culture in experimental design and interpretation, to explore how environmental factors, including culture, shape or interact with the development of structure and function of the healthy nervous system in such a way that several vocabularies of description—social, cultural, psychological, and biological—can coexist (Kirmayer 2006; Langlitz, this volume; Lock & Nguyen, 2010).

Historians of science trace historical trajectories of the conceptual construals, interpretive contexts, and experimental set-ups common to contemporary neuroscience (Foucault, 1973; Hacking, 2002; Young, 1995). Historical analysis will thus show how particular problems such as the criminal brain, posttraumatic stress disorder, the risky teen, or the empathic female become questions for the neurosciences and how particular methodologies are valued over others as more objective. Critical neuroscience will yield important insights from the history of concepts, practices, and objects of scientific inquiry, to understand how technologies, political, and moral contexts converge to give rise to diagnostic categories, how aspects of the self have come to be objectified and considered in certain contexts as clearly reducible to the brain (Vidal, 2002, 2009) and how scientific objectivity itself developed as an epistemic virtue (Daston & Galison, 2007). Longue durée analysis can additionally serve to interrogate the air of radical departure that surrounds much of the rhetoric around neuroscience (Borck & Hagner, 2001). Unpacking these histories might help to gain distance from the inflated, spectacular, and brain-centric rhetoric which parts of the neuroindustry seem to dictate (Stadler, this volume).

Conclusion

We have sketched a picture of a critical neuroscience that probes the extent to which claims *about* neuroscience do in fact match neuroscience's real world (social) effects. It sets out to analyze the allure and functions of the *neuro* in the broader scheme of intellectual and political contexts including the rise in recent years of a new (neuro) biologism in many academic disciplines and popular culture at large. Our aim is to contribute these observations from the human sciences to neuroscience so as to demonstrate the contingencies of neuroscientific findings and, at the same time, to open up new experimental and interpretive possibilities.

Assembling and broadening ontological landscapes of behavioral phenomena requires us to move beyond the tenacious nature–nurture distinction when conceptualizing phenomena such as addiction, adolescence, autism, or depression. Instead, critical neuroscience will work with concepts such as "cultural biology" and "local biology" which bring to the fore the co-constitutive relationship between the brain and its context. The "endorphin-challenged alcoholic," the "neurological adolescent," or the "female brain" are richly situated and sustained in a habitat made up of interactions between institutional, cultural, and neuronal infrastructures. Such a framework poses intellectual challenges to cognitive and clinical neuroscience—challenges that must be taken up, especially as the notion of neuroplasticity or the field of cultural neuroscience open up potential to investigate brain–environment interactions. We emphasize the need to rethink the conception and location of these borderlines at the skull or the skin in a way that troubles the arbitrary distinctions and moves beyond biological determinism and social constructionism. If fMRI can show that cultural upbringing modulates brain activity or new biotechnologies permit us to tinker with the brain and cognition, it is apt for neuroscience to acknowledge that our brains are represented in terms of cultural categories and that our brains also do "cultural work" in distinguishing what is natural, who is healthy, different, normal, or rational (Lock, 2001; Lock & Nguyen, 2010).

The chapters in this volume undertake initial explorations of the discursive space that is opened up once the outworn distinctions and dualisms are surpassed, and once open-minded interaction between practitioners from different methodological universes is enabled. The critical ethos we invoke, therefore, is not one that rejects but one that aims to elicit change: both in how social phenomena are explored within neuroscience, and in how the social implications of neuroscience are analyzed. The conceptual changes involved in studying the situated brain in its context, the pedagogical initiatives that bring multiple traditions of scholarship into contact, and the calls for contestation in neuroscience funding and application, all disturb boundaries—between the brain and its environment, between disciplinary vocabularies and methodologies, and between science and society. These very interruptions will provoke us to imagine the brain in different terms and to probe its functions in alternative ways. Such changes—towards which we sense an increasing openness among neuroscientists and social scientists alike—will, we believe, open up potential for a more realistic picture of the function of neuroscience in society while simultaneously commenting on the broader socio-political changes in contemporary societies that steer its developments, for better or for worse.

Acknowledgements

We gratefully acknowledge Max Stadler's important contribution to the development of the ideas expressed in this chapter. We are also thankful to Allan Young, whose comments on an earlier draft of this chapter proved immensely valuable. Our discussions and joint work with Saskia K. Nagel were important during the early stages of developing the approach described here. Lutz Fricke, Jan-Christoph Heilinger and John Protevi also made several helpful suggestions.

References

Abi-Rached, J. M., & Rose, N. (2010). The birth of the neuromolecular gaze. *History of the Human Sciences, 23*(1), 1–26.

Anderson, W. (2009). From subjugated knowledge to conjugated subjects: science and globalisation, or postcolonial studies of science? *Postcolonial Studies, 12*(4), 389–400.

Beck, U. (1997). *The reinvention of politics: Rethinking modernity in the global social order.* Cambridge: Polity Press.

Beck, U. (1995). *Ecological politics in an age of risk.* Cambridge: Polity Press.

Beck, U., Giddens, A., & Lash, S. (1994). *Reflexive modernization: Politics, tradition and the aesthetic in the modern social order.* Cambridge: Polity Press.

Beddington, J., Cooper, C. L., Field, J., Goswami, U., Huppert, F. A., Jenkins, R., & Thomas, S. M. (2008). The mental wealth of nations. *Nature, 455,* 1057–1060.

Bennett, C., Miller, M., & Wolford, G. (2009). Neural correlates of interspecies perspective taking in the post-mortem Atlantic salmon: An argument for multiple comparisons correction. *NeuroImage, 47*(Supplement 1), S125. doi:10.1016/S1053-8119(09)71202-9.

Borck, C., & Hagner, M. (2001). Mindful practices: On the neurosciences in the twentieth century. *Science in Context, 14*(4), 615–641.

Brigandt, I., & Love, A. (2008). Reductionism in biology. In *Stanford encyclopedia of philosophy.* Retrieved from http://plato.stanford.edu/entries/reduction-biology/ on October 19, 2010.

Brinkmann, S. (2008). Changing psychologies in the transition from industrial society to consumer society. *History of the Human Sciences, 21*(2), 85–110.

Butler, J. (2009). *Frames of war: When is life grievable?* London: Verso.

Campbell, N. (2010). Towards a critical neuroscience of "addiction". *BioSocieties, 5*(1), 89–104.

Chiao, J. Y., Iidaka, T., Gordon, H. L., Nogawa, J., Bar, M., Aminoff, E., & Ambady, N. (2008). Cultural specificity in amygdala response to fear faces. *Journal of Cognitive Neuroscience, 20*(12), 2167–2174.

Choudhury, S., Gold, I., & Kirmayer, L. (2010). From brain image to the Bush doctrine: critical neuroscience and the political uses of neurotechnology. *American Journal of Bioethics: Neuroscience, 1*(2), 17–19.

Choudhury, S., & Gold, I. (in press). Mapping the field of cultural neuroscience. *BioSocieties.*

Choudhury, S., & Kirmayer, L. (2009). Cultural neuroscience and psychopathology: Prospects for cultural psychiatry. *Progress in Brain Research, 178,* 263–279.

Choudhury, S., Nagel, S. K., & Slaby, J. (2009). Critical neuroscience: Linking neuroscience and society through critical practice. *BioSocieties, 4*(1), 61–77.

Cohn, S. (this volume). Disrupting images: Neuroscientific representationsin the lives of psychiatric patients.

Cooter, R. (2010). Neuroethical brains, historical minds, and epistemic virtues. Talk given at the WTCHOM Symposium (Feb. 19, 2010): *Neuroscience and Human Nature.*

Cooter, R. (2007). After death/after-"life": The social history of medicine in post-postmodernity. *Social History of Medicine, 20*(3), 441–464.

Cooter, R., & Stein, C. (2010). Cracking biopower. *History of the Human Sciences, 23.* doi: 10.1177/0952695110362318

Daston, L., & Galison, P. (2007). *Objectivity*. New York: Zone Books.

De Vries, R. (2007). Who guards the guardians of neuroscience? Firing the neuroethical omagination. *EMBO Reports, 8*, 1–5.

Dumit, J. (this volume). Critically producing brain images of mind.

Dumit, J. (2004). *Picturing personhood: Brain scans and biomedical identity*. Princeton: Princeton University Press.

Emery, N. J., Clayton, N. S., & Frith, C. D. (2010). Introduction. In Social intelligence: From brain to culture. *Philosophical Transactions of the Royal Society B, 362*, 485–488.

Fine, C. (2010). *Delusions of gender: How our minds, society and neurosexism create difference*. New York: Norton.

Fish, E. W., Shahrokh, D., Bagot, R., Caldji, C., Bredy, T., Szyf, M., & Meaney, M. J. (2004). Epigenetic programming of stress responses through variations in maternal care. *Annals of the New York Academy of Sciences, 1036*, 167–180. doi:10.1196/annals.1330.011

Fleck, L. (1979). *Genesis and development of a scientific fact*. Chicago: University of Chicago Press. (Original work published 1935).

Foresight Mental Capital and Wellbeing Project (2008). Final project report. The Government Office for Science, London.

Forman, P. (2010). (Re-)cognizing postmodernity: Helps for historians–of science especially. *Berichte zur Wissenschaftsgeschichte, 33*(2), 1–19.

Foucault, M. (1973). *Madness and civilization: A history of insanity in the age of reason*. New York: Vintage.

Fricke, L., & Choudhury, S. (2011). Neuropolitik und plastische Gehirne: Eine Fallstudie des adoleszenten Gehirns. *Deutsche Zeitschrift für Philosophie, 59*(3), 391–402.

Frith, U., & Frith, C. D. (2010). The social brain: Allowing humans to boldly go where no other species has been. *Philosophical Transactions of the Royal Society B, 365*, 165–176.

Gallagher, S. (this volume). Scanning the lifeworld.

Gallagher, S. (2003). Phenomenology and experimental design toward a phenomenologically enlightened experimental science. *Journal of Consciousness Studies, 10*, 85–99.

Gallagher, S., & Zahavi, D. (2008). *The phenomenological mind: An introduction to phenomenology and cognitive science*. London: Routledge.

Garner, A., & Hardcastle, V. G. (2004). Neurobiological models: An unnecessary divide–Neural models in psychiatry. In J. Radden (Ed.), *The philosophy of psychiatry: A companion* (pp. 364–380). Oxford: Oxford University Press.

Geuss, R. (1981). *The idea of a critical theory: Habermas and the Frankfurt School*. Cambridge: Cambridge University Press.

Giddens, A. (1991). *Modernity and self-identity: Self and society in the late modern age*. Stanford: Stanford University Press.

Giroux, H. A. (2007). *University in chains: Confronting the military-industrial-academic complex*. Boulder: Paradigm.

Greenberg, G. (2010). *Manufacturing depression: The secret history of a modern disease*. London: Bloomsbury.

Greenslit, N. (2002). Pharmaceutical branding: Identity, individuality, and illness. *Molecular Interventions, 2*, 342–345.

Hacking, I. (2007). Kinds of people: Moving targets. *Proceedings of the British Academy, 151*, 285–318.

Hacking, I. (2002). *Historical ontology*. Cambridge, MA: Harvard University Press.

Hacking, I. (1999). *The social construction of what?* Cambridge, MA: Harvard University Press.

Hacking, I. (1995). The looping effect of human kinds. In D. Sperber & A.J. Premack (Eds.), *Causal cognition* (pp. 351–383). Oxford: Oxford University Press.

Han, S., & Northoff, G. (2008). Culture-sensitive neural substrates of human cognition: A transcultural neuroimaging approach. *Nature Reviews: Neuroscience, 9*(8), 646–654.

Hartmann, M. (this volume). Against first nature: Critical theory and neuroscience.

Heinemann, L. V., & Heinemann, T. (2010). Optimize your brain! Popular science and its social implications. *BioSocieties, 5,* 291–294.

Healy, D. (2004). *Let them eat Prozac: The unhealthy relationship between the pharmaceutical industry and depression.* New York: New York University Press.

Henrich, J., Heine, S. J., & Norenzayan, A. (2010). The weirdest people in the world? (target article, commentaries & response by authors). *Behavioral and Brain Sciences, 33*(2/3), 61–83; 111–135.

Honneth, A. (2009). *Pathologies of reason: On the legacy of critical theory.* (J. Ingram, Trans.). New York: Columbia University Press. (Original work published 2007).

Hyman, S. E. (2007). Can neuroscience be integrated into the DSM-V? *Nature Reviews Neuroscience, 8*(9), 725–732.

Hyman, S. E., & Malenka, R. C. (2001). Addiction and the brain: The neurobiology of compulsion and its persistence. *Nature Reviews Neuroscience, 2,* 695–703.

Illes, J., Moser, M. A., McCormick, J., Racine, E., Blakeslee, S., Caplan, A., & Weiss, S. (2010). Neurotalk: Improving the communication of neuroscience research. *Nature Reviews Neuroscience, 11*(1), 61–69.

Insel, T. R., & Quirion, R. (2005). Psychiatry as a clinical neuroscience discipline. *Journal of the American Medical Association, 294,* 2221–2224.

Jasanoff, S. (2003). Technologies of humility: Citizen participation in governing science. *Minerva, 41*(3), 223–244.

Joyce, K. A. (2008). *Magnetic appeal: MRI and the myth of transparency.* Ithaca: Cornell University Press.

Kant. I. (1992). *The conflict of the faculties.* (Mary I. Gregor, Trans.). Lincoln, NE: University of Nebraska Press. (Original work published 1798).

Kendler, K. S. (2008). Explanatory models for psychiatric illness. *The American Journal of Psychiatry, 165*(6), 695–702.

Kerr, A., & Cunningham-Burley S. (2000). On ambivalence and risk: Reflexive modernity and the new human genetics. *Sociology, 34*(2), 283–304.

Kirmayer, L. J., & Gold, I. (this volume). Re-socialising psychiatry: Critical neuroscience and the limits of reductionism.

Kirmayer, L. J. (2006). Beyond the "new cross-cultural psychiatry": Cultural biology, discursive psychology and the ironies of globalization. *Transcultural Psychiatry, 43,* 126–144.

Knorr-Cetina, K. (1999). Epistemic cultures: How the sciences make knowledge. Cambridge, MA: Harvard University Press.

Kuhar, M. (2010). Contributions of basic science to understanding addiction. *BioSocieties, 5*(1), 25–35.

Langlitz, N. (this volume). Delirious brain chemistry and controlled culture: Exploring the contextual mediation of drug effects.

Latour, B. (2005). *Reassembling the social: An introduction to actor-network-theory.* Oxford: Oxford University Press.

Latour, B. (2004). Why critique has run out of steam: From matters of fact to matters of concern. *Critical Inquiry, 30,* 225–248.

Latour, B. (1993). *We have never been modern.* (C.Porter, Trans.). Cambridge, MA: Harvard University Press.

Leshner, A. I. (2001). Addiction is a brain disease. *Issues in Science and Technology, 2001,* April 9. Retrieved from http://www.issues.org/17.3/leshner.htm# (2010, November 16).

Logothetis, N. K. (2008). What we can do and cannot do with fMRI. *Nature, 453*(7197), 869–878.

Lock, M. (2001). Containing the elusive body. *The Hedgehog Review, 3*(2), Summer 2001, 65–78.

Lock, M. (1993). *Encounters with aging: Mythologies of menopause in Japan and North America.* Berkeley: University of California Press.

Lock, M., & Kaufert, P. (2001). Menopause, local biologies, and cultures of aging. *American Journal of Human Biology, 13,* 494–504.

Lock, M., & Nguyen, V.-K. (2010). *An anthropology of biomedicine.* Chichester/Oxford: Wiley-Blackwell.

Malabou, C. (2008). *What should we do with our brain?* New York: Fordham University Press.

Margulies, D. (this volume). The salmon of doubt: Six months of methodological controversy within social neuroscience.

McGoey, L. (2009). Pharmaceutical controversies and the performative value of uncertainty. *Science as Culture, 18*(2), 151–164.

McGowan, P. O., Sasaki, A., D'Alessio, A. C., Dymov, S., Labonté, B., Szyf, M., & Meaney, M. J. (2009). Epigenetic regulation of the glucocorticoid receptor in human brain associates with childhood abuse. *Nature Neuroscience, 12*(3), 342–348.

Meaney, M. J., & Szyf, M. (2005). Maternal care as a model for experience-dependent chromatin plasticity? *Trends in Neurosciences, 28,* 456–463.

Mirowski, P., & Sent, E. (2005). The commercialization of science and the response of STS. In E. Hackett, O. Amsterdamska, M. Lynch & J. Wajcman (Eds.), *Handbook of science and technology studies.* Cambridge, MA: MIT Press.

Mitchell, S. D. (2009). *Unsimple truths: Science, complexity, and policy.* Chicago: University of Chicago Press.

Moreno, J. D. (2006). *Mind wars: Brain research and national defense.* Washington, D.C.: Dana Press.

Noë, A. (2009). *Out of our heads: Why you are not your brain and other lessons from the biology of consciousness.* New York: Hill and Wang.

Nowotny, H. (2003). Democratising expertise and socially robust knowledge. *Science and Public Policy, 30*(3), 151–156.

Oullier, O., & Sauneron, S. (2010). *Improving public health prevention with behavioural, cognitive and neuroscience.* Centré d"Analyse Stratégique, French Government. http://www.strategie.gouv.fr/IMG/pdf/NeuroPrevention_English_Book.pdf

Pickersgill, M. (2009). Between soma and society: Neuroscience and the ontology of psychiatry. *BioSocieties, 4,* 45–60.

Protevi, J. (2009). *Political affect: Connecting the social and the somatic.* Minneapolis: University of Minnesota Press.

Raikhel, E. (this volume). Radical reductions: Neurophysiology, politics and personhood in Russian addiction medicine.

Rasmussen, N. (2010). Maurice Seevers, the stimulants and the political economy of addiction in American biomedicine. *BioSocieties, 5*(1), 105–123.

Ratcliffe, M. (2009). Understanding existential changes in psychiatric illness: The indispensability of phenomenology. In M. Broome & L. Bortolotti (Eds.), *Psychiatry as cognitive neuroscience: Philosophical perspectives* (pp. 223–244). Oxford: Oxford University Press.

Ratcliffe, M. (2008). *Feelings of being: Phenomenology, psychiatry, and the sense of reality.* Oxford: Oxford University Press.

Raz, A. (this volume). Critical neuroscience: From neuroimaging to tea leaves in the bottom of a cup.

Richardson, R. C. (2007). *Evolutionary psychology as maladapted psychology*. Cambridge, MA: MIT Press.

Rizzolatti, G., & Craighero, L. (2004). The mirror-neuron system. *Annual Review of Neuroscience, 27*, 169–192.

Roepstorff, A. (2004). Mapping brain mappers, an ethnographic coda. In R. Frackowiak, et al. (Eds.), *Human Brain Function* (2nd ed.), (pp. 1105–17). London: Elsevier.

Rose, N. (2010). "Screen and intervene": Governing risky brains. *History of the Human Sciences, 23*(1), 79–105.

Rose, N. (2007). *The Politics of life itself: Biomedicine, power, and subjectivity in the twenty-first century*. Princeton: Princeton University Press.

Rose, N., & Miller, P. (1992). Political power beyond the state: Problematics of government. *British Journal of Sociology, 43*(2), 173–205.

Rouse, J. (2002). *How scientific practices matter*. Chicago: Chicago University Press.

Rouse, J. (1996). *Engaging science: How to understand its practices philosophically*. Ithaka, NY: Cornell University Press.

Rouse, J. (1987). *Knowledge and power: Toward a political philosophy of science*. Ithaka, NY: Cornell University Press.

Seligman, R., & Kirmayer, L. J. (2008). Dissociative experience and cultural neuroscience: Narrative, metaphor and mechanism. *Culture, Medicine & Psychiatry, 32*, 31–64.

Singh, I., & Rose, N. (2009). Biomarkers in psychiatry. *Nature, 460*(9), 202–207.

Slaby, J. (2010). Steps towards a critical neuroscience. *Phenomenology and the Cognitive Sciences, 9*, 397–416.

Stadler, M. (this volume). The neuromance of cerebral history.

Strathern, M. (2004). Laudable aims and problematic consequences, or: The 'flow' of knowledge is not neutral. *Economy and Society, 33*(4), 550–561.

Vidal, F. (2009). Brainhood: Anthropological figure of modernity. *History of the Human Sciences, 22*(1), 6–35.

Vidal, F. (2002). Brains, bodies, selves, and science: Anthropologies of identity and the resurrection of the body. *Critical Inquiry, 28*(4), 930–974.

Vul, E., Harris, C., Winkielman, P., & Pashler, H. (2009). Puzzlingly high correlations in fMRI studies of emotion, personality, and social cognition. *Perspectives on Psychological Science, 4*, 274–290.

Weisberg D. S., Keil, F. C., Goodstein, J., Rawson, E., & Gray, J. R. (2008). The seductive allure of neuroscience explanations. *Journal of Cognitive Neuroscience, 20*(3), 470–477.

Wartenburg, T. (1990). *The forms of power*. Philadelphia: Temple University Press.

Watters, E. (2010). *Crazy like us: The globalization of the American psyche*. New York: Free Press.

Weaver, I. C., Cervoni, N., Champagne, F. A., D'Alessio, A. C., Sharma, S., Seckl, J. R., & Meaney, M. J. (2004). Epigenetic programming by maternal behavior. *Nature Neuroscience, 7*, 847–854.

Wexler, B. E. (2006). *Brain and culture: Neurobiology, ideology, and social change*. Cambridge, MA: MIT Press.

Wise, N. M. (2006). Thoughts on the politicization through commercialization of science. *Social Research, 73*, 1253–1272.

Young, A. (1995). *The harmony of illusions: Inventing post-traumatic-stress-disorder*. Princeton, NJ: Princeton University Press.

Zahavi, D. (2004). Phenomenology and the project of naturalization. *Phenomenology and the Cognitive Sciences, 3*, 331–47.

Zhu, Y., Zhang, L., Fan, J., & Han, S. (2007). Neural basis of cultural influence on self-representation. *Neuroimage, 43*, 1310–1316.

2

The Need for a Critical Neuroscience
From Neuroideology to Neurotechnology

Steven Rose

The Rise of Neuroscience

Neuroscience is one of the hottest fields of research within the life sciences, and its theoretical claims, research findings and technological prospects have implications that extend far beyond the internal debates within the discipline. The scope of my discipline's claims impinges on psychology, philosophy, and public social policy. Our science—or more accurately our technoscience—has an annual budget that runs into hundreds of millions of dollars, provided by State funding agencies like NIH (National Institutes of Health), charities like Wellcome, biotech companies, Big Pharma, and, of course, the military. On the back of such funding, we are offering not just to explain the human mind and its elusive properties, from memory to consciousness, but also to provide technologies to cure brain and mind diseases and enhance human happiness; indeed to use these technologies to control and manipulate the mind.

It is precisely for this reason that neuroscience has become too important to be left to the neuroscientists. Whilst "critical theory" has a specific meaning within the social sciences and humanities as deriving from the work of the Frankfurt School, as a neuroscientist I use it here in a more general sense as emphasizing the need for a critical examination of the philosophical, ideological, and methodological underpinnings both of the neurosciences as currently practiced, and of the emergent neurotechnologies. In this chapter I will discuss what we as neuroscientists know about the brain, what we might know, and what I will maintain we can't know. But as the neurotechnologies proceed apace, irrespective of the science, I will also refer briefly to some of the prospects and perils of the new neurotechnologies (Rees & Rose, 2004).

Neuroscience itself is an uneasy alliance of many subdisciplines: neurogenetics, neuroanatomy, molecular neurobiology, neurophysiology, neuropharmacology, neuro-imaging, and cognitive neuroscience, to name but a few. Half a century ago, passionate

Critical Neuroscience: A Handbook of the Social and Cultural Contexts of Neuroscience, First Edition.
Edited by Suparna Choudhury and Jan Slaby.

to study the brain, I began my graduate research in a gloomy red brick building in southeast London—the Maudsley Institute of Psychiatry. There, in the biochemistry department I was rapidly disabused of any idea that my research might lead to a greater understanding of how the brain could be "the organ of mind"—and still less that it might provide any help for the situation of the hospital's patients who I could dimly see through my laboratory windows. Neurochemistry meant grinding rats' brains up and extracting their enzymes; neuroanatomy was about cutting thin slices and staining them to be viewed under the microscope; neurophysiology was sticking minute electrodes into nerve cells and checking their electrical responses.

To articulate the thought that this might tell one anything about so-called "higher nervous functions" was strictly out of bounds. We were no further forward than when the great neurophysiologist Charles Sherrington, in his 1937 Gifford lectures, discussed "Man on his Nature". Despite his marvelous metaphor of neural activity within the brain as an "enchanted loom" he was clear that although science could describe how information entered the brain from the sense organs, and how it left it down motor nerves, what went on within the mass of cells in the cortex to generate mind or consciousness was a mystery that physiology could not penetrate. All that was clear to him was that in some way brain acted "in collaboration" with psyche. Even when, as a post-doc in the 1960s, I was captivated by the idea of researching something less grandiose than the totality of brain–mind relationships but merely the biochemical processes occurring when animals learn and remember—which has been my subsequent life's work—I was warned off by my two successive Nobel professors, Hans Krebs and Ernst Chain, as it was deemed a project not suitable for a respectable neurochemist to work on. Even a dozen or so years ago I heard a young American physiologist describe the study of consciousness as a "CLM"—a career limiting move. No topic for a young and ambitious neuroscientist, best left for those old enough to be experiencing the "philosopause" said to affect scientists who had run out of research steam.

The Limits to Reductionism

How times have changed! What was once dangerous territory is now the hottest theme in brain research. Two technologies above all have come to symbolize today's neuroscience: genetics, with its capacity to "construct" laboratory species, from Caenorhabditis elegans and Drosophila to mice with specific genes added or deleted; and brain imaging, which provides the false color images of the brain that grace most popular books and articles on the brain. Both encourage reductionist thinking, and both have pitfalls that reveal the limitation of such thinking. Thus the belief that genes "for" particular bioprocesses can be slotted in and out at will is based on the implicit assumption that the genes are unitary and independent. The very existence of an entity such as "a gene" has had to be greatly revised in the light of modern molecular biology. At the least, how a gene is expressed is contingent to varying degrees on all the other genes in the genome with which it and its gene products interact, and also on the plasticity and self-organizing capacity of living systems during development. Take out a gene coding for a protein believed to be essential for some physiological function and frequently, to the researcher's surprise, the animal shows "no phenotype"—that is, no observable physiological response. The system has simply reorganized itself in compensation. Alternatively, a wide range of body processes may be affected, indicating that the gene product has major pleiotropic functions. Newer technologies (time-and site-dependent temporary knockdowns) may overcome such limitations, but are subject to other caveats. Indeed one lesson from the gene knock-out and knock-in studies which has also caused some consternation is how dependent their effects are on the genetic background of the mice which have been manipulated and the specific conditions and experience of the laboratories in which they have been bred.

Other caveats apply to the interpretation of data from brain imaging systems. Conceptually and technically, "mapping the brain" is a task that is orders of magnitude harder than sequencing the genome. The genome is a linear and stable sequence, the brain a dynamic structure organized in three dimensions of space and one of time. However, the power of informatics is making possible a human brain project modeled on the human genome project, though more informally organized. The idea is to produce a brain-gene map, in which all the genes expressed in the brain are localized, and from which the mind can be read off. How such a map may change our concept of how the brain works is, however, another matter. Identifying "sites" or "genes" "for" particular brain processes or mental attributes ignores both the complexity and dynamism of the brain.

It is the advent of brain imaging, coupled with informatics, which has technically driven such proposals. Placing subjects into a functional magnetic resonance imager (fMRI) and asking them to think of God or contemplate moral dilemmas identifies regions of the brain that show increased blood flow compared with those in the control group. In such studies, blood flow is taken as a surrogate measure for neural activity. Another technique, magnetoencephalography, which measures the fluctuating transient magnetic fields around the head, offers a millisecond by millisecond record of the brain's activity during such thought processes. Reciprocally, focusing an intense magnetic beam through the skull onto specific brain regions can influence thoughts and emotions. The mathematical manipulations which lead to the identification of these brain regions are disguised by the dramatic false-color representations describing the latest aspect of human nature to be thus given a specific site within the brain.

I do not at all wish to diminish the insights into brain processes that neuroimaging can provide, nor its clinical utility. But the dramatic images may hide as much as they reveal. At best they provide a correlative indication of those regions of the brain that are active when the brain's owner is engaged in some mental activity; they do not mean that these regions are therefore the "sites" of such mental activity. A recent re-evaluation of some well-publicized claims to have identified specific brain sites refers to them as "voodoo correlations"(Vul, Harris, Winkielman, & Pashler, 2009).[1]

What unites the various disciplines that address the brain is a common interest in the structure and functioning of the 1.5 kilos of tissue inside the human skull, or their

---

[1] This was the title of the paper as circulated on the web prior to publication. It generated an intense web-based controversy and several efforts at rebuttal. The final title of the paper was modified to be less inflammatory: "Puzzlingly high correlations in fMRI studies of emotion, personality, and social cognition."

non-human animal or machine analogues. However, and despite such technological advances, the neurosciences lack a unifying "brain theory" within which their diverse findings could be assimilated and integrated. The 30,000 or so neuroscientists who meet annually at the US Society for Neuroscience gatherings predominantly talk past one another.

To take an example from my own field; learning and memory is researched by neuroscientists ranging from cognitive psychologists to molecular geneticists, but it is hard to recognize that they are studying the same topic (Rose, 2003). The very terms learning and memory, used across the range of disciplines, do not necessarily refer to the same phenomena or processes. Molecular biologists study animal models of memory that typically involve training rats or mice in well-controlled learning paradigms and investigating the effects of manipulating biochemical variables on either learning or recall. They speak of short- and long-term memory phases lasting minutes to hours, periods of "learning," "consolidation," and even "reconsolidation" each associated with specific biochemical processes and brain regions. Yet one of the many paradoxes of brain processes is that those brain regions that seem necessary for learning to occur, and in which one can document quite precise molecular and structural changes in response to the animal's novel experience, seem however not to be required for the subsequent expression of that memory in terms of a learned response. Memory does not "reside" in any specific brain region, though many are required for its expression—a localization problem that has dogged neuroscience ever since the days of phrenology.

Cognitive neuroscientists approach the study of learning and memory very differently. Absent are the phases of consolidation. They speak instead of working memory, evoked within seconds, and reference memory. Memory requires the dynamic reactivation of many brain regions—including even primary sensory ones—when individuals are called upon to remember. Books written by molecular biologists on memory rarely reference those written by psychologists, and the compliment—or lack thereof—is routinely returned. So if we can't even agree on what we mean when we say we are researching memory, or how it should be studied, what hope is there for a unified theory?

This lack of unifying theory is partially disguised by a largely shared commitment to what one of its exponents has called a "ruthless reductionism" (Bickle, 2006)—the type of reductionism sometimes referred to as physicalist, and characterized long ago by Marx and Engels as "mechanical." In such a reductionism, mind is an epiphenomenal product of brain, much as it was to the nineteenth-century mechanical materialists like Moleschott ("the brain secretes thought like the kidney secretes urine; genius is a matter of phosphorus") and Huxley ("mind is to brain as the whistle to the steam train"). But what were then merely provocations have become full-fledged research programs in the twenty-first century. In all cases, mind reduces to brain, as in the title of a recently formed "Society for cellular and molecular cognition."

On Consciousness

Recall that while for NIH the 1990s were the "decade of the brain," for many neuroscientists we are currently in the "decade of the mind." This programmatic agenda has been articulated by the new neurophilosophers, notably in the US by Dennett and the Churchlands, with their robust dismissal of mind language as mere folk psychology to be replaced by the rigors of computational neuroscience (Churchland, 2002; Dennett, 1991), a project shared by many leading neuroscientists. More reflective neuroscientists love to quote Emily Dickinson's 1862 poem "The brain is wider than the sky."

> The brain - is wider than the sky—
> For—put them side by side—
> The one the other will contain
> With ease—and You—beside—
> The Brain is deeper than the sea—
> For—hold them—Blue to Blue—
> The one the other will absorb—
> As Sponges—Buckets—do
> The Brain is just the weight of God—
> For—Heft them—Pound for Pound—
> And they will differ—if they do—
> As Syllable from Sound—

Consciousness theorist Gerald Edelman (2004) employs the poem as a frontispiece to one of his books before asserting "you are your brain (plus free will);" neurobiologist Eric Kandel comfortably agrees (Kandel, 2006). For neurophysiologist Semir Zeki, it is the brain rather than the mind which has "knowledge" and "acquires concepts" (Zeki, 2009). Zeki has even imaged "romantic love," seeing human characteristics with brain correlates as a universal outside society and culture. Larry Young extends Zeki's brain localization of romantic love and reprises Moleschott when, in a recent essay in *Nature*, he argues that human love (by analogy with the mating practices of voles) depends on a polymorphism in the *AVPR1A* gene (Young, 2009). Francis Crick is in robust Alice in Wonderland mode: "You're nothing but a bunch of neurons" before going on to speculate that "free will" is located in the anterior cingulate gyrus (Crick, 1994). Gall and Lombroso redux.

The problem with this reductionism is to equate a part with a whole—an error I was fully guilty of when many years ago I wrote a book incautiously called *The Conscious Brain*. But it simply won't do. For sure, the brain is "the organ of mind"—always bearing in mind (!) that brains are in bodies, which have their physiological role to play. There are, it is chastening to note, as many nerve cells in the gut as there are in the brain. However, it is not brains that have concepts or acquire knowledge or have "free will," it is people, using their brains. To paraphrase the anthropologist Tim Ingold (2000), I need legs to walk, but I don't say "My legs are walking." Similarly, I need my brain to think, but it is I, not my brain, which does the thinking. Indeed Zeki gives the game away when he quotes Kant as saying "The Mind does not derive its laws … from nature but prescribes them to her" and goes on to say "he might as well have been writing about the brain." No, indeed; the mind may need the brain, but it is not reducible to it, and we neuroscientists need to recognize our limitations.

Of course, such reductionism is not confined to my trade, but it is currently rampant amongst neuroscientists. The linguistic reduction of minds to brains, the attribution of a higher order phenomenon to a lower order, albeit necessary, component of the phenomenon is of course not unusual. Genes—or even DNA sequences—are "selfish," molecules have cognition, brains are conscious. Brain banks are still regularly offered "psychopathic brains" and in the psychiatric hospital where I did my PhD people spoke of "schizophrenic urine" meaning, of course, urine samples provided by people diagnosed as suffering from schizophrenia.

When challenged, the users of such phrases will of course readily insist that they are only using a convenient linguistic shorthand, but such shorthand is not innocent; they help shape our thinking. And it would be a mistake to assume that reductionism in all its forms is to be adjured. For most laboratory workers, a methodological reductionism is the only way to design a successful experimental strategy. Holding parameters constant whilst manipulating variables singly is a tried and tested procedure. Isolating individual aspects or components of a phenomenon is a surer way to get to Stockholm than emphasizing the blooming, buzzing confusion of the real world within which components, variables, and processes are embedded. As I have written about at length elsewhere (Rose, 2005), reductionism becomes a problem when it is seen as the only game in town or when a methodological approach tips over into a full-blown philosophical commitment—as in the quotes from Crick and others earlier—and above all when it becomes ideological and impinges on medical or public policy.

The neuroscientific reach into the mind has by now gone beyond even love and religious experience to approach what many consider humans' most enigmatic attribute, that of consciousness itself. Consciousness studies no longer inhabit a borderland between the speculations of theoretical physicists and New Age "mysterians." These days, ambitious young neuroscientists employ all the armory that brain imaging and computer simulation can provide, even while the still proliferating books on mind and consciousness are mainly written by their seniors. One consequence has been that where in the past philosophers of mind pondered the problems of qualia and first versus third person experience without feeling the need to relate them to findings from the neurosciences, this is no longer adequate. Philosophers—at least in the US—are beginning to enter the labs to observe the scientists at work. But the confidence, even hubris, of neuroscientists that their accounts of brain functioning will explain the mind can indicate a failure by the neuroscientist to understand what the philosopher is saying, as in the case of the public debate between the neurochemist Jean-Pierre Changeux and the hermeneutic philosopher Paul Ricoeur (Changeux & Ricoeur, 2000).

The truth is that in order to approach consciousness as a neuroscientist, one first has to strip the term of any of its richer meanings. It isn't just Freudian consciousness with its contrasting subconscious that goes, but also Marxian class consciousness, feminist consciousness, and race consciousness. As feminist sociologist of science Hilary Rose has pointed out (Rose, 1999), consciousness in this neuroscientific sense has been taken out of history and culture; there is no possibility of understanding the extraordinary transitions in consciousness that have occurred through, for instance, the emergence of the women's movement in the 1970s. Instead, consciousness is simply what happens when you are awake, the obverse of being asleep. It is no more than mildly ironic that one of the early figures in the emergence of modern consciousness studies was an anaesthetist, Stuart Hameroff. Consciousness is a "dimmer switch" (Susan Greenfield); it reduces to mere "awareness." As awareness is akin to perception and perception can be studied via the visual system, consciousness modelers like Francis

Crick and Christof Koch are up and away (Crick, 1994; Koch, 2007). But the essential human meanings embedded in our being conscious have somehow been lost in this reduction.

This is why, despite the confidence with which my colleagues cut the Gordian knots tied through centuries of philosophical debate, I would want to argue, without disrespect to Emily Dickinson, that she is wrong, that the mind is wider than the brain. Until neuroscience can respond to the meditations of St Augustine, 1600 years ago, when he wonders how the brain/mind can encompass vast regions of space and time, past, present, and future abstract thoughts and numbers, logical propositions, false arguments, and the idea of God, then we need to show a little more humility. A critical theory must thus examine the metaphysical and ideological context in which such reductionism has become the dominant mode within neuroscience. Its roots can be traced back beyond the Cartesian birth of Western science through classical debates between materialist/dualist/neutral monist accounts of mind, from Aristotle to Spinoza. I argue that whilst methodological reductionism is an essential experimental tool for the natural sciences, including biology, it is inadequate and flawed when its explanatory power is over-extended as in the quotations of the preceding paragraphs. Instead, it is necessary to contest such reductionism with a more integrative understanding of embodied brains and embedded bodies, to provide an evolutionary and developmental perspective firmly located within a biosocial framework.

The Autopoietic View

What might this mean in practice? We biologists enjoy quoting the evolutionary geneticist Theodosius Dobzhansky's memorable assertion "Nothing in biology makes sense except in the light of evolution." That is surely true, but insufficient. That the human mind and human nature have been shaped by evolutionary pressures is of course not in question. Humans are long-lived social animals whose offspring are born neotenous, requiring several years of care-giving before they can live independently. These parameters must play a central part in the formation of the human mind. Living in groups requires learning social skills—adjusting an individual's ways of being and thinking to the needs of others—a theme currently being actively explored by a variety of researchers. A new field, "social neuroscience" is emerging, stimulated by the discovery by neurophysiologists of so-called "mirror neurons" that are active both when the individual performs a particular act or watches others doing the same— allegedly the neural base for empathy. Empathy—or at least mirror neurons—are present in humans' nearest evolutionary neighbors. The social nature of human existence must also have driven the evolution of mind and consciousness. Thus, evolution has ceased to be seen as an entirely biological process, and many now speak of the emergence of modern humans as a co-evolutionary process, involving both biology and culture.[2] Such an argument insists on the inseparability of human biology from human culture, not as a matter of arbitrary partitioning of such-and-such a

---

[2] See for instance *A Mind so Rare: The evolution of human consciousness* by Merlin Donald (2001).

percent genes and such-and-such environment, but of the continual interplay between both during development. Humans are biosocial creatures.

However, the possibility of an empirically based evolutionary psychology has been sullied by its hijacking by a group of self-proclaimed evolutionary psychologists, the more recent avatars of 1970s sociobiology. Evolutionary psychology (EP) bases itself not just on the assumption that human nature is an evolved property, but on the profoundly un-Darwinian assertion that this—by contrast with the rest of nature—was fixed in the Pleistocene period and there has not been enough evolutionary time for human nature to change subsequently. Thus, it is not just that the demands of social living may have been one factor in the evolution of morality, but humanity is, according to the evolutionary psychologist Marc Hauser endowed with a universal set of moral principles, independent of culture or social context (Hauser, 2006). Also prominent amongst these apparently fixed human characteristics are the expression of so-called basic emotions (Ekman, 2003), racial preferences, and gender relations. Male preferences for mating with younger women of defined body shape, and female for richer, older, more powerful males, do little other than repeat in contemporary language Darwin's own assertions in *The Descent*. EP has been subject to severe criticism. Scholars across the disciplines, through the humanities to social and life scientists, have challenged its theoretical base and empirical adequacy; I won't expand further here (see Rose & Rose, 2000).

To EP theorists, the human mind is "massively modular," consisting of a large number of semiautonomous, innate components (Barkow, Cosmides, & Tooby (1992) liken it to a Swiss Army knife). However, not only is this claim disputed by those who argue that the mind's specificities are formed during development through interaction with the infant's social environment (Karmiloff-Smith, 2000), but brain imaging studies also find no evidence for such mental modularity. The complexity of the brain, with its 100 billion nerve cells, and 100 trillion internal connections, still defies comprehension. Twenty-two thousand genes cannot begin to specify in any more than generalities the pattern of these connections, which are shaped by the activities of the developing child.

For this reason we cannot understand the adult organism unless we add to Dobzhansky's dictum the concept of development. Humans are, as everyone now knows, 98.45% genetically identical to chimpanzees—a difference which implies no more than some 70,000 adaptively significant amino acid differences between the two species—a tiny number when one recalls that the 100,000 or so different proteins present in the body comprise at least 10 million different sequences. Yet no one would mistake a human for a chimp phenotype. The differences arise through the regulatory sequences that control the expression of genes during development, a process profoundly affected by the dynamic environmental context in which humans, above all, develop. The 22,000 genes in the human genome have to be transcribed and edited in multiple ways to generate those 100,000 proteins, to say nothing of the combinatorial explosion required to generate the 100 billion neurons in the cortex with their 100 trillion synaptic connections, continuously being made and remade not only during development, but in adult life as well. Furthermore, organisms—and here I speak of all living organisms, not just humans—are not simply the passive expression of the interplay between genes and environment during development, but are constantly acting on, choosing, and transforming those environments. In that sense, organisms are actors on the world—for humans perhaps one might call this agency.

It is this process that Humberto Maturana and Francisco Varela called autopoiesis (Maturana & Varela, 1992) and Susan Oyama, Paul Griffiths, and Russell Gray, Developmental Systems Theory (2003). In this view the living world consists of processes not things; organisms are constantly both *being* one thing and *becoming* another, as when a baby, born with a suckling reflex, becomes transformed over months from a suckler to a chewer, involving quite different muscles and nerves. The dichotomies between genes and environment are replaced by those between specificity—the persistence of form, of memory, despite the dynamic turnover of all body constituents—and plasticity, the capacity of the organism to adapt its own structure, physiology, and chemistry to short- or long-term changes in its environment. This dynamism is important; one of the most common but misleading terms in the biology student's lexicon is homeostasis—Shannon's term for Claude Bernard's concept of the stability of the body's internal environment. But such stability is achieved by dynamic responses; stasis is death, and homeodynamics needs to replace homeostasis as the relevant concept.

What I have hinted at so briefly in the preceding paragraph is a description of living processes that applies universally, even should life be discovered on Mars. But who we are as humans today, and how our brains and minds are constituted, is inseparably a product both of evolution and development and the culture and history within which we are embedded—that is, we are inexorably biosocial organisms. The brains and minds of twenty-first century people differ not just from those of our Pleistocene ancestors, but even from those of our great-grandparents. We are wired differently, and a neuroscience that fails to take this into account fails at the first hurdle in its attempts, for instance, to reduce consciousness to mere neural activity.

The Cerebroscope

So let me try a thought experiment, and tackle what philosopher David Chalmers calls "the hard problem," the division between objective, third person, and subjective, first person, knowledge, and experience (Chalmers, 1996). Chalmers, Colin McGinn, and others worry about qualia—that is, those aspects of subjective experience such as the sensation of seeing red. How, they ask, following earlier generations of philosophers, can brain stuff—the neural firing or whatever—which an outsider can measure "objectively," generate such a first-person, subjective experience? Experiencing the redness of red seems to belong to an entirely different universe—or at least an entirely different language system—to statements about neural firing.

It may be because I am philosophically tone-deaf, but I have never found this a very troubling question. It is surely clear that, granted enough knowledge of the visual system, we can in principle and to some extent in practice identify those neurons which become active when "red" is perceived. (Indeed in animal experiments such neurons have already been identified.) This pattern of neural activity translates into the seeing of red, and seeing red is simply what we call in mind language the phenomenon that we call in brain language the activity of a particular ensemble of neurons. This doesn't seem harder to understand than the fact that we call a particular small four-legged furry mammal "cat" in English and "gatto" in Italian; the two terms refer to the same object in different and coherent, but mutually translatable, language. No problem. Does another person experience red when these neurons fire in exactly the same way as I do? Probably not identically, because of the unique wiring of every individual's brain, and in any case it is an unanswerable and not very interesting question. However, assuming we are brought up in the same cultural context, we both agree on what we mean by red, and we must be satisfied with that.

But can we go further? Let us imagine that we have all the techniques and information-processing power that neuroscientists can dream of, and consider a hypothetical machine—let's call it a cerebroscope (a term I believe was invented many years ago by the information scientist—and explicitly Christian anti-determinist—Donald Mackay (1982))—that can report the activities at any one time of all the 100 billion neurons in the brain. Define this activity at any level—molecular, cellular, systems—that seems appropriate. Now consider a trivial day-to-day experience; I am standing at a bus-stop and I see a familiar (at least to Londoners) red bus coming towards me that we call subjectively "seeing a red bus coming towards me."

The cerebroscope will record and integrate the activity of many neurons in the visual cortex, those that are wavelength sensitive and report red, those that are motion sensitive that report directional movement, edge detecting neurons, neurons tuned to binocularity, all of which combine, via some solution to the binding problem, to create an image of an object of a given shape and volume, moving towards me with a speed I can estimate from the rate of change the image subtends on my retinae. Acoustic information is also bound in so I can register the engine noise of the approaching bus. But hold on—how do I know that the noise is that of an engine, or the object a bus? There must be bound in with all the sensory information some other neural activity which scans and extracts the recognition memory which defines the object as a bus, and the noise as that of an engine. Perhaps the memory involves inferior temporal cortex, and the naming of the bus will engage Broca's area as we identify it as a "bus".

But let's go one step further. Is seeing this bus a good thing or a bad thing? If I am on the pavement waiting for it, a good thing; if I am crossing the road and it is coming at me fast, a dangerous thing. There is affect associated with these images. Amygdala and ventromedial frontal cortex are involved. Then I must decide how to act—do I prepare to enter, or jump out of the way—perhaps right parietal cortex and frontal lobe engagement? The appropriate muscles must be engaged, blood circulation adjusted, and so forth. The cerebroscope will enable an observer to record all this activity over the few seconds during which I am perceiving the bus and acting on my perception, and such an observer is entitled to say that the sum total of this activity represents, in brain language, my mental processes of seeing and evaluating the bus. So, once more, what's the problem?

Consider the process I have just described from the other end. Suppose the cerebroscope stores all this information in its gigaterabyte information processor. Then at some later time, an experimenter asks the machine to present the data and translate it back into mind language, that is, to deduce from the neural activity the thought and action processes that it represents. Could it interpret all the data and print out a statement saying "what the person, Steven Rose, associated with this brain, is experiencing is a red bus coming towards him and that he is in danger of being run over by it?"

The answer seems to me to almost certainly be no. The interpretation of the firing pattern of any particular neuron is very much dependent on its history. Plasticity during development may mean that even the wavelength to which any particular neuron is sensitive may vary from individual to individual, so what ends up as one person's "red" neuron may in another person's brain respond to blue and not to red. Even more sure is that whatever the pattern of neural firing and connectivity in my inferotemporal cortex that corresponds to *my* recall or recognition memory of a bus, it will not be the same as the pattern in *yours*, even though the outcome—recognizing a bus—will be the same in both cases. This is because your experience and my experience of buses, and how we each store that experience are inevitably both different and unique to each of us. So for the cerebroscope to be able to interpret a particular pattern of neural activity as representing my experience of seeing the red bus, it needs more than to be able to record the activity of all those neurons at this present moment, over the few seconds of recognition and action. It needs to have been coupled up to my brain and body from conception—or at least from birth, so as to be able to record my entire neural and hormonal life history. Then, and only then, might it be possible for it to decode the neural information.

The hypothetical cerebroscope could only do so, however, if there were a one-to-one relationship between the history and the present state of my neurons and my mental activity. And this we simply do not know. There may be an indefinite number of histories of neurons from conception to the present which could be interpreted as meaning the experiencing of a red bus coming towards me—and equally there might be an infinite number of experiences that could be inferred from any particular pattern. Even more fundamentally though, the embodied brain and embedded body mean that our consciousness of—the meaning of—our experiences are not entirely brain properties. (Leave aside for now the relatively trivial point that how we experience depends not only on neural firing but on our hormonal state and general physiology.) Consider that we, as humans, are embedded in a mesh of history, society, and culture. The meaning of any experience is then not "in the brain" but in a mind which is an open system, depending to be sure on the brain, but not isolable within it. That is, the mind is wider than the brain. This is not a dualist position, but a rejection of the philosophy of a mechanical materialism that constantly seeks to reduce higher order phenomena to lower ones.

As a Christian and troubled by the problem of free-will and determinism, Mackay was interested in a further question. Could the information obtained by his hypothetical cerebroscope enable an observer to predict what the observed person would do next? He wondered what would happen if the cerebroscope was constructed so as to continually report back to a person the state of his or her own brain—and therefore mind—so as to predict that person's future thoughts and actions. But, as he pointed out in an argument against determinism and in favor of some form of "free will," this raises a paradox, for the act of reporting back to a person will change the state of his or her brain in unpredictable ways, and hence the predicted outcome would itself be modified. (In a hugely simplified manner this is of course just what the biofeedback devices, which are supposed to help people reduce stress or learn to meditate, can do.)

Thus, Mackay concluded, even if we knew everything we needed to know subjectively or mentally about the "objective" state of our brain at any one time, our actions would not, therefore, be determined. I don't really think that this provides the solution to the "free will paradox," but it does explore the limits to how any understanding of the brain might help us to understand the mind.

On Neurotechnology

However theoretically inadequate neuroreductionism may be, it has and will continue to be used to provide an ideological justification for social theories with major policy implications. It is not my intention to belittle the advances in diagnostics and treatment that neuroscience is making possible, from new approaches to Alzheimer's disease to prosthetics for damaged sensory and motor systems (though in many cases the practice lags well behind the hype, and the results are often no better than earlier technologies) especially in the available drug treatments for conditions like depression. Other contributors to this book discuss these issues more fully and I have also written more about these issues elsewhere (Rose, 2006). Other examples include IQ theory with its accompanying claims of genetically determined group differences in intelligence between races, classes, and genders (see Rose, 2009). The essential point is, that claims as to the neurogenetic base of such DSM-IV categories as attention deficit hyperactivity disorder, oppositional defiance disorder, antisocial behavior, and others serve first to reify, then to locate and fix, socially defined forms of undesirable thoughts and behaviors in a causally directional manner within the individual, within his or her brain and genes, rather than in a relationship between the individual and their economic, social, and cultural environment.

Reductionist neuroscience first locates a "problem" and then offers to generate powerful neurotechnologies, framed within this reductionist theoretical framework, to fix it. Gene scanning to detect potentially "antisocial" genes such as predispositions to drug abuse, pharmacological agents used for social control purposes, such as Ritalin for ADHD, brain imaging to detect potential psychopathic behavior, or "terrorist thought patterns" are gaining credence or are a prospect for the immediate future. Brain imaging is already admissible in court in India and even in the US in at least one well publicized case, in spite of the fact that the interpretation of its images is often flawed.[3] Further down the road are the various DARPA-funded projects to direct and alter thought processes and behaviors by focused magnetic pulses (transcranial brain stimulation).

So What should Critical Neuroscience Do?

So what should a critical neuroscience—in my sense of the word—be doing? First, it needs to analyze and make transparent the metatheoretical and ideological under-pinning of the current neuroscientific enterprise. Second, it needs to scrupulously

---

[3]   See Editorial in *Nature Neuroscience, 11,* 1231, (2008) doi :10.1038/nn1108–1231.

unpick the empirical claims made by neuroscientists offering to "explain" memory, intelligence, love, or consciousness and to "locate" them in specific brain sites, neuronal ensembles, or molecular processes. Third—and this is the harder part—it needs to offer a credible alternative to the "ruthless reductionism" that dominates neuroscientific thought and practice, without collapsing into what Richard Dawkins once memorably referred to as a "holistier than thou" rejection of what reductionism has to offer. Fourth, it needs to work to help integrate neuroscientific understandings into the many rich and varied discourses on human thought and action. Fifth, it needs to keep a very wary eye on the developing neurotechnologies with their power to intrude and intervene in the fundamental processes and freedoms of civil society. And finally, it needs to do all these without simply becoming professionalized into a new academic discipline, speaking only to itself without engaging either working neuroscientists, or, more importantly, the wider civil society. If it can achieve these goals, it will truly justify its self-designation as "critical."

Acknowledgements

This chapter draws in part on material from my book *The 21st Century Brain: Explaining, amending and manipulating the mind*, (2006) and on a more recent text by Hilary Rose and myself (2009): The changing face of human nature, *Daedalus, 136*, (3).

References

Barkow, J., Cosmides, L., & Tooby, J. (Eds.). (1992). *The adapted mind: Evolutionary psychology and the generation of culture*. Oxford: Oxford University Press.

Bickle, J. (2006). Ruthless reductionism in recent neuroscience. *Systems, Man, and Cybernetics, Part C: Applications and Reviews, IEEE Transactions, 36*, 134–140.

Chalmers, D. (1996). *The conscious mind: In search of a fundamental theory*. Oxford: Oxford University Press.

Changeux, J-P., & Ricoeur, P. (2000). *What makes us think?* Princeton, NJ: Princeton University Press.

Churchland, P. M. (2002). *Brain-wise: Studies in neurophilosophy*. Cambridge, MA: Bradford Books, MIT Press.

Crick, F. (1994). *The astonishing hypothesis: The scientific search for the soul*. New York: Simon and Schuster.

Dennett, D. (1991). *Consciousness explained*. New York: Little, Brown.

Donald, M. (2001). *A mind so rare: The evolution of human consciousness*. New York: Norton.

Edelman, G. M. (2004). *Wider than the sky: The phenomenal gift of consciousness*. London: Allen Lane.

Ekman, P. (2003). *Emotions revealed: Understanding faces and feelings*. New York: Henry Holt.

Hauser, M. D. (2006). *Moral minds: How nature designed our universal sense of right and wrong*. London: HarperCollins.

Ingold, T. (2000). Evolving skills. In H. Rose & S. Rose (Eds.) *Alas poor Darwin: Arguments against evolutionary psychology* (pp. 225–246). London: Cape.

Kandel, E. (2006). *In search of memory: The emergence of a new science of mind.* New York: Norton.

Karmiloff-Smith, A. (2000). Why babies' brains are not Swiss army knives. In H. Rose & S. Rose (Eds.) *Alas poor Darwin: Arguments against evolutionary psychology* (pp. 144–156). London: Cape.

Koch, C. (2007). *The quest for consciousness: A neurobiological approach.* New York: Roberts.

Mackay, D. (1982). Our selves and our brains: Duality without dualism. *Psychoneuroendocrinology, 7,* 285–294.

Maturana, H., & Varela, F. (1992). *The tree of knowledge: The biological roots of human understanding.* Boston: Shambhala.

Oyama, S., Griffiths, P., & Gray, R. (2003). *Cycles of contingency: Developmental systems and evolution.* Cambridge, MA: MIT Press.

Rees, D., & Rose, S. (Eds.) (2004). *The new brain sciences: Perils and prospects.* Cambridge: Cambridge University Press.

Rose, H. (1999). Changing constructions of consciousness. *Journal of Consciousness Studies, 6,* 251–258.

Rose, H., & Rose, S. (2009). The changing face of human nature. *Daedalus, 138,* (3)11.

Rose, H., & Rose, S. (Eds.) (2000). *Alas poor Darwin: Arguments against evolutionary psychology.* London: Cape.

Rose, S. (2009). NO: Science and society do not benefit. *Nature, 457,* 786–788.

Rose, S. (2006). *The 21st century brain: Explaining, mending and manipulating the mind.* London: Vintage.

Rose, S. (2005). *Lifelines: Biology beyond determinism.* London, Vintage.

Rose, S. (2003). *The making of memory: From molecules to mind* (2nd ed.). London: Vintage.

Vul, E., Harris, C., Winkielman, P., & Pashler, H. (2009). Puzzlingly high correlations in fMRI studies of emotion, personality, and social cognition. *Perspectives on Psychological Science, 4*(3), 319–324.

Young, L. (2009). Being human: Love: Neuroscience reveals all. *Nature, 457,* 148.

Zeki, S. (2009). Splendors and miseries of the brain: Love, creativity and the quest for human happiness. Oxford: Wiley-Blackwell.

3

Against First Nature
Critical Theory and Neuroscience

Martin Hartmann

In the following reflections my basic intention is to focus on the degree to which the specific conceptual apparatus developed by various authors in the tradition of critical theory helps to articulate a critical stance towards some of the methodologies, procedures, and practices of the present-day neurosciences. To this, I should add several warnings. To begin with, I do not pretend to offer a general critique of the neurosciences, since critical theory, in the narrower sense this term has acquired in the context of the so-called Frankfurt School, has concentrated only on specific aspects of the sciences criticized while completely omitting or ignoring others. Critical theory, therefore, has not so much developed a critique of various scientific *practices*, as of the theoretical and methodological models fueling these practices.

It is an easily overlooked fact that critical theory, certainly at its inception, put as much weight on its being critical as on its being an explicatory *theory* in opposition to other scientific theories. Consequently, what was at the core of its enterprise was a more or less close analysis of the methodological self-understandings of the sciences variously called positivistic, scientistic, or, more recently, naturalistic. These self-understandings, to be sure, were seen as linked to specific practical efforts or interests, a fact ignored by some sciences and highlighted by others (namely by critical theory itself). Nevertheless, what is truly at stake in critical theory is what one might call the rationality of the scientific stances analyzed, a rationality that includes the way the sciences themselves conceptualize the distinction between theory and practice.

In its programmatic writings, early critical theory should to a large extent be seen as a criticism of certain self-understandings of science, a point of great importance since the clarification of its intellectual and practical sources impinges on the question of the very possibility of a critical theory. "Critical theories," says Raymond Geuss in his *The Idea of a Critical Theory*, "have cognitive content, i.e. they are forms of knowledge" (1981, p. 2). In this, however, these theories are not only rivals of competing claims to knowledge, but attempt to develop a specific kind of knowledge

Critical Neuroscience: A Handbook of the Social and Cultural Contexts of Neuroscience, First Edition.
Edited by Suparna Choudhury and Jan Slaby.
© 2012 John Wiley & Sons, Ltd. Published 2016 by John Wiley & Sons, Ltd.

that helps to free agents from socially-induced forms of coercion and also helps to understand why competing claims to knowledge cannot adopt this same emancipatory role—or even participate in the repressive mechanisms being incriminated. It is part of the Marxist and Hegelian heritage of critical theory to insist on the practical relevance of theory and to be interested in the causal mechanisms that prevent agents from realizing a given potential for emancipation. In addition, it is more or less agreed that the basic distortions of rationality must be ascribed to the effects of a capitalist economy that pervades all areas of life, including the realm of science and scientific methodology, yet somehow manages, in Honneth's words, to remove "from recognition those social conditions through which this [capitalist] system is at the same time structurally produced" (Honneth, 2008, p. 796).

Another point that needs to be mentioned is, that there is as yet no critical theory of the neurosciences. Given the aforementioned elements of a critical theory of the sciences even in their narrow form, this cannot come as a surprise. A critical theory of the neurosciences would have to be aware of the methodological presuppositions of the neurosciences in order to fairly evaluate them from its own standpoint (assuming that this standpoint is itself coherently articulated). Furthermore, it would have to prove that these presuppositions generate scientific results that can fuel large scale processes of social repression and it would have to show what causal role capitalism plays in bringing about this particular mode of the relationship between science and the world outside science. Lastly, it would have to supply ethical, political, sociological, and psychological knowledge allowing us to overcome the repressions diagnosed (but who exactly is "us"?).

It should be obvious that this is a lot to achieve, even for a philosophical and social scientific tradition that has continuously stressed the necessity of interdisciplinary work in scientific research. Habermas has, to be sure, launched an attack on the objectifying tendencies of the natural sciences and attempted to defend our phenomenological sense of freedom against naturalizing interpretations of the very phenomenon; but it is hard to identify the points at which his criticism deserves to be called critical in the narrow sense of the Frankfurt School. Whilst part of the core business of critical theory is to attempt to distinguish a non-positivistic, non-scientistic, or non-naturalistic perspective on human beings from positivistic, scientistic, or naturalistic approaches to human beings, the question of the coercive character or effect of the neurosciences' scientific outlook has not been adequately posed. This, I believe, is essential if one wants to develop a specific critical theory approach to the neurosciences. In addition, I maintain that part of the problem of Habermas' approach to the neurosciences is that his (neo-Kantian) acceptance of the dualism of nature and culture, or, for that matter, body and mind, does not allow for a more radical critique of naturalism in its various guises.

Having these warnings in mind, what can critical theory tell us about contemporary neuroscience? Here, I will introduce briefly some of the main criticisms that critical theory has launched against competing scientific claims to knowledge, in order to explore later to what extent these criticisms can be applied or extended to the neurosciences of today. I will maintain that the traditional form of the critique of positivism cannot be continued in the face of typical insights and methods of the neurosciences. In the last part of the chapter, I will sketch a modified version of the traditional critique that is sensitive to the particular claims to knowledge raised by the neurosciences.

The Failures of Traditional Theory

I will begin with Max Horkheimer's famous essay "Traditional and Critical Theory," published in 1937, still considered to be one of the founding documents of the Frankfurt School. In this essay Horkheimer distinguishes between two types of theory, namely traditional and critical theory. In it, he aims not just to delineate two different but legitimate approaches to the study of social or non-social objects but to present critical theory as better theory, largely meaning that it has developed the means to analyze its own practical preconditions and effects where traditional theory has not.

Traditional theory, according to Horkheimer, can be characterized as positioning itself outside practical, political, economic, or social contexts. Thus, it identifies itself as an independent, "self-sufficient" form of gathering knowledge about whatever object is studied, a form that follows its own laws and scientific procedures. The aim of this self-sufficient theory is to gather "propositions (*Sätze*) about a subject" where the propositions are "so linked with each other that a few are basic and the rest derive from these" (Horkheimer, 1972, p. 188). Moreover, these propositions are to be applied to observable facts (*Sachverhalte*), which serve as corroborating or falsifying evidence for the explanatory implications of the propositional system. These facts are seen as merely given, or as there to be observed, even if it is granted that the way they impinge on the observer is conceptually mediated and that they can be seen as the result of human intervention. "Subject and object," says Horkheimer, "are kept strictly apart. Even if it turns out that at a later point in time the objective event is influenced by human intervention, to science this is just another fact" (Horkheimer, 1972, p. 229). If the logical structure of the propositional system is coherent, it enables the observer to infer probable conclusions from given observations and this is what is meant to explain an event, be it natural or cultural.

Probably the most important aspect of traditional theory as defined in this way is its abstinence from any sort of practice outside the scientific stance itself. Traditional scientists are not interested in the practical effects and presuppositions of their research even though they may very well admit that they exist. If they turn up in a traditional scientist's work he or she will "factualize" them, as Horkheimer suggests, and treat them as just another set of facts to be studied and observed (and not to be changed). In this sense the scientists, as it were, place themselves outside history, which appears to them to be the step required to generate law-like propositions. In this way, even social or cultural events appear to be naturalized in this stance, since they follow a necessity of their own that only the detached outside view of the scientist can detect.

Critical Theorists adopt a different stance towards the objects they study, considering themselves to be part of a movement—in contrast to the traditional theorist—and readily accepting the fact that "the scientific calling (*Beruf*) is only one, non-independent, element in the work or historical activity of man" (Horkheimer, 1972, p. 198). It is integral to their work as theorists, to generate knowledge that allows them and others to change reality, to improve it and thereby to render it less unjust and oppressive. Compared to the traditional theorist, the Critical Theorist does not study different objects but the perspective he or she adopts towards these objects is radically different as it is one of engagement and involvement with these very objects.[1] While this difference appears to be merely one of subjective attitude quite a lot follows from it in terms of the larger scientific stance. Thus, the Critical Theorist does not only accept the fact that his or her own perspective is driven by practical interests, but generalizes this insight to science per se. The idea, for example, that subjective differences of perception can be overcome through improved (experimental) methods of observation is taken, in Horkheimer's essay from 1937, "The Latest Attack On Metaphysics," as belonging to the "passing world of liberalism" which stipulates a "harmonious relation of individuals" and rejects the possibility of "theoretical differences which rest on historically conditioned antagonisms of interest" (Horkheimer, 1972, pp. 147–148). The basic conceptual and methodological tools of science cannot in any way be isolated from social, political, or economic interests.

Defining this contextual aspect of science away or simply ignoring it, as the traditional theorist does, amounts to a willful distortion of the larger context that fuels the construction of facticity within any scientific endeavor. The very desire to construe science in terms of facts is not itself a neutral fact to be analyzed by, say, a philosophy of science; instead it should be seen as resulting from a desire to eliminate all kinds of subjective influence on scientific results, a desire that does not admit its own subjective (practical) sources. This denial, to be sure, does have practical effects and it is the Critical Theorist who can spell them out. The claim of "Traditional and Critical Theory" is that traditional theory serves as the justification of a science that has as its main business the "manipulation (*Handhabung*) of physical nature and of specific economic and social mechanisms" (Horkheimer, 1972, p. 194). Horkheimer readily admits that modern science has more or less achieved this goal, but the problem is that it has not enabled those involved in this progress to see it as resulting from their own work and their own reasonable interventions. This, in fact, is part of the naturalizing effect of traditional theory: in postulating law-like generalizations on the basis of objectifying observations, the role of conscious human interventions into the natural and social world drops out of the picture. As a consequence, the world as it is or has become is experienced by the subjects as alien and driven by forces outside their own control.

The Failures of Positivism and Naturalism

Let me now move on to a second approach to the sciences that has been promulgated within the context of the Frankfurt School. In his writings of the 1960s, particularly *Theory and Practice* and *Knowledge and Human Interests*, Jürgen Habermas has adopted the general frame of Horkheimer's criticism of traditional theory but has added important modifications to it. While Horkheimer located the practical basis of critical theory (and in some sense of theory in general) in the sphere of work—"the

---

[1] Horkheimer, 1972, p. 209: "Its [Critical Theory's] opposition to the traditional concepts of theory springs in general from a difference not so much of objects as of subjects."

goals of human activity, especially the idea of a reasonable organization of society that will meet the needs of the whole community, are immanent in human work" (Horkheimer, 1972, p. 213)—Habermas has dropped this Marxist frame and placed what he then called "emancipatory interest" outside the sphere of the developing productive forces. This difference is important in many respects that cannot be commented upon in this context.

I would like to stress one point; as I have indicated, Horkheimer never denied the progressive impact modern science had in at least potentially improving the living conditions of the masses. The possibility of an "association of free men," he claims, "can be shown to be real even at the present stage of productive forces" (Horkheimer, 1972, p. 219). Habermas also admits that modern natural science has a progressive side; after all, Descartes and his followers battled the prejudices and dogmatisms of their time and used the insights of natural science as moral guides. An experience-based study of nature should not only generate knowledge with respect to nature but also knowledge with respect to how humans should behave according to nature. Reason itself demanded, as it were, the turn to nature and the natural sciences as part of its struggle against prejudice and (religious) ideology. Positivism in Habermas' sense, however, has successfully destroyed the liaison between naturalism and reason as it disenchanted nature and raised serious doubts about the possibility of deriving normative conclusions from the knowledge of causal laws (Habermas, 1973, p. 258). Habermas treats positivism as a method of knowledge generation that became influential in the nineteenth century and has continued way into the twentieth century. The basic task of this method was to "describe" reality as adequately as possible without taking account of the constitutive role the knowing subject plays in this process (Habermas, 1971). The hard sciences were seen as paradigmatic models of knowledge generation, a doctrine Habermas also called "scientism." Positivism and scientism therefore can be seen to be heirs to traditional theory that have not only forgotten, but also actively shaken off their practical moorings and any interest in the social context within which scientists develop their theories.

This, of course, does not mean that these doctrines no longer have an interest-based background. The science they legitimize is a science that helps to subjugate nature following the dictates of a thoroughly instrumental rationality. However, the problem with this sort of doctrine is that the turn to nature and the natural sciences is no longer guided by a reason that is able to specify norms or goals of action independently of the precepts of instrumental rationality itself. In fact, the rationality linked to technical forms of domination of nature absolutizes itself and thus rejects the existence of any form of non-instrumental rationality. This is what Habermas calls "technocratic consciousness:" the inability to differentiate between practical and technical rationality and the assumption that matters of social or cultural dimension can be dealt with in a technological perspective. In the technocratic perspective, human beings lose their character as beings who "live together and discuss matters with each other" as they are addressed as beings "who manipulate" (*hantieren*) (Habermas, 1973, p. 255). This means that the technocratic elite does not address human beings as capable of influencing their fate, or shaping and steering the direction of history, but as beings implementing the pre-given norms of technological rationality. In this sense, the technocratic perspective itself articulates and implements a notion of scientifically induced domination and coercion.

Horkheimer's notion of traditional theory and Habermas' notion of positivism (and scientism) refer not only to the methods of the natural sciences but also, and probably even more so, to traditional or positivistic theories of the human sciences following the lead of the natural sciences. Nevertheless, the notion of naturalism that has recently been Habermas' main target cannot as easily be adopted by the human sciences as it is more aggressive in its denial of the core material of these sciences. Unfortunately, Habermas' notion of naturalism is complex and not always very clear, so I will introduce several distinctions into the debate that cannot be found in Habermas. For the sake of argument, I will differentiate between epistemic naturalism, interventionist naturalism, ethical naturalism, and normative naturalism. As will become clear, Habermas concentrates largely on epistemic naturalism and for this reason we will have to transcend his approach if we want to develop a stronger basis for a critical theory of the neurosciences.

Epistemic naturalism considers the natural sciences as the only avenue to truth. Typical of this naturalism are the following elements:

(a) What we can know about mental events reduces to what our observational instruments allow us to know about them. Non-observational properties of mental events cannot be treated as existing.
(b) Events are causally linked by natural laws that can be detected by the methods of the natural sciences. This leads to some form of acceptance of determinism.
(c) Consciousness of free will or of responsible agency should be treated as a form of reality-distortion. Neither free will nor responsible agency (as linked to the notion of free will) exists; at least their existence cannot be proven by scientific methods. These distortions may serve evolutionary purposes, but if so they can be causally explained.

It should be obvious that these observations about naturalism lead back to some of the models of positivism and scientism just discussed. Epistemic naturalism breaks with all attempts to understand human behavior in terms of non-natural categories such as mind, free will, intentionality, responsibility, and so forth; or it naturalizes these categories if it proves possible to treat them exhaustively in an observer's perspective.[2] If this turns out to be possible we can conclude that these mental categories are not just in need of an organic basis (a point hardly denied by any philosopher), but that the naturalized perspective adopted towards them generates all the knowledge we need about them. Habermas, by contrast, claims that it is not possible to reduce our self-understanding to the objectifying perspective of the natural sciences. His main argument is complicated but it runs roughly along the following lines: to consider ourselves free agents implies considering ourselves to be involved as participating subjects in those reflective processes that lead us to judge what to do next in a given situation. No attempt to prove that something else has lead to "my" decision, for example some brain mechanism unknown to me, can convince me of the

---

[2] For a much more sophisticated account of this variant of naturalism see Kitcher (1992).

obsolescence of my concept of freedom. The grammar of this concept forces me, as it were, to resist attempts at naturalistic reductions and explains why we often do indeed continue to use the vocabulary of intentionality, responsibility, and free will.

This is not to say that the physicalist perspective is wrong, it is just to say that it cannot claim to explain all of our behavior. We must retain both perspectives, the physicalistic and mentalistic "language game," without wanting to reduce one to the other. This attack on naturalism, of which I have only mentioned very few aspects, concentrates on the epistemological and methodological question of how we, as humans, should understand and conceptualize ourselves, of how we can gain knowledge about ourselves. The naturalistic perspective is identified by Habermas as the third-person perspective of an observer who does not participate as a first person in the processes analyzed.

Apart from this epistemic naturalism, Habermas does mention something I will treat as a variant of naturalism; Habermas calls this variant "practical" but I will call it interventionist. In *The Future of Human Nature* (2003) Habermas hypothesizes about the possible impact of genetic engineering. If parents were ever able to select from genetic pools some of the basic characteristics of their children, this would amount to a form of reifying the child, treating it "as if disposing over an object" (Habermas, 2003, p. 51). The "unavailabilities" (*Unverfügbarkeiten*) of the "natural lottery" (born with blue eyes) are sharply distinguished from humanly induced unavailabilities (born with blue eyes because it's the favorite color of my parents), which tend to instrumentalize human beings on the basis of biologically available data without taking into consideration the communicative capacities which they will develop later.

Unfortunately, it is not easy to see how these variants of naturalism relate to one another. Briefly, what Habermas suggests is that the problems generated by epistemic naturalism form the basis of the problems generated by interventionist naturalism. In other words, what is wrong with interventions such as genetic engineering is that they would rob us (as future carriers of free will) of just those natural unavailabilities that we need as beings endowed with free will. If a naturalistic conception of the self, as propounded by variants of epistemic naturalism, disempowers us as responsible and free agents, the manipulated causalities of genetic engineering do just the same even though they spring from attempts to actively shape our genetic inheritance.

Interventionist naturalism gives rise to a third type, or variant, of naturalism that I call ethical naturalism. To understand what this is let us look back to Horkheimer's version of traditional theory. Horkheimer locates the coercive aspect of traditional theory in the fact that human beings are seen in the light of seeming natural causalities, and are thereby robbed of their pre-existing capacity to submit natural fate to humanly invented goals. Habermas, on the other hand, locates the coercive aspect of naturalism—in its interventionist variant—in the very desire to submit natural fate to human goals.[3] Manipulating the natural lottery destroys what is now called the

---

[3] Horkheimer's criticism of traditional theory stresses its tendency to submit the sphere of human action to the strict regularities of causality which allow for lawlike generalizations by the natural sciences. Critical Theory, by contrast, wants to broaden the basis for human interventions into seemingly natural processes in order to extend the realm of human freedom. For this reason Horkheimer and other members of the early Frankfurt School have been blamed for wanting to reduce the realm of nature impervious to the demands of reason, to a minimum (see Theunissen, 1981, p. 14).

"practical unavailability of subjective nature" seen as a necessary horizon of the decisions of a free will (Habermas, 2008, p. 203).

While I am not able to comment on this theory of free will, I do want to stress that the notion of certain aspects of subjective nature that are not to be manipulated, opens the door to "ethical" naturalism. Since we are, as Adorno says (finding Habermas' approval), "part of nature" we may very well naturalize our self-understanding to some extent in order to battle varieties of a rampant idealism or uncontrolled subjectivism. However, since we can in principle manipulate our nature (at least in the thought experiments envisioned by Habermas) all this ethical naturalism demands is that we should not do so; doing so would amount to destroying something about our concept of freedom that is, so Habermas claims, inextricably linked to it, namely its need for a basis in non-manipulable conditions (Habermas calls his concept of freedom a concept of "conditioned" freedom). That there is something about our subjective nature we should not manipulate thus derives, more or less, from our concept of freedom (from its "grammar" as Habermas is fond of saying); this is why epistemological considerations appear to be the normative precondition for the criticism of the more practical forms of intervention into human nature.

Clearly there is a certain tension in Habermas' approach between epistemic and interventionist naturalism; I want to explore this tension to indicate where it is necessary to move beyond Habermas' position. While epistemic naturalism, if taken seriously, deprives us of our capacity to conceptualize ourselves as free and responsible agents, interventionist naturalism seems to greatly enhance our manipulative power over nature on the basis of the results of scientific research. But how can the disempowering and the empowering perspectives be reconciled? One possible answer, not given by Habermas, is to point out that the empowering perspective is based on something we might call normative naturalism, a variant of naturalism. This treats the findings of the natural sciences as sure indicators of what to do, as they reveal to us what William Casebeer calls "natural ethical facts" (Casebeer, 2003). If this is the way to find out what we should do, however, it is clear that only those who know enough about these "natural" matters can really tell us what to do. Consequently, interventionist variants of naturalism imply what one might call a naturalistic expertise or advisory position that somehow attempts to disprove other methods and procedures of gaining knowledge about what to do.

In this sense, processes of empowerment and disempowerment go hand in hand, as is often the case. While scientists seemingly have the means to discern the facts, others lack these means and must therefore be advised as to what follows from these facts. The presupposition that something does actually follow from these facts only reflects the instrumental perspective of the scientific stance itself which should be emphasized much more than the debilitating effects of extending the objectifying third-person perspective to first-person lifeworld matters.[4]

The neurosciences, or so I will claim, are characterized by a coalition of interventionist and normative naturalism where the proposal for intervention often follows the generation of apparent facts about nature. The point of these reflections is that instead of talking about two distinct perspectives—namely a subjective, involved

---

[4]   See Wellmer, (2008, p. 11) for a similar reminder of the instrumentalist stance of the natural sciences.

first-person perspective and a distanced, objective third-person perspective—we should stress the grounding of third-person observation in a practical attitude of intervention guided by seemingly natural norms. The first- and third-person observational stances are not as distinct as Habermas takes them to be. If we apply these insights to the case of the booming neurosciences we might say that they gain an interventionist moment through the knowledge they generate about the brain. This, one might add, holds true not just for instances of brain lesion with severe impact on specific mental capacities, but also for the general workings of the brain that apparently allow for practical measures to be taken (just consider the field of neuroenhancers which I shall deal with later).

The large mass of scientific influence exerted on the public cultures of our day thus rests on arguments of the following shape: "If *a* is the case (about, say, the brain) *b* should follow." My claim is that the scientifically induced construction of a realm of normative facticity should be the target of a critical theory of the neurosciences. The leading idea behind such a critical theory would be to detect at what points neuroscientific approaches to human behavior rest the persuasiveness of their arguments on models of first nature that carry normative weight. This may appear like a return to older criticisms of positivism and traditional theory but I think there is a difference of approach. While positivism in the strict sense denies that normative arguments can be derived from factual statements, recent naturalistic models of human behavior again and again derive normative conclusions from their findings in a surprisingly straightforward manner. And it is the seeming facticity of a first nature that justifies the oughts at the base of these naturalistic models.

Defending the realm of freedom against naturalistic encroachment does not so much require us to praise a sphere of "subjective nature" that is not to be manipulated in order to rescue the first-person lifeworld perspective from a debilitating third-person influence; and neither does it require us to emphasize the unavoidable "complementarity" (a favorite phrase of Habermas) of the first-person and third-person perspectives. Rather it requires us to take a very close look at what appears to be natural, in order to analyze the interested or practical basis of the construction of this appearance. This also means that we accept the fact that the lifeworld is not beyond the influence of political and societal forces that shape the way the world is interpreted even in the first-person perspective—a fact that Habermas himself recognized when, earlier in his career, he talked of the "colonization" of the lifeworld by economic or political imperatives. Stressing the historicity of the lifeworld, and the instrumentality of many branches of the natural sciences, allows us to take a closer look at the mechanisms responsible for the construction of a normatively relevant facticity in scientific discourse. What this means will be the subject of the next sections.

Is Neuroscience Positivistic—or Naturalistic?

It should be obvious that traditional, positivistic, technocratic, and naturalistic methodologies and outlooks share many characteristics. However, I want to highlight only specific aspects. The most important characteristic of traditional theory, according to Horkheimer, was that it took a non-intervening, "neutral" stance towards its objects and submitted them to conceptually guided lawlike generalizations. The reality of human creativity in history was not denied, but it was seen as a phenomenon with causal antecedents and effects that could be studied in an objective scientific perspective. While it is not entirely clear whether Horkheimer considered this mode of science to be coercive as such, it is clear that he considered it to be affirmative of reality, as possible improvements to it were not part of the scientific endeavor. Since injustice and coercion were real enough in capitalism, any seemingly neutral scientific stance was treated as compromised.

While traditional theory may have had emancipatory effects due to its close, though unrecognized, connection to the more or less progressive movement of productive forces, positivism and technocracy have abandoned all (explicit) normative ambitions. They reduce rationality to instrumental rationality and adopt the norms generated by the technological apparatus itself (technocracy). Human beings are thereby classified as beings capable of manipulating nature and other humans, but not as communicative beings able to critically scrutinize and modify the ends seemingly inherent in technology. They are treated as technologically active, as it were, and this type of activity and its concomitant rationality is extended to other forms of, say, moral or political activity, activities which cannot in the same sense be considered rational as they are not open to observation-based research.

Critical theory has always struggled against the domination of the general sciences by the methods and epistemology of the natural sciences, so its quarrel with naturalism appears just to be the last round in a longstanding battle. However, as I have indicated, its treatment of naturalism is ambiguous. Naturalism is justified as long as it does not "colonize" the field of knowledge about humans as such. Epistemic naturalism renders our notion of responsibility obsolete (as long as we link it to the existence of a free will) and hereby destabilizes our self-understanding; interventionist naturalism destroys part of the "natural" conditions for being able to assume responsibility in the first place. We cannot assume responsibility, Habermas suggests, for decisions our parents have made without in any way asking for our opinion. This then constitutes a bad form of naturalizing, or objectifying, ourselves. Freedom-based, good forms of (ethical) naturalism will prevent such forms of coercive intervention into our subjective nature.

However forceful one finds these arguments, I have added that Habermas' criticism of naturalism does not pay enough attention to the interest-based generation of "natural" facts and their ever closer connection to normative propositions. The perspective of the natural sciences per se is accepted or taken at face value; it is only criticized where it attempts to replace a first-person or lifeworld vocabulary.[5] A critical theory of the neurosciences will have to take a look at the way in which arguments, which one might call *normative first nature arguments*, structure many approaches of

[5] Habermas and his followers do admit that scientific explanations are shot through with non-neutral lifeworld categories. In other words they accept a practical basis of scientific language. However, in stressing the irreplaceable role of lifeworld or participant language in science, their main interest is still to defend the participant language against attempts to wholly naturalize it. In addition, the extent to which the relevant lifeworld categories are ideological or under the influence of socially powerful regimes of justification and interpretation is inadequately dealt with. For an interesting but depoliticized conception of the relation between lifeworld and science, see Wingert (2007).

the neurosciences. In this sense, findings about the brain are not just treated as findings about the natural preconditions of mental life, but as powerful guides to the way this life should function. This may not in all cases amount to a naturalistic reduction of the mental categories at hand, but it certainly amounts to the thesis that whatever purposes they serve for human organisms, they ought to pay attention to the "demands" of the brain and the specific requirements of its functioning.

In what follows I will analyze briefly how neuroscience fares when confronted with the normative apparatus developed by critical theory. I will not be able to give neuroscience a fair treatment but I hope it will become clear to what extent it makes sense to submit neuroscience to what Bernard Williams, in *Truth and Truthfulness* (2002), has called a "Critical Theory Test."

The general heading for the following reflections could be: does neuroscience ignore its practical basis? Well, it does and it doesn't. Let me once more take up Horkheimer's characterization of traditional theory: "In traditional theoretical thinking, the genesis of particular objective facts, the practical application of the conceptual systems by which it grasps the facts, and the role of such systems in action, are all taken to be external to the theoretical thinking itself" (Horkheimer, 1972, p. 208). Now while I do not claim that the neurosciences consider these practical aspects to be internal to their conceptual apparatus, I would maintain that most neuroscientists are fully aware of the practical impact that their science might have. In fact, part of their often provocative structure (what Habermas calls their debilitating effects on our self-understanding) rests on the fact that the neurosciences deal with the neurological underpinnings of our usual mental faculties and do not restrict their research to studying causally determined natural events as if they were unrelated to our first-person perspective on the world. As Wellmer has said, they build a bridge between the neurological basis of consciousness and consciousness as it presents itself to subjective experience and it is this that is part of their special scientific status (see Wellmer, 2008, p. 11).

In this sense their results cannot be kept at bay and seem to flow more or less freely into our usual mentalistic vocabulary. Whether we deal with paraplegia, Alzheimer's or Parkinson's disease, or Attention Deficit Hyperactivity Disorder (ADHD), many neuroscientists suggest that their research might have an impact on improving the treatment of these "diseases." They thus realize that what they do is part of the "manipulation of physical nature." Many even go so far as to use their insights as instruments of critique of given social practices. For example, take the German debate concerning recommendations by neuroscientists to get rid of the notion of personal guilt as a basis of the penal system. If it can be shown that humans have no free will it makes absolutely no sense to base judicial judgments on the presupposition of such a notion. The neuroscientist turns out to be a reformer of institutions and self-understandings and is quite confident in the presentation of his or her case. What this illustrates is that a more or less explicit ethics of neuroscience has developed. In other words, there is a tendency to accept the functional description of biological phenomena as a basis for normative oughts (see Wingert, 2006, pp. 246–248).

While traditional theory, according to Horkheimer, was not aware of its practical basis or even consciously ignored it, recent neuroscience is much less reticent with respect to this question and has adopted some elements that Horkheimer had hoped to reserve for critical theory. Yet Horkheimer had more in mind when insisting on the practical basis of science. Recalling the practical basis of science meant recalling the fact that, as he puts it in "The Latest Attack on Metaphysics," "the world of perception is not merely a copy nor something fixed and substantial, but, to an equal measure a product of human activity" (Horkheimer, 1972, pp. 157–158).[6] This means that not only are the objects studied shaped by human activity but that the scientific way of observing these objects is too. On both counts neuroscience tends to falter severely.

Take the various steps that constitute the construction of ADHD. The German neuroscientist Gerhard Roth, in his book *Fühlen, Denken, Handeln* (2002), first claims that ADHD exists, that it is a factually given disorder. He then moves on to present the elements that constitute this disorder (usually early on in childhood): lack of attention, physical restlessness, and lack of control of spontaneous impulses. In a third step the disorder is naturalized, that is, it is treated as being based on genetic defects. To be sure, it is admitted that there is some sort of motherly (sic!) misconduct that might influence the conditioning of this disorder but, if I read Roth correctly at this point, the idea is that personal, social, or cultural factors can only reinforce existing genetic deficiencies, and cannot originally produce them. Fourthly, it is suggested that there are palpable deficits in specific brain regions that are causally responsible for the onset of ADHD. Despite the fact, then, that many brain researchers conceptualize the brain as an open and plastic structure that interacts with its environment and thereby takes on individual characteristics, there is still a strong tendency to downplay this plasticity and emphasize early processes of lifelong determination. But why should this be the case given that the brain itself—or what we know about it at this stage—does not preclude assumptions about its regenerative and open structure? I will come back to this point.

What about lack of attention, restlessness, and deficits in impulse control? Are these transcultural deficits? Or could we say that they are only seen as deficits within a specific cultural context that treats certain ways of behaving as hyperactive and thus helps to construct the deficit in the first place (Hacking, 1999)? Roth does not reflect upon this aspect of the practical basis of the objects studied by him. Neither does he reflect upon the prescientific basis of his own conceptual apparatus. This is probably the central deficiency of neuroscience in general—call it, its hidden hermeneutics. Neuroscientists interpret the data they acquire while presenting them as allowing factual descriptions of an intersubjectively valid sort. In addition, they suggest that these data allow for normative conclusions and hereby treat them as the "natural ethical facts" already mentioned. Typical arguments of this sort run as follows: facts about human nature or about specific aspects of human nature which we can detect using the means of natural science, are facts about entities that have a *telos*, say survival or wellbeing or homeostasis. From this seemingly incontrovertible fact we draw conclusions about requirements that need to be fulfilled if the specific telos of the entity at stake is to be achieved or realized.

In this functionalist sense Casebeer, for example, derives moral demands concerning our relationship to others from needs we have as the biological entities we are. "Almost

[6] See also "Traditional and Critical Theory" (Horkheimer, 1972, p. 200) "Even the way they [men] see and hear is inseparable from the social life-process as it has evolved over the millennia."

all of our functional needs," writes Casebeer, "can be satisfied only by working with others" (Casebeer, 2003, p. 60). For this reason, obligations towards others can be derived from facts about our nature. In other words, we are by nature social beings and it is from this fact alone that we gain the knowledge concerning how to interact with others. It is important to add that what is natural in this sense must not be treated as an a priori given; on the contrary it can be accepted as the result of experience-based, falsifiable scientific research. However, whatever normative authority the natural has rests on its facticity and it is this proposition that is relevant to the two claims I want to make in the final part of my discussion.

Flexible Capitalism and Neuroscience

Let me follow up on some of the themes just discussed. I maintain that the discourse of facticity surrounding neuroscientific research does not only disempower us epistemically (as Habermas thinks) but also practically. What we are to do depends upon what we know about our brain or about the facts of the brain that are seen to have an inherent normativity of their own; in this sense we are to react to this knowledge or to these facts as the brain is interpreted as centrally relevant to almost all human endeavor. At a profound level, much of neuroscientific research places us in a passive relation to the knowledge or to the facts about our brain, even though this knowledge may require us to do something about our brain or to restructure relevant institutions and practices (see Farah, 2005). This passive stance even survives acceptance of the insight that our knowledge of the brain may change according to scientific standards. At least, this seems to be the case given the constellation within which the neuroscientific agenda is presently set. Put differently, whether we construct the brain as command center (as was done in the past) or as non-hierarchical neuronal network, both models could be inscribed into a larger discourse explicating the practical conclusions that follow from these "facts" about the functioning of the brain.

Let me clarify what I mean by this; in doing so I will situate neuroscientific work more in the social, economic, and political context in which it actually takes place (see the essays in Karafyllis & Ulshöfer, 2008 for a similar perspective). Earlier I suggested that neuroscience does not reflect the complexity of its practical basis; so far, however, I have not indicated whether this leads to coercive mechanisms of self-understanding. The problem at this point is, of course, that all variants of critical theory assume a notion of freedom from coercion or domination as central to their critical enterprise. Yet it is this very notion of freedom that is denied by the neurosciences (in fact, if the possibility of freedom or free will is at the normative heart of many critical projects, a hard-boiled naturalist might call "critical neuroscience" an oxymoron). Neuroscientists themselves will, of course, explain their position in neurological terms but we do not have to accept this as the only position one can take towards this question. Consequently, I will accept for the moment what the French philosopher Catherine Malabou, in her book *What Should we do with our Brain?*, calls the "political and ideological" construction of neuronal man (Malabou, 2008, p. 13). The idea behind this formula is that most of the basic categories of the neurosciences "mirror" political or social categories that structure contemporary ways of understanding ourselves (and, one might add, vice versa). This is, of course, a complex claim that cannot be fully substantiated in the space of these pages. A few suggestive examples will have to suffice.

Take the picture of the brain that has more or less replaced the explicitly politicized picture of the brain as a command center (this is, perhaps, exaggerated for the picture of the brain as a deterministic process is still alive). If I am not fully mistaken (see Rosenfield & Ziff, 2008), the picture of the brain presently emerging is one of a network of non-hierarchical decentered structures which take on a more definite form through evolutionary pressures (from inside and outside); this helps to select the synaptic junctions that are to become more or less stable (Changeux' "learning by selection").[7] The brain is not, according to this picture, a preprogrammed mechanism with innate reaction models to outside stimuli. (Incidentally, this is another respect in which brain research seems to break with central tenets of (older) critical theory.) Horkheimer saw positivism as a methodology that explains human behavior in terms of causal laws. Yet as far as I can see, neuroscience construes the brain more and more as an active organism that shapes its environment and is shaped by it. In a certain sense, then, the brain is less deterministic in its working than is often supposed.

What about the new picture emerging of the brain as a non-hierarchical network of variable synaptic junctions? Is this picture simply neutral; does it objectively reflect scientific findings? In "Traditional and Critical Theory," Horkheimer emphasizes the "influence of the current social situation on change in scientific structures" and mentions the example of the slow acceptance of the Copernican worldview (Horkheimer, 1972, p. 195). What then is the social situation into which neuroscience carries its message about the network brain? Malabou mentions the parallel between models of contemporary "new" and flexible capitalism, which frequently use the network metaphor and refer also to the decentered non-hierarchical structure of the firm and the network model of the brain (see also Baumeler, 2008). The basic imperative of this new capitalism is flexibility which means that only those economic agents are valued who are willing to adapt quickly to new circumstances, resist settling permanently, create networks of contacts that allow them to move from one place to another with minimal friction, and who, if higher up in the seemingly non-existent hierarchy, do not rely on vertical commands but on the willingness of network co-workers to accept responsibilities on a voluntary and self-organized basis (see Boltanski & Chiapello, 2006).

It will seem strange to move from this description of a new economic and political order to recent descriptions of brain research. To be sure, management advice literature frequently makes use of "neuronal" metaphors and freely assimilates elements of scientific discourse for its own ends. But what about the neuroscientific images of human behavior themselves? In what sense can one say that they reflect or mirror political or social discourses? I would maintain that they are shaped by and partly shape these images, by naturalizing a stance of passivity towards the mechanisms of brain functioning that are allowed to generate norms of behavior. In this sense,

---

[7]   See Changeux, Heidmann, & Patte, (1984).

models of the brain as either command center or as non-hierarchic synaptic network do not differ very much from one another if the command center is taken to issue its orders autonomously or prereflectively. Consider the norm of flexibility as just sketched where to be flexible seems to presuppose an activity of the agent who is the center of all his movements, self-responsible, individualized, and unsettled. But underneath this activistic image of human behavior lies, of course, the norm of adaptation to ever-changing circumstances. If the agent is really flexible he or she will be able to detect situational requirements and adapt his or her behavior immediately to the new circumstances of action. The agent is thus not truly free as he or she is more or less forced to be flexible and this inevitably tinges the agent's perspective with a rather large measure of passivity.

My thesis is this: the role of what we might want to call social, economic, or political circumstances of action in everyday life (including their specific pressures and constraints) is played by the brain in neuroscientific discourse. This is what one might call the naturalization of social categories that combine, in an intricate and often paradoxical manner, elements of an enhancement of individual responsibility with neo-Darwinian elements of passivity and adaptability in the light of specific circumstances of action (Hartmann & Honneth, 2006). In other words, the typical neuroscientific discourse runs along the following lines: given what we (now) know about the brain we (as individuals, as schools, as universities, as political bodies) should restructure our institutional frameworks along the lines of what the brain, in its functional requirements, demands. It should be obvious that this description combines elements both of losing responsibility and of gaining new kinds of responsibility. Much brain research indulges in proving our notion of responsibility for what we do as wrong or as mere illusion.

From a social-psychological perspective this might explain in part the popularity of the neurosciences (for another, less sociological explanation see Weisberg et al., 2008). Given the pressures of a neoliberal order of individualized responsibility, it might come as a relief to hear that the increasing knowledge about our brain lets us off the hook a little. At the same time, however, a new responsibility emerges, namely a naturalized responsibility for our brain and its specific functionings (that is a responsibility without, in some sense, full responsibility). Foucaultians would certainly want to interpret this phenomenon as part of the biologization of power that is typical of modern societies (see Rose, 2007). The forceful social and political reformism of much contemporary neuroscience (and not just in its popularized versions) cannot be explained without this larger context in mind. The idea is that something about our brain—about its first nature—allows us to draw conclusions about what we should do.

The question arises, nevertheless, whether force or unrecognized repression is involved in these processes. In answer, it might have been easier to point to the liaison between neuroscientists and their pharmaceutical sponsors which structures much research in this field and reflects, among other factors, the rise of the "entrepreneurial university" (Etzkowitz, 2005). It is to be expected that these co-operative ventures generate their own constraints and obligations, but certainly little research has been done in this field concerning the neurosciences. One could also point to the pressure exerted on contemporary subjects to enhance and improve their mental and physical capacities in order to cope better with the demands of a thoroughly flexibilized capitalism. To mention just one example, Martha Farah warns that "the freedom to remain unenhanced [by not using Ritalin for example] may be difficult to maintain in a society where one's competition is using enhancement. American courts have already heard cases brought by parents who were coerced by schools to medicate their children for attentional dysfunction" (Farah, 2005, p. 37).

What we are presently witnessing is the rise of a form of "neurocapitalism" that binds pharmaceutical interests, neuroscientifc research, and the pressures of a growing commodification of the self together in a rather unhealthy alliance (Jokeit & Hess, 2009). Yet if we admit that the pressures exerted here are more or less open and straightforward—certainly in courtrooms in which parents defend their children against coercive tranquilization—it still remains to be seen whether I have detected elements in the system of flexible capitalism that render unrecognizable, in Honneth's words, those social conditions through which the capitalist system is structurally produced. The closest I have come is probably my analogy of the brain in neuroscience to the seemingly heteronomous economic pressures of contemporary capitalism that force us to flexibly adapt to changing circumstances. The more concrete entanglement of economic interest and neuroscience shows that what is actually at stake here is more than mere analogy.

Neuroscientific research seems to reveal, or is treated as if it reveals, natural facts about the brain that help naturalize some of the economic practices of contemporary society. If capitalism has, as the old Marxist vernacular put it, become second nature to us and could at least in principle be changed, we might add that elements of a theory of first neuronal nature at present help support the economic status quo. It is as if the brain is made for flexible capitalism, as if it has a more or less natural response to the demands of the present. In adopting Malabou's phrase about the "political and ideological" construction of neuronal man, I do not just mean that the categories neuroscientists use to structure and set up their research, reflect political, economic, or social categories. I also mean that the results of their research have an impact on contemporary capitalist societies in justifying the processes of institutional restructuring according to the seemingly natural demands of the brain's functional mechanisms. In that sense neuroscience helps to depoliticize these suggested institutional reforms and that is part of what ideologies do.

There is no need to assume that the strange but powerful coalition between neuroscience and the given economic system will continue to exist forever. Perhaps the model of the cultural brain will slowly gain ground and bring an end to it, as it will no longer allow us to simply naturalize "facts" about the brain (see Wellmer, 2008, p. 12). Yet if—as Dewey once said—it is the task of all the intelligent activities of men "no matter whether expressed in science, fine arts, or social relationships" to convert "causal bonds, relations of succession, into a connection of means-consequence, into meanings" (Dewey, 1991, p. 277), I cannot see that neuroscience has been very successful in generating meaningful results. It is still much too involved in drawing up causal bonds for us by holding us responsible for something for which we cannot—or so the story goes—be fully responsible, namely our brain and its specific functionings.

References

Baumeler, C. (2008). Technologies of the emotional self: Affective computing and the "Enhanced second skin" for flexible employees. In N. C. Karafyllis & G. Ulshöfer (Eds.), *Sexualized brains: Scientific modeling of emotional intelligence from a cultural perspective* (pp. 179–190). Cambridge, MA: MIT Press.

Boltanski, L., & Chiapello, E. (2006). *The new spirit of capitalism*. London: Verso Books.

Casebeer, W. (2003). *Natural ethical facts: Evolution, connectionism and moral cognition.* Cambridge, MA: MIT Press.

Changeux J. P., Heidmann T., & Patte P. (1984). Learning by selection. In P. Marler and H. Terrace, (Eds.), *The biology of learning* (pp. 115–139). Berlin: Springer-Verlag.

Dewey, J.(1991). *Experience and nature*. In J. A.Boydston & J. Gouinlock, (Eds.), *J. Dewey, The later works, 1925-1953 (Vol. 1)*. Carbondale and Edwardsville: Southern Illinois University Press.

Etzkowitz, H. (2005). The rise of the entrepreneurial university. *International Journal of Contemporary Sociology, 42*(11), 28–43.

Farah, M. J. (2005). Neuroethics: The practical and the philosophical. *Trends in Cognitive Science, 9*(1), 34–40.

Geuss, R. (1981). *The idea of a critical theory: Habermas and the Frankfurt School.* Cambridge: Cambridge University Press.

Habermas, J. (2008). "I myself am part of Nature"–Adorno on the intrication of reason in Nature: Reflections on the relation between freedom and unavailability. In *Between naturalism and religion: Philosophical essays* (pp. 181–208). Cambridge: Polity Press.

Habermas, J. (2003). *The future of human nature*. Cambridge: Polity Press.

Habermas, J. (1973). Dogmatism, reason, and decision: On theory and praxis in our scientific civilization. In *Theory and Practice* (pp. 253–282). London: Heinemann.

Habermas, J. (1971). *Knowledge and human interests*. London: Heinemann.

Hacking, I. (1999). *The social construction of what?* Cambridge, MA: Harvard University Press.

Hartmann, M., & Honneth, A. (2006). Paradoxes of capitalism. *Constellations, 13*(3), 41–58.

Honneth, A. (2008). Critical theory. In D. Moran (Ed.), *The Routledge companion to twentieth century philosophy* (pp. 784–813). London: Routledge.

Horkheimer, M. (1972). Traditional and critical theory. In *Critical theory: Selected essays* (pp. 188–243). New York: Continuum.

Horkheimer, M. (1972). The latest attack on metaphysics. In *Critical theory: Selected essays* (pp. 132–187). New York: Continuum.

Jokeit, H., & Hess, E. (2009). Neurokapitalismus. *Merkur, 63*(721), 541–545.

Karafyllis, N. C., & Ulshöfer, G. (Eds.), (2008). *Sexualized brains: Scientific modeling of emotional intelligence from a cultural perspective.* Cambridge, MA: MIT Press.

Kitcher, P. (1992). The Naturalists Return, *Philosophical Review, 101*(1), 53–114.

Malabou, C. (2008). *What should we do with our brain?* New York: Fordham University Press.

Rose, N. (2007). *The politics of life itself: Biomedicine, power, and subjectivity in the twenty-first century.* Princeton and Oxford: Princeton University Press.

Rosenfield, I., & Ziff, E. (2008). How the mind works: Revelations. *The New York Review of Books, 55*(11), 62–65.

Roth, G. (2002). *Fühlen, Denken, Handeln. Wie das Gehirn unser Verhalten steuert.* Frankfurt am Main: Suhrkamp Verlag.

Theunissen, M. (1981). *Kritische Theorie der Gesellschaft. Zwei Studien.* Berlin and New York: Walter de Gruyter.

Weisberg, D. S., Keil, F. C., Goodstein, J., Rawson, E., & Gray, J. R. (2008). The seductive allure of neuroscience explanations. *Journal of Cognitive Neuroscience, 20*(3), 470–477.

Wellmer, A. (2008). "Bald frei, bald unfrei"–Reflexionen über die Natur im Geist. *WestEnd, 5*(2), 3–21.

Williams, B. (2002). *Truth and truthfulness.* Princeton and Oxford: Princeton University Press.

Wingert, L. (2007). Lebensweltliche Gewissheit versus wissenschaftliches Wissen? *Deutsche Zeitschrift für Philosophie, 55*(6), 911–927.

Wingert, L. (2006). Grenzen der naturalistischen Selbstobjektivierung. In D. Sturma, (Ed.), *Philosophie und Neurowissenschaften* (pp. 240–260). Frankfurt am Main: Suhrkamp Verlag.

4

Scanning the Lifeworld
Toward a Critical Neuroscience of Action and Interaction[1]

Shaun Gallagher

A recent report published in *Neuron*, a leading journal of neuroscience, by researchers at Japan's ATR Computational Neuroscience Laboratories (Miyawaki et al., 2008) has been the basis for a claim that new technology able to analyze signals in the brain "can reconstruct the images inside a person's mind and display them on a computer monitor." Although claims made in the actual research paper were much more modest, in the media the standard, optimistic predictions were quick to come. "These results are a breakthrough in terms of understanding brain activity. In as little as 10 years, advances in this field of research may make it possible to read a person's thoughts with some degree of accuracy." And again:

> The researchers suggest a future version of this technology could be applied in the fields of art and design—particularly if it becomes possible to quickly and accurately access images existing inside an artist's head. The technology might also lead to new treatments for conditions such as psychiatric disorders involving hallucinations, by providing doctors a direct window into the mind of the patient. … In the future, it may also become possible to read feelings and complicated emotional states.[2]

Similar kinds of claims have been made about advancing brain imaging technology by others[3] but, at best, the technology described may allow a scientist to make

---

[1] The author thanks the Zentrum für Literatur- und Kulturforschung (ZfL) in Berlin for support as Visiting Researcher in 2008 and 2009 to complete this chapter. Special thanks to Sabine Flach and Jan Georg Söffner at the ZfL. An earlier version of the chapter was presented as a paper at the UCLA conference on critical neuroscience in January 2009.

[2] All of these quotations were reported on the science blog, *Pink Tentacle* – http://www.pinktentacle. com/2008/12/scientists-extract-images-directly-from-brain/

[3] Chris Frith has claimed that we may someday be able to read mental states off brain scans (see Frith & Gallagher, 2002; also in Gallagher, 2008d). Also Elger et al. (2004) declare that, "within the foreseeable future, it will be possible to explain and predict psychological processes such as sensations, emotions, thoughts, and decisions on the basis of physiochemical processes in the brain" (Habermas, 2007, p. 14).

Critical Neuroscience: A Handbook of the Social and Cultural Contexts of Neuroscience, First Edition.
Edited by Suparna Choudhury and Jan Slaby.
© 2012 John Wiley & Sons, Ltd. Published 2016 by John Wiley & Sons, Ltd.

inferences about whether a person is experiencing one sample of a certain pre-delineated set of stimuli.

These kinds of claims—coupled with all the other claims made about what neuroscience is capable of showing about human experience—amount to a super claim that the richness of human experience, informed by emotion, memory, imagination, and diverse perceptual encounters is entirely reducible to brain events. Furthermore, they imply that in principle there is a clean translation possible from measurable processes in the brain to the fullness of the meaningful, personal, and interpersonal experience of the lifeworld. These claims are the target of what seems to me a justifiable critique of neuroscience from the perspective of Frankfurt School critical theory. Axel Honneth has recently written:

> Surrounding the current discussions concerning the results and social implications of brain research, it has often been remarked that the strictly physio-biological approach employed in this sphere betrays a reifying perspective. The argument goes that by presuming to explain human feelings and actions through the mere analysis of neuron firings in the brain, this approach abstracts from all our experience in the lifeworld, thereby treating humans as senseless automatons and thus ultimately as mere things. ... [T]he fact that the neuro-physiological perspective apparently does not take humans' personal characteristics and perspectives into account is thus conceptualised as an instance of reification.
>
> (Honneth, 2008, p. 20)

While endorsing this critique of reification and reductionism, I want to suggest that the relationship between critical theory and cognitive neuroscience is a two way street.

- Critical theory can certainly take aim at the reifying and reductionistic tendencies of cognitive neuroscience (see Choudhury, Nagel, & Slaby, 2009 and this volume).
- But also, cognitive neuroscience and cognitive science more generally may be able to tell us things about human behavior that need to be accommodated by critical theory, or that can even support the aims of critical theory.

I want to pursue the second of these proposals in this chapter, but in doing so it will become clear that if cognitive neuroscience is to inform critical theory, it already needs to be a critical neuroscience, that is, a cognitive neuroscience that is non-reifying and non-reductionistic. In regard to this project I want to suggest two things. First, that a cognitive neuroscience informed by phenomenological insights about embodied, enactive, and situated cognition can be non-reductionistic in a way that is not subject to the particular critique mentioned by Honneth. In other words, a phenomenologically informed neuroscience can also be a critical neuroscience. And second, that cognitive neuroscientific studies of agency and social cognition, in particular, can reveal aspects of human relations important for improving the kind of actions and communicative practices that are championed by critical theory. I think there are clear but implicit connections between questions of social cognition and questions of agency, intention formation, and free will, but I will not argue for these connections here. Instead I will focus on the fact that for both agency and social cognition the relevant phenomena are not reducible to brain processes alone, but involve larger pragmatic and social interactions in the lifeworld.

Agency and Free Will

There are a number of things to say about the kind of claims made in regard to fMRI reconstructions of mental images found in studies by Miyawaki and others. From a philosophical perspective it is not clear what neuroscientists mean by "images inside a person's mind" (and here Miyawaki is not alone in using this terminology; see for example, Damasio, 1999). Many philosophers would consider this a return to an eighteenth-century epistemological vocabulary. But even current terminology is not settled or uncontentious. Neuroscientists, perhaps oblivious to ongoing philosophical debates about representationalism (see Dreyfus, 2002; Gallagher, 2008e; Hutto, 2008; Ramsey, 2007), often use terms like "representation" without clearly defining what they mean. These are not just terminological squabbles; they register serious conceptual issues about exactly what one is imaging. There are also methodological considerations that should limit claims about what brain imaging actually shows. As Overgaard (2004) points out, brain imaging does not give us a direct snapshot of anything like an image in the mind. The brain imager does not see the brain state itself, but can only work with statistically massaged data based for example on BOLD (blood flow) signals; data that is manipulated in contrastive analyses which are never perfect. Nor are scientists able to access the subject's mental experience in any direct way; at best they are working with the subject's report or with interpretations of overt behavior. From this indirect view of the brain and mediated report on experience, conclusions about "images inside a person's mind" seem several steps removed (Figure 4.1).

Debates within neuroscience on methodological questions ranging from statistical analysis (Vul, Harris, Winkielman, & Pashler, 2009), to the use of concepts much too general to inform interpretations of brain-imaging studies (see Legrand & Ruby, 2009), to important issues regarding ecological validity in brain-imaging experiments must qualify any quick conclusions about what is being captured in the scanner. On any reading, however, it is clearly not the fully embodied and environmentally situated experience of an active agent in his or her lifeworld.

A good example of how experimental data can lead to serious confusion—with moral and legal implications in the broader contexts of situated action—can be seen in the use made of the Libet experiments (Libet, Gleason, Wright, & Pearl, 1983) to argue against the notion of free will. Philosophers, psychologists, and neuroscientists have argued that consciousness does not control behavior, that our sense of agency for action is retrospective and purely epiphenomenal, and that there is therefore no such thing as free will (Prinz, 2001; Wegner, 2002). Part of the argument is based on the Libet experiments (in which the subject is required simply to flick their wrists or finger whenever they please) and on the idea that whilst we have a sense or feeling of freely deciding, in fact what we are going to do is already determined by brain processes (see Libet, 2000).

The focus of the Libet experiments, however, is on approximately 500 ms of neuronal processing, and the question is whether we can say that what counts as free will is contained in this very short time span. I have argued elsewhere that Libet's experiment is about motor control mechanisms rather than free will, precisely because it focuses on control of bodily movement rather than engaged action in the world

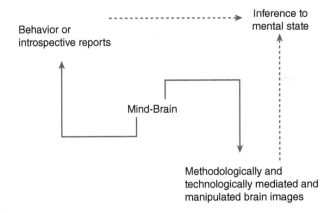

Figure 4.1   Indirect measures of mind and brain.

(Gallagher, 2005, 2006). This, however, may be puzzling for philosophers since the standard understanding of agency and mental causation has been framed precisely in terms of control of bodily movement, at least since the time of Descartes. For Descartes, if we will an action, it produces a movement in the brain which then produces a movement in the body; such actions, he suggests, "terminate in our body, as when from our merely willing to walk, it follows that our legs are moved and that we walk" (Descartes, 1649/1989, §§ xli, xliii, xviii). One need only compare Descartes' view to that of recent philosophers' statements about mental causation to see what the long-standing, standard view is. Here is just one example:

> In the case of normal voluntary action, movements of the agent's body have amongst their causes intentional states of that agent which are "about" just such movements. For instance, when I try to raise my arm and succeed in doing so, my arm goes up—and amongst the causes of its going up are such items as a desire of mine *that my arm should go up*. The intentional causes of physical events are always "directed" upon the occurrence of just such events, at least where normal voluntary action is concerned.
>
> (Lowe, 1999, pp. 235–36)

Raising one's hand is a favorite example in the philosophical literature. Lowe is in good company; similar statements can be found in many contemporary philosophers (Frankfurt, 1978; Proust, 2003; Searle, 1984). Moreover, neuroscientists follow this standard view. Haggard and Libet (2001), for example, frame the problem in the same way, referring to it as the traditional concept of free will: "how can a mental state (my conscious intention) initiate the neural events in the motor areas of the brain that lead to my body movement?" (p. 47).

Against this standard view, I suggest we think of the consciousness that pertains to action not as a consciousness of deciding to move one's body—indeed, as neuropsychology and phenomenology suggest, one's consciousness of normal bodily movement is minimal and recessive (Jeannerod & Pacherie, 2004). For most intentional action when I decide to act, I do not first decide, for example, to locate my hand and then decide to move it in a specific way. We do not ordinarily think about flicking our wrists in the larger contexts of action. The processes that Libet studied are automatic body-schematic processes—processes that we are not normally aware of in the details of our movements. Libet's experiments attend to a reflective consciousness of bodily movement that normally does not exist in situated action.

To be sure, complexities tied to embodied and situated action have not been carried over into neuroscientific experiments. But some experiments do start to point to a broader action arena, and they do so in a way that begins to show that the sense of agency is more complex than Libet or Wegner suggest. I'll discuss just one of several experiments (Farrer & Frith, 2002; see also Chaminade & Decety, 2002; Farrer et al., 2003) that begin with the phenomenological distinction between "sense of agency" (SA: the experience of being the author of one's action) and "sense of ownership" (SO: the experience that one's body is moving),[4] and then attempt to identify the neural correlates of the sense of agency. The experimental design in Farrer & Frith (2002) is as follows:

> Subjects manipulated a joystick [to drive a coloured circle moving on a screen to specific locations on the screen]. Sometimes the subject caused this movement [on the screen] and sometimes the experimenter. This paradigm allowed us to study the sense of agency without any confounding from the sense of ownership. To achieve this, subjects were requested to execute an action during all the different experimental conditions. By doing so the effect related to the sense of ownership (I am performing an action [I am moving]) would be present in all conditions and would be cancelled in the various contrasts.
>
> (Farrer & Frith, 2002, p. 597)

Why does SO remain constant while SA changes? The experimenters understand SA to be something more than an experience generated by motor control processes; rather, it is tied to the intentional aspects of the action, that is, to the perceptual monitoring of what I am accomplishing in the world rather than to motor control. It is not about moving the joystick or moving one's body; it's about doing something on the computer screen. Results show that when subjects feel that someone else is controlling the action on the screen (no SA) the right inferior parietal cortex is activated. By contrast, when they feel that they are controlling the action on the screen, the anterior insula is activated bilaterally.

One thing that seems important for the proper interpretation of these results is the phenomenological distinction between a feeling of agency for bodily action, generated in efferent processes, and the correlated perceptual intentionality of accomplishing something in the world. The experimenters do not keep this distinction as conceptually clear as they should, and when they attempt to explain why the anterior insula is involved in SA they focus on bodily movement and motor control rather than the intentional aspect of what I am accomplishing in the world. They also point out that the anterior insula involves the integration of three kinds of self-specifying signals generated in self-movement:

---

[4] See Gallagher (2000) for this distinction. SA and the SO are difficult to distinguish in normal action, but they can easily be distinguished in involuntary movement or reflex. In the case of involuntary movement, for example, if someone is manipulating my body I have the experience of my body moving (SO), but not the sense that I am the author of the movement (SA).

- somatosensory signals (sensory feedback from bodily movement, e.g. proprioception);
- visual and auditory signals containing ecological information about movement; and corollary discharge associated with efferent motor commands that control movement.

Ecological information, of course, is tied to non-conscious or possibly pre-reflective monitoring of my relation to the environment, and this is certainly an important element in SA. It is likely, however, that in some cases an even more explicitly conscious perceptual monitoring is ongoing in action. Searle (1983) calls this "intention-in-action," and it is certainly one of the contributories to a full SA, along with more deliberated intentions formed prior to the action (Gallagher, 2010; Pacherie, 2006, 2007).

The argument here is based on a complex phenomenological account of SA which does not just depend on the efferent signals of motor control mechanisms; it also depends on pre-reflective perceptual monitoring of what one is accomplishing in the world, as well as on reflective, prospective, and retrospective deliberations about means and ends. Beyond this relatively narrow phenomenology, and beyond the brain-based and cognitive processes just mentioned, embodied action is enacted in a world that is both physical and social. At the most relevant pragmatic level, experience reflects perceptual and affective saliences, as well as the effects of physical and social forces, including constraints, affordances, and normative elements of the subject's social and cultural milieu. SA will accordingly be modulated by this larger context.

More generally, conceptions of agency, intention, and free will are best conceived in terms that integrate all of these aspects. What I freely decide to do is not about bodily movements—and it is certainly not reducible to 300 ms of brain activity—it is about my pragmatic or socially defined actions. Intentions often get co-constituted in interactions with others—indeed, some kinds of intention may not be reducible to processes that are contained exclusively within one individual. In this case, free will and sense of agency are matters of degree: they can be won or lost, enhanced or reduced by physical, social, economic, and cultural factors, including our own communicative and narrative practices.

The point I want to make here is that this kind of interchange between phenomenology and neuroscience points in the direction of a more complex picture involving not just brain processes "in the head," but certain physical and social aspects of the environment. In effect, even if we cannot "PET" or "fMRI" the lifeworld, as phenomenologists define it and critical theorists analyze it, the lifeworld, and what Habermas (2007) calls "the participant's involvement in shared lifeworld practices," need to be taken into consideration in the interpretation of experimental results. This, accordingly, would be a non-reductionist use of neuroscience.

In this respect, Habermas sets the discourse of neuroscience and the discourse of responsible agency in opposition; he contends that "the constellations of conditions that render actions intelligible and explainable differ in kind, conceptually, from the constellations of events linked by laws of nature" (2007, p. 17). I would argue, in fact, that we should not see these as completely distinct realms, even if we need different vocabularies to explain them. A breakdown at the level of neuronal processing is never just a brain event, since the brain is embodied and the body is embedded in an environment that is physical, social, cultural, and so forth. To say that one acts the way one does because 300 ms before each bodily movement the brain engages in preparatory processes amounts to a completely inadequate explanation of intentional action; but neither can we claim to have the whole picture by pointing exclusively to conscious intention formation or the enabling and constraining factors involved in social structures. Non-neural factors have an effect on neural factors, and vice versa, since the system in question is brain–body–environment and is organized dynamically across time. Disruptions in any part of the system—in the brain or in those aspects that involve conscious deliberation and amounts of time in excess of 300 ms, and even in social settings—can lead to disruptions in action. If, as Habermas and many others suggest, the languages of neuroscience and freedom/responsibility are irreducible to each other, that should not be a problem since we have both languages and we can say more with both than we can with only one.

Theory of Mind

To pursue the idea that a non-reductive neuroscience can contribute to critical theory, I want to show how a phenomenologically informed account of the neuroscience of social cognition can reframe certain aspects of a critical theory of communicative action. This project is larger than I can outline here, but I will focus on one important controversy in social neuroscience and try to show the implications of a particular interpretation for critical theory.

Let me first set the stage for the current debate on social cognition. Theory of mind (ToM) is one way to name the standard theories that dominate this broad debate in philosophy of mind, psychology, and social neuroscience. Under this heading there are two main contenders. The first is theory theory (TT), so called because it proposes that our understanding of others is based on a theory, namely, folk psychology—that is, the general, commonsense understanding that we have about human behavior. The idea defended by TT is that in understanding others, we use folk psychology to make inferences about their mental states (typically identified as propositional attitudes like belief and desire). Alternatively, simulation theory (ST) contends that we have something better than theory; we have direct access to our own mind and are capable of using it as a model to simulate the mental states of others. We do this explicitly (consciously) by introducing pretend mental states (pretend beliefs, pretend desires) into the mechanisms of our own minds, and then projecting the results to the minds of others. Alvin Goldman offers a concise three-step formula for this procedure.

1. First, the attributor creates in herself pretend states intended to match those of the target. In other words, the attributor attempts to put herself in the target's "mental shoes."
2. The second step is to feed these initial pretend states [for example, beliefs], into some mechanism of the attributor's own psychology … and allow that mechanism to operate on the pretend states so as to generate one or more new states [decisions].

3. Third, the attributor assigns the output state to the target ... [we infer or project the decision to the other's mind]

(Goldman, 2005, pp. 80–81)

Let me add two complications to this basic account. First, because it is left unexplained how we might accomplish the first step without already having the knowledge that ST is meant to explain, many theorists (including Goldman himself, 2006) now adopt a hybrid approach that combines TT and ST. Specifically, one appeals to folk psychology in order to understand what the target's "mental shoes" look like. Second, theorists of ST have recently developed an implicit version of this approach that makes use of the neuroscience of mirror neurons, a topic that I shall return to shortly.

Notably, both TT and ST share three basic assumptions.

-   **Mentalizing supposition**: we understand others to be other minds that are inaccessible. Mindreading involves an attempt to *explain* or *predict* their behavior on the basis of their mental states—the beliefs or desires they have, or as John Flavell recently put it, "the inner world inhabited by beliefs, desires, emotions, thoughts, perceptions, intentions and other mental states" (2004, p. 274).
-   **Spectatorial supposition:** for the most part we are observers who take a third-person stance toward others. Descriptions in TT or ST picture the subject as standing back observing the actions of others and trying to interpret them from that stance.
-   **Universal supposition**: mindreading is our primary and pervasive way of understanding others (starting sometime around the age of three or four years, based on data from traditional false-belief tests).[5]

Two things are clear from this set of suppositions. First, whatever our ordinary, usual way of understanding others actually is, it must enter into and constrain our communicative practices. Second, if our ordinary, usual way of understanding others is best described by either TT or ST, or by some hybrid version, then there are already some issues of concern to critical theorists. Honneth, following a long tradition in critical theory, objects to reifying practices that involve "detached observation" of one subject by another:

> Here the subject is no longer empathetically engaged in interaction with its surroundings but is instead placed in the perspective of a neutral observer, psychically and existentially untouched by its surroundings. The concept of "contemplation" [observation] thus indicates not so much an attitude of theoretical immersion or concentration as it does a stance of indulgent, passive observation
>
> (Honneth, 2008, p. 24)

---

[5] I have a large collection of statements endorsing this universal (or at least close to universal) supposition (see Currie & Sterelny, 2000, p. 145; Goldman, 2002, pp. 7–8). The most recent endorsement comes from Peter Carruthers: "Human beings are inveterate mindreaders. We routinely (and for the most part unconsciously) represent the mental states to (sic) the people around us ... We attribute to them perceptions, feelings, goals, intentions, knowledge, and beliefs, and we form our expectations accordingly. While it isn't the case that all forms of social interaction require mindreading ... it is quite certain that without it, human social life would be very different indeed" (Carruthers, 2009, p. 121).

Yet, if ToM genuinely captures our natural capacities for social cognition, taking a detached, observational stance may be unavoidable.

Similarly, ST throws up further problems from the perspective of critical theory. One criticism, targeting an earlier version of ST (the argument from inference by analogy) was voiced by both Max Scheler (1923/1954) and Gilbert Ryle (1949). According to the latter, for example, "the observed appearances and actions of people differ very markedly, so the imputation to them of inner processes closely matching [one's own or] one another would be actually contrary to the evidence" (Ryle, 1949, p. 54). In other words, people are diverse and it is somewhat presumptuous to reduce this diversity to something that can be easily modeled by one's own first-person experience, as ST would suggest. In a broader view this is an important point for critical theory, but again I will return to this later.

Before we leave these standard accounts, we need to say something about the recent development of neural ST. Neural ST conceives of simulation as a subpersonal process. This is an approach that has gained more ground in recent years by appealing to neuroscientific evidence involving subpersonal activation of the mirror (neuron) system.[6] Mirror neurons (MNs) in the pre-motor cortex, including Broca's area, are said to be activated both when the subject engages in specific instrumental actions and when the subject sees someone else engage in those actions (Rizzolatti, Fadiga, Gallese, & Fogassi, 1996; Rizzolatti, Fogassi, & Gallese, 2000). In broad terms, one's motor system resonates when one encounters another person. The claim made by ST is that these subpersonal mechanisms constitute a simulation of the other's intentions (Gallese, 2001; Gallese & Goldman, 1998). The hypothesis is this: understanding others is achieved by simulating the other's action "with the help of a motor equivalence between what the others do and what the observer does" (Gallese, 2001, p. 39). This is a subpersonal process generated by "automatic, implicit, and non-reflexive simulation mechanisms" (Gallese, 2005, p. 117).

Neural ST understood in these or in similar terms has been the growing consensus. Indeed, use of the term "simulation" has become the standard way of referring to mirror system activation. Goldman (2006) distinguishes between simulation as high-level mindreading and simulation as low-level mindreading where the latter is "simple, primitive, automatic, and largely below the level of consciousness" (p. 113), the prototype for which is "the mirroring type of simulation process" (p. 147; see also Jeannerod & Pacherie, 2004). That MN activation is a simulation not only of the goal of the observed action but of the intention of the acting individual, and is therefore a form of mindreading, is suggested by research that shows MNs discriminate identical movements according to the intentional action and the simple pragmatic contexts in which these movements are embedded (Fogassi et al., 2005; Iacoboni et al., 2005; Kaplan & Iacoboni, 2006).

---

[6]   For example, Fadiga, Fogassi, Pavesi, & Rizzolatti (1995); Rizzolatti, Fadiga, Gallese, & Fogassi (1996); Rizzolatti, Fogassi, & Gallese (2000). I leave aside the recent criticisms that raise questions about the existence of MNs in humans (see Dinstein, Thomas, Behrmann, & Heeger, 2008; Hickok, 2009). This is clearly an empirical question with implications for ST. But even if we assume that there are MNs in the human brain, the more philosophical question is whether they can be considered what Oberman & Ramachandran (2008) call "simulator neurons."

An Alternative Theory

For present purposes I want to focus on some important criticisms of neural ST, but before I do that let me outline an alternative theory that emphasizes embodied interaction.[7] Interaction theory (IT) offers an alternative account of social cognition that opposes the three basic suppositions made in ToM approaches. In opposition to the *mentalistic supposition*, which treats the other as a Cartesian mind that is hidden away, IT maintains that we have an enactive perceptual access to the other's intentions and emotions via their embodied actions, movements, gestures, facial expressions, and so forth. In opposition to the *spectatorial supposition* that takes third-person observation to be our normal stance toward others, IT holds that in our everyday encounters with each other we are primarily interacting in second-person relations where the task is not explanation but continuing participatory interaction and pragmatic doing. In opposition to the *universal supposition*, IT holds that there are many kinds of human relations, but interaction rather than mindreading characterizes most of our encounters, while the use of theory or simulation for this purpose is a relatively rare occurrence.

IT points to three broad kinds of capacities for understanding others. The first consists of a set of sensory-motor capacities included under the concept of *primary intersubjectivity*, a term originating with Colwyn Trevarthen (1979) working in developmental psychology. These basic sensory-motor capacities, some of which are found in infants from birth, are geared to interaction with others. They include the capacity for neonate and very early imitation (Gallagher & Meltzoff, 1996; Meltzoff & Moore, 1977, 1994). Young infants are also able to parse the surrounding environment into those entities that perform human actions and those that do not (Johnson, 2000; Johnson, Slaughter, & Carey, 1998; Legerstee, 1991; Meltzoff & Brooks, 2001). For the infant, the other person's body presents opportunities for action and expressive behavior—opportunities that the infant can pursue through imitation. There is, in this case, a common bodily intentionality that is shared by infant and caregiver. In addition, the ability to detect correspondences between visual and auditory information that specify the expression of emotions starts as early as five to seven months old (also Hobson, 1993, 2002; Walker, 1982), and at nine months, the ability to follow the other person's gaze (Senju, Johnson, & Csibra, 2006). Such perceptual abilities serve affective coordination between the gestures and expressions of the infant and those of caregivers with whom they interact. Infants "vocalise and gesture in a way that seems 'tuned' [affectively and temporally] to the vocalisations and gestures of the other person" (Gopnik & Meltzoff, 1997, p. 131). Infants, then, are not simply observing others; they are interacting with them from the very beginning.

In primary intersubjectivity, then, our preliminary access to others is based on these innate or early developing capacities manifested at the level of perceptual experience— we "see", in the other person's bodily movements, facial expressions, gestures, eye direction, and so forth, what they intend and what they feel—and we react to them. Infants as young as six months perceive grasping as goal directed; infants at 10–11

---

[7] I've outlined the phenomenological critique of TT and ST elsewhere (Gallagher, 2001, 2004, 2005, 2007, 2008a, 2008b).

months are able to parse some kinds of continuous action according to intentional boundaries (Baird & Baldwin, 2001; Baldwin & Baird, 2001; Woodward & Sommerville, 2000). By the end of the first year of life, infants have a non-mentalistic, perceptually-based embodied understanding of the intentions and dispositions of other persons (Allison, Puce, & McCarthy, 2000; Baldwin, 1993; Johnson, 2000; Johnson et al., 1998).

These primary capacities do not disappear with maturity; they become more nuanced. Thus, Scheler (1923/1954), Wittgenstein (1967, §229), and others have described the everyday phenomenology of perceiving others as a direct perception of their feelings and intentions (see Gallagher, 2008a for more on this concept of direct perception). As we'll see, the best way to characterize this ability to see meaning in the other person's actions and expressions is to say that social perception is enactive—that is, it is geared to interaction with others.

A second set of capacities belongs to what Trevarthen calls secondary intersubjectivity (Trevarthen & Hubley, 1978). Expressions, intonations, gestures, and movements, along with the bodies that manifest them, do not float freely in thin air; we find them situated in the world, and infants soon start to notice how others interact with things in the environment. In such interactions the child looks to the body and the expressive movement of the other to discern the intention of the person or to find the meaning of some object. Around the age of 9–12 months, the infant goes beyond person-to-person immediacy and enters contexts of joint attention (Phillips, Baron-Cohen, & Rutter, 1992; Reddy, 2008)—shared situations—the pragmatic and social situations in the everyday lifeworld where we learn what things mean and what they are for. The child can understand that the other person wants food or intends to open the door; that the other can see him (the child) or is looking at the door. They begin to see that another's movements and expressions often depend on meaningful and pragmatic contexts and are mediated by the surrounding world.

There are two aspects involved in secondary intersubjectivity that are important to distinguish (see Gallagher, 2009). The first pertains to social cognition—understanding the other person. As we interact with others we learn their intentions and we gain understanding of them through their behavior towards us, towards things in the surrounding world, and through the richly pragmatic and social contexts of such interactions. The second aspect is what De Jaegher and Di Paolo (2007) call "participatory sense making." As we interact with others we not only gain an understanding of them, we gain an understanding of the world that we share with them. Our attention to objects in the world around us changes when others are present—even if our attention is not explicitly guided by others (Becchio, Bertone, & Castiello, 2008). Empirical studies show that when we see another person's face simply looking towards or away from an object, we evaluate the object looked at as more valuable than the object not looked at. An emotional expression on the face results in a stronger effect (Bayliss, Frischen, Fenske, & Tipper, 2007; Bayliss, Paul, Cannon, & Tipper, 2006).

Brain imaging studies show that what we see other people do primes our system for action with objects. Motor-related areas of the brain—dorsal premotor cortex, the inferior frontal gyrus, the inferior parietal cortex, the superior temporal sulcus—are activated not only when we see someone reach for an object, but simply if we see them gaze at an object (Friesen, Moore, & Kingstone, 2005; Pierno, Becchio, Tubaldi,

Turella, & Castiello, 2008; Pierno et al., 2006). Thus, objects surrounding us take on meaning within the context of our shared projects. We begin to make sense of the world through our participation with others in pragmatic contexts, and the shared lifeworld starts to open up precisely in such participation.

The actions of others are always framed in pragmatic and socially defined contexts. It follows that there is not one, uniform way in which we relate to others, but that our relations are mediated through the various pragmatic (and ultimately, institutional) circumstances of our encounters. In understanding others, then, the world itself does some of the work—sometimes the actual physical situation, or location; sometimes the institutional setting and the various social roles played by individuals. We come to understand how things work and how contexts can inform the emotions, intentions, and thoughts of others. In this very real sense, a complete explanation of social cognition is not possible simply in terms of neuronal processes; since it clearly involves others in extra-neural contexts and worldly events, it is not reducible to the worldless realm of brain events. The lifeworld cannot be reduced to appropriate scanner size.

A third kind of capacity is required to account for the more nuanced and sophisticated understandings we attain as adults. These are communicative and narrative competencies that allow us to fill in and properly frame the interactive contexts that help to make sense of people's actions (Gallagher & Hutto, 2009; Hutto, 2008). Capacities involved in communicative and narrative competencies are not only essential for a more complete account of social cognition, but are also of great importance to critical theory. They can offer something more positive to the kind of critical theory that develops around the concept of communicative practice. Although there is much more to say about this, particularly in regard to narrative competency, for the purposes of this chapter I will focus on issues more pertinent to critical neuroscience. Specifically, I will focus on some questions about IT that involve the neuroscience of the mirror system and the concept of simulation.

Neural Simulation or Enactive Perception?

At the neuronal level, the mirror neuron (MN) system may (or may not[8]) underlie some of the capacities of primary intersubjectivity. For this reason it might seem that the idea of an implicit neural simulation would help to support interaction theory, or that IT is just a version of ST. There are, however, several things wrong with thinking of MN activation as a simulation process (see Gallagher, 2007, 2008c). First, the meaning of the term "simulation" as defined by ST involves two essential aspects that are simply missing in the activation of MNs. According to that definition, simulation (1) involves pretense and (2) has an instrumental character.

For example, in Goldman's explanation: simulation involves "pretend states," by which he means "some sort of surrogate state, which is *deliberately adopted* for the sake of the attributor's task ... In simulating practical reasoning, the attributor *feeds* pretend

---

[8]   There is still some controversy about whether there is good scientific evidence for MNs in humans (see, for example, Dinstein, Thomas, Behrmann, & Heeger, 2008); and if there are MNs in humans, there are still debates about whether they have anything to do with social cognition (see Hickok, 2009).

desires and beliefs into her own practical reasoning system" (2002, p. 7; see Adams, 2001; Bernier, 2002). The aspect of pretense is part of what distinguishes simulation from a TT model or a simple practice of reasoning (see Fisher, 2006). The claim for pretense is found even in subpersonal accounts; as Gallese puts it, "our motor system becomes active *as if* we were executing that very same action that we are observing" (2001, p. 37). Likewise for Gordon (2005, p. 96) the neurons that respond when I see your intentional action, respond "*as if* I were carrying out the behaviour ..."

Despite these claims, it is difficult to see how pretense can be involved at the subpersonal level, in the neuronal processes themselves. Activation of MNs per se cannot represent or register pretense in the way required by ST since, as is often claimed, MNs are "neutral" in regard to the agent (Gallese, 2005; Hurley, 1998; Jeannerod & Pacherie, 2004); that is, they are activated both when one engages in intentional action and when one sees someone else engage in intentional action. Accordingly, in MN activation there is no first- or third-person specification, in which case, it is not possible for them to register "my" intentions as pretending to be "your" intentions. There can be no "as if" of the sort required by ST because there is no "I" or "you" represented. Even if some aspect of MN activation did distinguish between my action versus my observation of your action (see Gallagher & Zahavi, 2008), more than that would be needed for pretense.

With respect to characterizing simulation as instrumental, it is often described as a mechanism or model that we manipulate or control in order to understand something to which we do not have access (see Goldman's description above). Gordon locates instrumental control at the neuronal level by suggesting that on the "cognitive-scientific" model, "one's own behaviour control system is employed as a *manipulable model* of other such systems. (This is not to say that the 'person' who is simulating is the model; rather, only that *one's brain can be manipulated to model other persons*)" (2004, p. 1).

In this regard, however, if simulation is characterized as a process that I (or my brain) instrumentally use(s), manipulate(s), or control(s), then the implicit processes of motor resonance are not good examples of simulation. Certainly, at the personal level, we do not manipulate or control the activated brain areas—in fact, we have no instrumental access to neuronal activation. "Pace" Gordon, it's not clear where anything like manipulation comes into play. Indeed, in precisely the intersubjective circumstances that we are considering, these neuronal systems do not take the initiative; they do not activate themselves. Rather, they are automatically activated by the other person's action. The other person has an effect on us and elicits this activation. It is not I (nor my brain) who manipulates anything; it is the other who does something to us via a perceptual elicitation.

Perhaps, however, this objection targets an incremental version of ST that is not favored by neural simulationists (Gordon, 2008; Hutto, in press). Indeed, anticipating objections about the involvement of pretence and instrumentality, Goldman (2006; see also Goldman & Sripada, 2005) argues that the instrumental and pretence conditions are not necessary conditions for simulation, and that simulation involves something more minimal.

> We do not regard the creation of pretend states, or the deployment of cognitive equipment to process such states, as essential to the generic idea of simulation. The

general idea of simulation is that the simulating process should be similar, in relevant respects, to the simulated process. Applied to mindreading, a minimally necessary condition is that the state ascribed to the target is ascribed as a result of the attributor's instantiating, undergoing, or experiencing, that very state. In the case of successful simulation, the experienced state matches that of the target. This minimal condition for simulation is satisfied [in the neural model].

(Goldman & Sripada, 2005, p. 208)

Rizzolatti and Sinigaglia (2008; see also Flanagan & Johansson, 2003; Sinigaglia, 2009) also favor what they term the Direct Matching Hypothesis (DMH). There are a number of things we could say in response to this tactic (see Gallagher, 2008c), but for now let me just note that against this view, the minimal condition of matching cannot be the pervasive or default way of attaining an understanding of others. If simulation were the default mode of social cognition, and as automatic as mirror neurons firing, then it would seem that we would automatically go into the mental or motor state of the other person whenever we properly see their action, and we would not be able to attribute to ourselves a state different from the other person's mental or action state. But we do this all the time. Seeing you angry does not automatically make me angry—indeed, it may make me afraid. There are many cases of encountering others in which we do not adopt, or find ourselves in a matching state. When I see you trip and start to fall, I do not simulate your movement; I do not find myself starting to fall, rather I find myself trying to reach out to catch you—so my motor system is clearly in a different state from yours. Furthermore, consider the difficulties involved if we were interacting with more than one person, especially if in such interpersonal interactions the actions and intentions of each person are affected by the actions and intentions of the others (see Morton, 1996).

In regard to this, a study by Catmur, Walsh, and Heyes, (2007) demonstrates that learning can work against matching. The experimenters trained subjects to move their fingers in a manner incongruent with an observed hand, for example, they moved their little finger when they observed the movement of an index finger. After training, MEPs were greater in the little finger when index finger movement was observed. "The important implication of this result is that study participants who exhibited incongruent MEP responses presumably did not mistake the perception of index finger movement for little finger movement" (Hickok, 2009, p. 1236). In other words, the lack of matching in the motor system does not pre-empt some kind of recognition of what the other person is doing.

In another study Dinstein and colleagues (2008) show that in fact, in certain areas of the brain where MNs are thought to exist—specifically *anterior intraparietal sulcus* (aIPS)—areas activated for producing a particular hand action are not activated for perceiving that hand action in another. Thus, "distinctly different fMRI response patterns were generated by executed and observed movements in aIPS ... aIPS exhibits movement-selective responses during both observation and execution of movement, but ... the representations of observed and executed movements are fundamentally different from one another" (Dinstein et al., 2008).

The phenomenology and behavioral logic of these results are supported by the details of the scientific data on MNs. MN activation in monkeys does not always involve a precise match between motor system execution and observed action. Approximately 60% of MNs are "broadly congruent," which means there may be some relation between the observed action(s) and their associated, executed action, but not an exact match. Only about one-third of MNs show a one-to-one congruence (Csibra, 2005).[9]

In denying that MN activation is a form of simulation, I am not denying the possibility that MNs may be involved in our interactions with others and possibly contribute to our ability to understand others or to keep track of ongoing intersubjective relations. There is some evidence that the mirror system is involved when we perceive not only the other person's action, but when we perceive the other person being touched (Blakemore, Bristow, Bird, Frith, & Ward, 2005), when we perceive their emotional states (Jabbi, Swart, & Keysers, 2007), facial expressions (Van der Gaag, Minderaa, & Keysers, 2007), and emotional bodily movements (Pichon, de Gelder, & Grézes, 2008). In contrast to thinking of MNs as part of a simulation process, over and above perceptual processes, a more parsimonious interpretation of MN activation which is consistent with interaction theory, is possible. The line between neuronal activation in the visual system and neuronal activation of the mirror system is not a line that we should draw between perception and simulation. Rather, mirror resonance processes can be considered part of the neuronal processes that underlie the kind of enactive *perception* found in primary intersubjectivity.

The idea of an enactive social perception is consistent with the basic idea of enactive perception (Varela, Thompson, & Rosch, 1991) where the act of perception is defined not simply as a sensory activation but includes motor components. Perception is primarily "for action". In the case of intersubjectivity, perception is "for interaction". The corresponding phenomenology is this: in most cases, when I see the other's action or gesture or emotional expression, I "directly perceive" the meaning in the action or gesture or expression. I see the joy or I see the anger, I see what they must feel, or I see the intention in the face or in the posture or in the gesture or action of the other. I see it. I don't have to simulate it; and I see is as something I can respond to—I see it as "respondable to." As Newman-Norlund, Noordzij, Meulenbroek, and Bekkering suggest (2007, p. 55), it's likely that broadly congruent MN activation is preparation for a complementary action rather than a matching action. This kind of perception depends of course not just on MNs, but on a complex of subpersonal neuronal processes that include activation of sensory areas (for example, the visual cortex), association areas, as well as motor areas. These articulated neuronal processes, that may include activation of MNs, contribute as part of the neural correlates of a non-articulated immediate perception of the other person's intentional actions, rather than being a distinct process of simulating their intentions. If mirror activation is involved, by this interpretation it is not the initiation of simulation but is part of an enactive intersubjective perception of what the other is doing. That is, I see the other's

---

[9]　A recent study by van Schie et al. (2008) does find matching (of same-action-related neurons to observed action) in the automatic initial timeframe of congruent MN activation (taking place less than 100 ms after activation of visual cortex). But this study also shows that this initial matching does not register any of the important specifics of the action in relation to goal or intention, correctness or incorrectness, or any other parameters that are important for social cognition.

action as something I can respond to. Action perception in the interactive context is action priming—I perceive the action of the other as a social affordance.

Implications for Critical Theory

Getting the story (and the science) of intersubjectivity and social cognition right has important relevance to critical theory. Questions about free will and responsibility, and about action and communication in social contexts, for example, are directly linked to questions about intersubjectivity, since responsibility may be conceived of as answerability to others, and actions are frequently interactions. That I am responsible for my actions suggests that I should be able to give (to others, or to myself as another) reasons for acting the way I did where such reasons go beyond a list of mechanical causes.

Habermas draws a strict line between reasons, which he characterizes as "arguments that express positions persons take on validity claims," and causes, such as unconscious brain states. When reasons are assimilated to causes, as in the neuroscientific claim that free will is an illusion, scientific understanding itself, which depends on the evaluation of reasons and arguments, is undermined. Whether a strict line can be maintained between reasons and causes, or between the two respective language games, it certainly seems right to suggest that in doing science scientists, who themselves are not just brains, engage in the evaluation of reasons and arguments about what they should do (how they should act) to maintain the validity of their experimental procedures, for example.[10] It is also clear that their actions are—or presuppose—intersubjective interactions with others. Science itself is an intersubjective practice.

Habermas's own characterization of intersubjectivity, however, reflects an emphasis on propositional communication, described in terms of pragmatic speech acts:

> Thoughts, intentions, and experiences can be attributed only to persons, who themselves can develop as persons only in contexts of social interaction. It is in the course of their ontogenesis that children first learn to take up the pragmatic roles of speaker, hearer, and observer and relate to oneself in the corresponding ways.... [T]he intersubjective consti- tution of a mind that is intentionally oriented towards the world, communicates via propositional contents, and is responsive to rules and standards of validity
>
> (Habermas, 2007, pp. 24–25)

Habermas suggests that there is a logical gap between the other's mind and his or her behavior. At the same time he emphasizes the priority of second-person interaction over experiential aspects of mind—which have too little propositional content and are seemingly inaccessible:

---

[10]   I think Searle (2007) misses the point when he argues against Habermas that it is perfectly legitimate for scientists to take their own free will as a presupposition which their investigation in the end shows to be false. The issue is not about an inconsistency within the methodological boundaries of a particular experiment, but about the ability to do science in the first place. The situation is more like the scientist presupposing that he has a right arm, and by investigation demonstrating that to be false, but doing so only by using his right arm.

That explains why the intentionalist predicates with which a vocabulary must be equipped (if it is to be suitable for describing persons and their utterances) can be learned only performatively, through being practised by agents who relate to each other in interaction as second persons.

(Habermas, 2007, p. 35)

Trevarthen settled on his terminology of primary and secondary intersubjectivity after finding the concept of intersubjectivity developed in Habermas's writings on critical theory (personal communication). In Habermas, however, all of the important action, the "action oriented toward reaching understanding" is to be found in communicative practices. Habermas links the developmental psychology of Piaget and Kohlberg to his discussion of communicative action, but in this respect the earliest experiences of the infant have little or nothing to do with such action (see Habermas, 1990). He comes at the developmental questions from a perspective already informed by social theory and his interest in moral development, and perhaps for this reason he misses the importance of the embodied and enactive aspects of non-verbal interaction. For him, co-ordination of action is the result of reflective processes, but, from the perspective of IT, such processes are always informed by pre-reflective intersubjective experiences about which Habermas has little to say. Indeed, in these primary and secondary intersubjective experiences the lifeworld, understood as a background to communicative action (Habermas, 1990, p. 135), is established. Habermas's starting point is too late in developmental terms to give an account of how the lifeworld—the shared world in which we interact—comes to be the established background. Building on work by Selman (1980) and Flavell (1968), the interaction he describes is something of a hybrid of ST and TT, framed in terms of inferring the intentions of others through the rational adoption of different perspectives (1990, p. 142ff).

As indicated above, whatever our usual ways of understanding others actually are, whether or not these capacities for social cognition are innate or early developing, or relatively late developing, they must enter into and constrain, as well as enable our communicative practices. If TT or ST best describe social cognition, however, we would end up with a communicative practice in which the second-person participatory perspective is waylaid since, for TT especially, our relations to others are characterized in terms of the third-person observational perspective, and for ST, interaction is framed in terms of first-person mechanisms—something, as noted, that raises questions about the nature of diversity. Neither of these standard accounts provides an explanation of what Bruner and Kalmar (1998) have called the "massively hermeneutic" background (that is, the lifeworld) which forms a necessary backdrop for communicative processes. At best, they appeal to an already formulated folk psychology as the *de jure* theory, or as an ad hoc primer that lets simulation take the first step into someone else's shoes.

More importantly, when one ignores the various capacities for interaction found in primary intersubjectivity, communicative practices can only be conceived of as a collection of speech acts motivated by a formal a priori trust that speakers want to be understood—a formal-pragmatic conception of interaction in which one has to argue for the idea that illocutionary force is not just something extra added on to expressed propositions (Habermas, 1987, 2000).[11] In contrast, if one starts with the idea of communicative practices as a continuation of the enactive sensory-motor performances of primary intersubjectivity, communication is already for action. That is, we would not have to undertake a demonstration of "how communicative acts—that is, speech acts or equivalent non-verbal expressions—take on the function of coordinating action and make their contribution to building up interactions" (Habermas, 1987, p. 278). Rather, we would clearly see that speech acts emerge with just this function in already established interactions. The task would be to show how such communicative speech acts transform actions that are already non-propositionally, non-verbally, communicative.

In other words, with a detailed account of primary and secondary intersubjectivity—a set of capacities and practices that characterize not only young infants, but continue to characterize relations with others in maturity, albeit transformed in communicative and narrative practices—we already have an enactive account of interaction, a form of (pre-verbal) communication; and we already have the basis for the opening up of the intersubjective, experiential lifeworld in participatory sense making. An analysis of communicative practice at the level of speech acts doesn't get off the ground without this more basic framework.

Moreover, an important part of this detailed account is neuroscientific; at least it is if we want the well-rounded story and if we don't want to appeal to a concept of the lifeworld as a representational (or simulated) illusion. Getting the story right, even at the level of brain processes, seems important with respect to this very point of understanding what the lifeworld is—namely, a landscape for action and interaction rather than a place for the meeting of minds.[12] Intentional actions are already underway, at a pre-reflective level constrained and enabled by sensory-motor processes, as well as by social, cultural, and institutional forces, before we can turn them to account by means of reflective intention formation. Deliberative speech acts are already shaped by *im*plicit communicative processes of interaction and participatory sense making, processes that have woven together the shared lifeworld which makes those more *ex*plicit acts possible.

In this regard, Honneth's recent work may be more promising, for he does include reference to concepts of primary and secondary intersubjectivity. Honneth, reviewing the developmental studies of Mead and Piaget, which emphasize perspective taking, rightfully criticizes their lack of inclusion of an emotional dimension. Among others, he turns to Peter Hobson and Michael Tomasello for clarification on the importance of emotion in developing communicative abilities:

> The starting point of these investigations consists in the same transition from primary to secondary intersubjectivity that the cognitivist approaches also have in mind. These theories suggest that at the age of nine months a child makes several notable advances in its interactive behavior. It acquires the ability to point out objects to its attachment figure

---

[11] Habermas clearly puts the emphasis on speech acts. "If we were not in a position to refer to the model of speech, we could not even begin to analyse what it means for two subjects to come to an understanding with one another" (2000, p. 120).

[12] For the distinction between the landscape of action and the mentalistic—folk psychological—concept of the landscape of consciousness, see Bruner (1986).

by means of protodeclarative gestures and then to view these objects with this person. It can further make its attitude toward meaningful objects dependent upon the expressive behaviour with which this other person reacts to these objects. And, finally, the child appears, in doing what G. H. Mead [and I would add, everyone else–S.G.] calls "playing," gradually to grasp the fact that familiar meanings can be uncoupled from their original objects and transferred to *other* objects, whose new borrowed function can then be creatively dealt with.

(Honneth, 2008, p. 43)

Although Honneth emphasizes the importance of joint attention and emotional attachment as an essential aspect of this development, he leaves aside the full complement of what Buckner, Shriver, Crowley, & Allen (2009) have called the "swarm" of interactive capacities found in primary intersubjectivity. Interaction theory, integrating the rich set of capacities of primary and secondary intersubjectivity (as well as narrative competency), presents a fuller picture of what constitutes intersubjective interaction than that acknowledged by Habermas or yet by Honneth. In specific regard to the question of communicative behavior, we should not stop short of considering the contribution of early interaction capabilities, or the questions that concern the neuroscientific understanding of these capabilities.

If most of the actions that are the concern of critical theory are more contextually complex than the typical dyadic relationship that developmental psychology addresses, the question still remains: can we not, at least to some limited degree, use principles that pertain to dyadic and small group interactions to understand the larger and more complex events that involve not only people, but institutions, technologies, and cultural practices?[13] One such principle may be that interaction always adds up to more than what each individual brings to the encounter. Interaction itself generates meaning irreducible to any or all of what the participants intended (see De Jaegher & Di Paolo, 2007). This relates directly to the concept of participatory sense making in which meaning emerges from the interaction itself. This is certainly consistent with the view of critical theory, that to limit analysis to what individuals intend (or to what is happening in one brain) would be to ignore some very real and powerful aspects of interaction that may be productive or counterproductive to communicative practices.

Similarly, the importance that IT places on pragmatic and social contexts is clearly relevant to the aims of critical theory. What makes Habermas's concept of the ideal speech situation ideal (if not idealistic) is precisely that it tries to strip away contextual differences (differences in individual backgrounds, for example)—differences that actually do count in communicative practices. IT would side with the hermeneutical claim that real communicative situations are always biased. The only way to deal with

---

[13] On a related issue Joel Anderson, who translated Habermas's essay on free will, makes a suggestion that has recently been explored, independently, in Crisafi & Gallagher (2009) and Gallagher & Crisafi (2009) namely, that these kinds of institutions might be considered from the perspective of the extended mind hypothesis. "The notion of 'objective mind' (which stems from Hegel, where it is often translated as 'objective spirit') is used to refer to social institutions, customs, shared practices, science, culture, language, and so on—those entirely real parts of the human world that are neither held within one individual's mind nor physically instantiated independently from humans. In this sense, then, recent discussions within philosophy of mind and cognitive science regarding 'situated cognition' or the 'extended mind' are also about the 'objective mind'" (Translators footnote 5, in Habermas, 2007, pp. 42–43).

such differences is through continued interaction, and specifically in ways that recognize and acknowledge them. Understanding others is not a matter of simulating or matching our experience to theirs; rather, it involves understanding why such simulation, or the presumption of simulation, may blind us to diversity and may lead to distorted communicative practices. It is at this point that one would need to say more about narrative competencies, but that is a topic for another time.[14]

On standard ToM accounts the problem is often posed as follows: "To understand interactive minds we have to understand how thoughts, feelings, intentions, and beliefs can be transmitted from one mind to the other" (Singer, Wolpert, & Frith, 2004, p. xvii). But mental states do not fly through thin air between minds; nor are they simply replicated in matching brains or externalized in pure speech acts. Rather, human feelings, intentions, thoughts, and beliefs are deeply embedded in backgrounds and contexts, in embodied social interactions and communicative practices in the everyday lifeworld, and all of these phenomena are characterized by a great many differences that need to be recognized and acknowledged.

References

Adams, F. (2001). Empathy, neural imaging, and the theory versus simulation debate. *Mind and Language, 16*(4), 368–92.

Allison, T., Puce, Q., & McCarthy, G. (2000). Social perception from visual cues: Role of the STS region. *Trends in Cognitive Science, 4*(7), 267–278.

Baird, J. A., & Baldwin, D. A. (2001). Making sense of human behavior: Action parsing and intentional inference. In B. F. Malle, L. J. Moses, & D. A. Baldwin (Eds.), *Intentions and intentionality: Foundations of social cognition* (pp. 193–206). Cambridge, MA: MIT Press.

Baldwin, D. A. (1993). Infants' ability to consult the speaker for clues to word reference. *Journal of Child Language, 20*, 395–418.

Baldwin, D. A., & Baird, J. A. (2001). Discerning intentions in dynamic human action. *Trends in Cognitive Science, 5*(4), 171–8.

Bayliss, A. P., Paul, M. A., Cannon, P., & Tipper, S. P. (2006). Gaze cueing and affective judgments of objects: I like what you look at. *Psychon. Bull Rev, 13*, 1061–1066.

Bayliss, A. P., Frischen, A., Fenske, M. J., & Tipper, S. P. (2007). Affective evaluations of objects are influenced by observed gaze direction and emotional expression. *Cognition, 104*, 644–653.

Becchio, C., Bertone, C., & Castiello, U. (2008). How the gaze of others influences object processing. *Trends in Cognitive Sciences, 12*(7), 254–58.

Bernier, P. (2002). From simulation to theory. In J. Dokic & J. Proust (Eds.), *Simulation and knowledge of action* (pp. 33–48). Amsterdam: John Benjamins.

---

[14] For a proponent of what is sometimes referred to as a hermeneutics of suspicion, Habermas puts a great deal of trust in reason: "Whoever enters into discussion with the serious intention of being convinced of something through dialogue ... has to presume ... that the participants allow their 'yes' or 'no' to be determined solely by the force of the better argument" (1996, p. 367). Certainly this needs to be qualified by the particular differences of the participants—differences that can be communicated by narrative means. See Harrist & Gelfand (2005) for a more detailed analysis of this approach. As they point out, "Non-narrative theoretical accounts necessarily emphasise the bias built into the theoretical model ... Narrative accounts, on the other hand are flexible enough to allow emphasis on any given factor at any given moment" (p. 238).

Blakemore, S. J., Bristow, D., Bird, G., Frith, C., & Ward, J. (2005). Somatosensory activations during the observation of touch and a case of vision–touch synaesthesia. *Brain, 128*(7), 1571–83.

Bruner, J. (1986). *Actual minds, possible worlds.* Cambridge, MA: Harvard University Press.

Bruner, J., & Kalmar, D. A. (1998). Narrative and metanarrative in the construction of self. In M. Ferrari & R. J. Sternberg (Eds.), *Self-awareness: Its nature and development* (pp. 308–31). New York: Guilford Press.

Buckner, C., Shriver, A., Crowley, S., & Allen, C. (2009). How "weak" mindreaders inherited the earth. *Behavioral and Brain Sciences, 32,* 140–41.

Carruthers, P. (2009). How we know our own minds: The relationship between mindreading and metacognition. *Behavioral and Brain Sciences, 32,* 121–182.

Catmur, C., Walsh, V., & Heyes, C. (2007). Sensorimotor learning configures the human mirror system. *Current Biology, 17,* 1527–1531.

Chaminade, T., & Decety, J. (2002). Leader or follower? Involvement of the inferior parietal lobule in agency. *Neuroreport, 13*(1528), 1975–1978.

Choudhury, S., Nagel S., & Slaby, J. (2009). Critical neuroscience: Linking neuroscience and society through critical practice. *BioSocieties, 4*(1).

Crisafi, A., & Gallagher S. (2009). Hegel and the extended mind. *Artificial Intelligence & Society.* DOI 10.1007/s00146-009-0239-9.

Csibra, G. (2005). Mirror neurons and action observation. Is simulation involved? *ESF Interdisciplines.* http://www.interdisciplines.org/mirror/papers/.

Currie, G., & Sterelny, K. (2000). How to think about the modularity of mind-reading. *The Philosophical Quarterly, 50*(199), 145–160.

Damasio, A. R. (1999). *The feeling of what happens: Body and emotion in the making of consciousness.* New York: Harcourt.

De Jaegher, H., & Di Paolo, E. (2007). Participatory sense-making: An enactive approach to social cognition. *Phenomenology and the Cognitive Sciences, 6*(4), 485–507.

Descartes, R. (1989). *The passions of the soul.* Indianapolis: Hackett Publishing Company, (Original work published 1649).

Dinstein, I., Thomas, C., Behrmann, M., & Heeger. D.J. (2008). A mirror up to nature. *Current Biology, 18*(1), R13–R18.

Dreyfus, H. (2002). Intelligence without representation: Merleau-Ponty's critique of mental representation. *Phenomenology and the Cognitive Sciences, 1*(4), 367–383.

Elger, C. E., Friederici, A. D., Koch, C., Luhmann, H.,Von der malsburg, C., Menzel, R., & Singer, W. (2004). Das Manifest: Elf führende Neurowissenschaftler über Gegenwart und Zukunft der Hirnforschung. *Gehirn und Geist, 6,* 30–37.

Fadiga, L., Fogassi, L., Pavesi, G., & Rizzolatti, G. (1995). Motor facilitation during action observation: A magnetic stimulation study. *Journal of Neurophysiology, 73*(6), 2608–2611.

Farrer, C., & Frith, C. D. (2002). Experiencing oneself vs. another person as being the cause of an action: The neural correlates of the experience of agency. *Neuroimage, 15*(3), 596–603.

Farrer, C., Franck, N., Georgieff, N., Frith, C. D., Decety, J., & Jeannerod, M. (2003). Modulating the experience of agency: A positron emission tomography study. *Neuroimage, 18,* 324–333.

Fisher, J. C. (2006). Does simulation theory really involve simulation? *Philosophical Psychology, 19*(4), 417–432.

Flanagan, J. R., & Johansson, R. S. (2003). Action plans used in action observation. *Nature, 424,* 769–771.

Flavell, J. H. (1968). *The development of role-taking and communication skills in children.* New York: John Wiley & Sons.

Flavell, J. H. (2004). Theory-of-mind development: Retrospect and prospect. *Merrill-Palmer Quarterly, 50*(3), 274–290.

Fogassi, L., Ferrari, P. F., Gesierich, B., Rozzi, S., Chersi, F., & Rizzolatti, G. (2005). Parietal lobe: From action organization to intention understanding. *Science, 308,* 662–667.

Frankfurt, H. (1978). The problem of action. *American Philosophical Quarterly, 15,* 157–62.

Friesen, C. K., Moore, C., & Kingstone, A. (2005). Does gaze direction really trigger a reflexive shift of spatial attention? *Brain & Cognition, 57,* 66–69.

Frith, C., & Gallagher, S. (2002). Models of the pathological mind: An interview with Christopher Frith. *Journal of Consciousness Studies, 9*(4), 57–80.

Gallagher, S. (2010). Multiple aspects of agency. *New Ideas in Psychology.* (http://dx.doi.org/10.1016/j.newideapsych.2010.03.003). Online publication April 2010.

Gallagher, S. (2009). Two problems of intersubjectivity. *Journal of Consciousness Studies, 16*(6–8), 289–308.

Gallagher, S. (2008a). Direct perception in the social context. *Consciousness and Cognition, 17,* 535–543.

Gallagher, S. (2008b). Inference or interaction: Social cognition without precursors. *Philosophical Explorations, 11*(3), 163–73.

Gallagher, S. (2008c). Neural simulation and social cognition. In J. A. Pineda (Ed.), *Mirror neuron systems: The role of mirroring processes in social cognition* (pp. 355–71). Totowa, NJ: Humana Press.

Gallagher, S. (2008d). *Brainstorming: Views and interviews on the mind.* Exeter: Imprint Academic.

Gallagher, S. (2008e). Are minimal representations still representations? *International Journal of Philosophical Studies, 16*(3), 351–69.

Gallagher, S. (2007). Simulation trouble. *Social Neuroscience, 2*(3–4), 353–65.

Gallagher, S. (2006). Where's the action? Epiphenomenalism and the problem of free will. In W. Banks, S. Pockett, & S. Gallagher (Eds.), *Does consciousness cause behavior? An investigation of the nature of volition* (pp. 109–124). Cambridge, MA: MIT Press.

Gallagher, S. (2005). *How the body shapes the mind.* Oxford: Oxford University Press.

Gallagher, S. (2004). Understanding interpersonal problems in autism: Interaction theory as an alternative to theory of mind. *Philosophy, Psychiatry, and Psychology, 11*(3),199–217.

Gallagher, S. (2001). The practice of mind: Theory, simulation, or interaction? *Journal of Consciousness Studies, 8*(5–7), 83–107.

Gallagher, S. (2000). Philosophical conceptions of the self: Implications for cognitive science. *Trends in Cognitive Science, 4*(1), 14–21.

Gallagher, S., & Crisafi, A. (2009). Mental institutions. *Topoi, 28* (1), 45–51.

Gallagher, S., & Hutto, D. (2009). Understanding others through primary interaction and narrative practice. In J. Zlatev, T. Racine, C. Sinha & E. Itkonen (Eds.), *The shared mind: Perspectives on intersubjectivity* (pp. 17–38). Amsterdam: John Benjamins.

Gallagher, S., & Meltzoff, A. (1996). The earliest sense of self and others: Merleau-Ponty and recent developmental studies. *Philosophical Psychology, 9,* 213–236.

Gallagher, S., & Zahavi, D. (2008). *The Phenomenological Mind.* London: Routledge.

Gallese, V. (2005). "Being like me": Self-other identity, mirror neurons and empathy. In S. Hurley. & N. Chater, (Eds.), *Perspectives on imitation: From cognitive neuroscience to social science, Vol I* (pp. 101–118). Cambridge, MA: MIT Press.

Gallese, V. (2001). The "shared manifold" hypothesis: From mirror neurons to empathy. *Journal of Consciousness Studies, 8,* 33–50.

Gallese, V., & Goldman, A. I. (1998). Mirror neurons and the simulation theory of mind-reading. *Trends in Cognitive Sciences, 12,* 493–501.

Goldman, A. I. (2006). *Simulating minds: The philosophy, psychology and neuroscience of mind-reading.* Oxford: Oxford University Press.

Goldman, A. I. (2005). Imitation, mind reading and simulation. In S. Hurley & Chater, N. (Eds.), *Perspectives on imitation: From cognitive neuroscience to social science, Vol.II* (pp. 79–93). Cambridge, MA: MIT Press.

Goldman, A. I. (2002). Simulation theory and mental concepts. In J. Dokic & J. Proust (Eds.), *Simulation and knowledge of action* (pp. 1–19). Amsterdam/Philadephia: John Benjamins.

Goldman, A. I., & Sripada, C. S. (2005). Simulationist models of face-based emotion recognition. *Cognition, 94,* 193–213.

Gopnik, A., & Meltzoff, A.N. (1997). *Words, thoughts, and theories.* Cambridge, MA: MIT Press.

Gordon R. M. (2008). Beyond mindreading. *Philosophical Explorations, 11*(3), 219–222.

Gordon, R. M. (2005). Intentional agents like myself. In S. Hurley & N. Chater (Eds.), *Perspectives on imitation: From cognitive neuroscience to social science, Vol I* (pp. 95–106). Cambridge, MA: MIT Press.

Gordon, R. M. (2004). Folk psychology as mental simulation. In N. Zalta (Ed.), *The Stanford encyclopedia of philosophy.* (http://plato.stanford.edu/archives/fall2004/entries/folkpsych-simulation/).

Habermas, J. (2007). The language game of responsible agency and the problem of free will. *Philosophical Explorations, 10*(1), 13–50.

Habermas, J. (2000). *On the pragmatics of communication.* (M. Cooke, Trans.). Cambridge, MA: MIT Press.

Habermas, J. (1996). *Between facts and norms.* (W. Rehg, Trans.). Cambridge, MA: MIT Press.

Habermas, J. (1990). *Moral consciousness and communicative action.* (C. Lenhardt & S. W. Nicholsen, Trans.). Cambridge, MA: MIT Press.

Habermas, J. (1987). *The theory of communicative action, Vols 1 & 2.* (T. McCarthy, Trans.). Cambridge: Polity. (Original work published in 1984).

Haggard, P., & Libet, B. (2001). Conscious intention and brain activity. *Journal of Consciousness Studies, 8*(11), 47–63.

Harrist, S., & Gelfand, S. (2005). Life story dialogue and the ideal speech situation. *Theory & Psychology, 15*(2), 225–246.

Hickok, G. (2009). Eight problems for the mirror neuron theory of action understanding in monkeys and humans. *Journal of Cognitive Neuroscience, 21*(7), 1229–1243.

Hobson, P. (1993). The emotional origins of social understanding. *Philosophical Psychology, 6,* 227–49.

Hobson, P. (2002). *The cradle of thought.* London: Macmillan.

Honneth, A. (2008). *Reification: A new look at an old idea* (The Tanner Lectures on Human Values. University of California, Berkeley). Oxford: Oxford University Press. (Pagination is from electronic copy of original lectures available at http://www.tannerlectures.utah.edu/lectures/documents/Honneth_2006.pdf)

Hurley, S. L. (1998). *Consciousness in action.* Cambridge, MA: Harvard University Press.

Hutto, D. D. (in press). Action understanding: How low can you go? In G. Rizzolatti, C. Sinigaglia & R. Viale (Eds.), *Mirror neurons and social cognition.* Oxford: Oxford University Press.

Hutto, D. D. (2008). *Folk psychological narratives: The sociocultural basis of understanding reasons.* Cambridge, MA: MIT Press.

Iacoboni, M., Molnar-Szakacs, I., Gallese, V., Buccino, G., Mazziotta, J., & Rizzolatti, G. (2005). Grasping the intentions of others with one's own mirror neuron system. *PLoS Biology, 3*(79), 1–7.

Jabbi, M., Swart, M., & Keysers, C. (2007). Empathy for positive and negative emotions in the gustatory cortex. *NeuroImage, 34*(4), 1744–53.

Jeannerod, M., & Pacherie, E. (2004). Agency, simulation, and self-identification. *Mind and Language, 19*(2), 113–46.

Johnson, S. C. (2000). The recognition of mentalistic agents in infancy. *Trends in Cognitive Science, 4*, 22–28.

Johnson, S., Slaughter, V., & Carey, S. (1998). Whose gaze will infants follow? The elicitation of gaze-following in 12-month-olds. *Developmental Science, 1*(2), 233–38.

Kaplan, J. T., & Iacoboni, M. (2006). Getting a grip on other minds: Mirror neurons, intention understanding, and cognitive empathy. *Social Neuroscience,* (3–4), 175–83.

Legerstee, M. (1991). The role of person and object in eliciting early imitation. *Journal of Experimental Child Psychology, 51*, 423–33.

Legrand, D., & Ruby, P. (2009). What is self-specific? Theoretical investigation and critical review of neuroimaging results. *Psychological Review, 116*(1), 252–282.

Libet, B., Gleason, C. A.,Wright, E. W., & Pearl, D. K. (1983). Time of conscious intention to act in relation to onset of cerebral activity (readiness potential): The unconscious initiation of a freely voluntary act. *Brain, 106*, 623–42.

Libet, B. (2000). Do we have free will? In A. Freeman, K. Sutherland & B. Libet (Eds.), *The volitional brain* (pp. 45–58). Exeter: Imprint Academic.

Lowe, E. J. (1999). Self, agency and mental causation. *Journal of Consciousness Studies, 6*(8–9), 225–39.

Meltzoff, A. N., & Brooks, R. (2001). "Like Me" as a building block for understanding other minds: Bodily acts, attention, and intention. In B. F. Malle, L. J. Moses & D. A. Baldwin (Eds.), *Intentions and intentionality: Foundations of social cognition* (pp. 171–191). Cambridge, MA: MIT Press.

Meltzoff, A. N., & Moore, M. K. (1994). Imitation, memory, and the representation of persons. *Infant Behavior and Development, 17*, 83–99.

Meltzoff, A. N., & Moore, M. K. (1977). Imitation of facial and manual gestures by human neonates. *Science, 198*, 75–78.

Miyawaki, Y., Uchida, H., Yamashita, O., Sato, M., Morito, Y., Tanabe, H., & Kamitani, Y. (2008). Visual image reconstruction from human brain activity using a combination of multiscale local image decoders. *Neuron, 60*(5), 915–929.

Morton, A. (1996). Folk psychology is not a predictive device. *Mind, 105*(417), 119–37.

Newman-Norlund, R. D., Noordzij, M. L., Meulenbroek, R. G. J., & Bekkering, H. (2007). Exploring the brain basis of joint attention: Co-ordination of actions, goals and intentions. *Social Neuroscience, 2*(1), 48–65.

Oberman, L. M., & Ramachandran, V. S. (2008). Preliminary evidence for deficits in multisensory integration in autism spectrum disorders: The mirror neuron hypothesis. *Social Neuroscience, 3*(3–4), 348–55.

Overgaard, M. (2004). Confounding factors in contrastive analysis. *Synthese, 141*(2), 217–31.

Pacherie, E. (2007). The sense of control and the sense of agency. *Psyche, 13*(1). (http://psyche.cs.monash.edu.au/).

Pacherie, E. (2006). Towards a dynamic theory of intentions. In S. Pockett, W.P. Banks, & S. Gallagher (Eds.), *Does consciousness cause behavior? An investigation of the nature of volition* (pp. 145–167). Cambridge, MA: MIT Press.

Phillips, W., Baron-Cohen, S., & Rutter, M. (1992). The role of eye-contact in the detection of goals: Evidence from normal toddlers, and children with autism or mental handicap. *Development and Psychopathology, 4,* 375–83.

Pichon, S., de Gelder, B., & Grézes, J. (2008). Emotional modulation of visual and motor areas by dynamic body expressions of anger. *Social Neuroscience, 3*(3–4), 199–212.

Pierno, A.C., Becchio, C., Wall, M., Smith, A., Turella, L., & Castiello, U. (2006) When gaze turns into grasp. *Journal of Cognitive Neuroscience, 18,* 2130–2137.

Pierno, A.C., Becchio, C., Tubaldi, F., Turella, L., & Castiello, U. (2008) Motor ontology in representing gaze-object relations. *Neuroscience Letters, 430,* 246–251.

Prinz, W. (2001). Kritik des freien Willens: Bemerkungen über eine soziale Institution. *Psychologische Rundschau, 55*(4), 198–206.

Proust, J. (2003). How voluntary are minimal actions? In S. Maasen, W. Prinz & G. Roth (Eds.), *Voluntary action: Brains, minds, and sociality* (pp. 202–219). Oxford: Oxford University Press.

Ramsey, W. (2007). *Representation reconsidered.* Cambridge: Cambridge University Press.

Reddy, V. (2008). *How infants know minds.* Cambridge, MA: Harvard University Press.

Rizzolatti, G., Fogassi, L., & Gallese V. (2000). Cortical mechanisms subserving object grasping and action recognition: A new view on the cortical motor functions. In M. S. Gazzaniga (Ed.), *The new cognitive neurosciences* (pp. 539–52). Cambridge, MA: MIT Press.

Rizzolatti, G., Fadiga, L., Gallese V., & Fogassi, L. (1996). Premotor cortex and the recognition of motor actions. *Cognitive Brain Research, 3,* 131–141.

Rizzolatti, G., & C. Sinigaglia, C. (2008). *Mirrrors in the brain. How our minds share actions and emotions.* Oxford: Oxford University Press.

Ryle, G. (1949). *The concept of mind.* New York: Barnes and Noble.

Scheler, M. (1954). *The nature of sympathy.* (P. Heath, Trans.). London: Routledge and Kegan Paul. (Original work published 1923).

Searle, J. (2007). Neuroscience, intentionality and free will: Reply to Habermas. *Philosophical Explorations, 10*(1), 69–76.

Searle, J. (1984). *Minds, brains, and science.* Cambridge, MA: Harvard University Press.

Searle, J. (1983). *Intentionality: An essay in the philosophy of mind.* Cambridge, MA: MIT Press.

Selman, R. (1980). *The growth of interpersonal understanding.* New York: Academic Press.

Senju, A., Johnson M. H., & Csibra, G. (2006). The development and neural basis of referential gaze perception. *Social Neuroscience, 1*(3–4), 220–234.

Singer, W., Wolpert, D., & Frith, C. (2004). Introduction: The study of social interactions. In C. Frith & D. Wolpert (Eds.), *The neuroscience of social interaction* (xii–xxvii). Oxford: Oxford University Press.

Sinigaglia, C. (2009). Mirror in action. *Journal of Consciousness Studies, 16.*

Trevarthen, C. (1979). Communication and cooperation in early infancy: A description of primary intersubjectivity. In M. Bullowa (Ed.), *Before speech.* Cambridge: Cambridge University Press.

Trevarthen, C., & Hubley, P. (1978). Secondary intersubjectivity: Confidence, confiding and acts of meaning in the first year. In A. Lock (Ed.), *Action, gesture and symbol: The emergence of language.* San Diego, CA: Academic Press.

Van der Gaag, C., Minderaa, R., & Keysers, C. (2007). Facial expressions: What the mirror system can and cannot tell us. *Social Neuroscience, 2,* 179–222.

van Schie, H. T., Koelewijn, T., Jensen, O., Oostenveld, R., Maris, E., & Bekkering, H. (2008). Evidence for fast, low-level motor resonance to action observation: An MEG study. *Social Neuroscience, 3*(3–4), 213–228.

Varela, F. J., Thompson, E., & Rosch, E. (1991). *The embodied mind: Cognitive science and human experience.* Cambridge: MIT Press.

Vul, E., Harris, C., Winkielman, P., & Pashler, H. (2009). Puzzlingly high correlations in fMRI studies of emotion, personality, and social cognition. *Perspectives on Psychological Science, 4*(3), 274–290.

Walker, A. S. (1982). Intermodal perception of expressive behaviors by human infants. *Journal of Experimental Child Psychology, 33,* 514–35.

Wegner, D. (2002). *The illusion of conscious will.* Cambridge, MA: MIT Press.

Wittgenstein, L. (1967). *Zettel.* In G. E. M. Anscombe & G. H. von Wright, (Eds.). (G. E. M. Anscombe. Trans.). Berkeley: University of California Press.

Woodward, A. L., & Sommerville, J. A. (2000). Twelve-month-old infants interpret action in context. *Psychological Science, 11,* 73–77.

Part II
Histories of the Brain

5

Toys are Us
Models and Metaphors in Brain Research

Cornelius Borck

"A model … is something you hold in your head rather than in your hands."
(Ian Hacking, 1983, p. 216)

Regardless of its seemingly abstract philosophical nature, historically the relation of mind and body has found a surprising variety of forms in the changing "mirror of machines" (Meyer-Drawe, 1996)—perhaps because the "invention" of abstract theorizing and knowledge coincided with the construction of tools and machines. Friedrich Nietzsche brought this articulation of knowledge and machines famously to the fore in his *Birth of Tragedy*, when he lamented that with the invention of machines, the artful wisdom of the mythos, and the metaphysical consolation it offered, was lost and replaced with the "*deus ex machina*, namely the god of the machines and crucibles, … the belief in knowledge and a life led by science" (Nietzsche, 1872, § 17). He dated this shift to the transformation of tragedy from its early to the classical form in Greek antiquity; a period which also saw the rise of science and philosophy. Nietzsche wrote his essay during the Franco–Prussian war, a fact that may have influenced his analysis, because this was a time when modern science and the process of industrialization changed the world in hitherto unanticipated ways (Nietzsche reflected upon this in an introduction of the book which he added later).

At the same time, two young physician-scientists once again changed the relation between mind and brain, when they experimented with the application of electric currents to the cortex and observed its excitability. In 1870, Gustav Fritsch and Eduard Hitzig reported on their experiments with cerebrally induced contractions in dogs (Fritsch & Hitzig, 1870; Hagner, 1993). The electric excitability of the brain had been demonstrated almost a hundred years after Luigi Galvani had discovered animal electricity, even though his discovery of it in 1780 had sparked an intense debate on—and much research into—the nature of the nervous principle (Brazier, 1961). Applying electrical stimuli to different sites of the exposed surface of

Critical Neuroscience: A Handbook of the Social and Cultural Contexts of Neuroscience, First Edition. Edited by Suparna Choudhury and Jan Slaby.
© 2012 John Wiley & Sons, Ltd. Published 2016 by John Wiley & Sons, Ltd.

the cortex, Fritsch and Hitzig observed that the stimulation of areas in the frontal half of the cerebral hemispheres resulted in muscular contractions in the opposite side of the body. This single publication ended a long and puzzling debate about psychophysiological principles by demonstrating the brain to operate like an electrical apparatus (Young, 1970, pp. 224–240).

Their findings were confirmed and further developed by the British physiologist David Ferrier, when he established, in experiments with monkeys, the principle of cerebral localization—for example the functional specialization and strict topographical organization of the cortex (Ferrier, 1873, 1876). Seeing one of Ferrier's experimental monkeys with a paralyzed limb, Martin Charcot is reported to have remarked "C'est un malade!" (quoted from Viets, 1938, p. 482). Ferrier's experiments were, indeed, quickly transferred to human beings (Zago, Ferrucci, Fregni, & Priori, 2008). In effect, the replication of the stimulation experiment in humans resulted in a second metonymic replacement, namely the transformation of the human body into an electromechanical device, executing automatic movements of specific definition due to the precise electrical stimulation of the centers of control. This transformation was to happen a generation later, when neurosurgeons started to introduce cerebral stimulation into the operating theater.

Starting with a closer look at Wilder Penfield's exploration of the functional topography of the cerebral cortex and his use of instruments as technical (material) as well as heuristic (metaphorical) tools, this chapter compares and contrasts different strands in the employment of machines as cognitive tools in the neurosciences. Over the course of more than 200 years, a changing series of models and metaphors was brought forward and new research methods were developed to conceive of the brain; this created a dynamic exchange between models, metaphors, and research strategies for accommodating and generating new masses of data. Focussing on twentieth-century research, this chapter's objective is not to examine the neuro-sciences in their entirety or to trace every single analogy or metaphor, but rather to analyze the cognitive significance of this strategy and the epistemic implications of different approaches.

For a long time tools and instruments were simply not sophisticated or complex enough for use as compelling correlates; yet their very limitations could be employed for demonstrating the brain's functional superiority. Only with the arrival of automata, in particular the computer as a logical apparatus of calculation, did the relation between brain and machine change; what had once been a comparison across a generally shared understanding of differences turned into rivalry and competition. Now the metaphorical explanation of brains as machines acquired a material, concrete reality and the philosophical program of materialism turned into reductionism. As if to testify to its allegedly anti-human nature, humans soon lost the competition with machines in calculation capability and other supposed indicators of human intelligence. At this point, towards the end of the twentieth century, the emphasis shifted to other aspects of human nature that apparently escaped comparison with machines, such as empathy and the social brain. At the same time however, the arrival of functional imaging promised to discard functional comparisons in general and to conflate the technological visualization with a neurophysiological substrate. Machines may always have lacked something essentially human but the mobilization of machine metaphors operated on the crucial basis of a differentiality between men and machines. In the interplay of shortcomings, limitations, and transgressions, machine metaphors opened the space for critical humanism in brain theory that is overlooked or extinguished in the current identification of substrate and significance in functional imaging.

The Tape Recorder and the Electrode

Realizing that, because of the brain tissue's insensitivity to pain, brain surgery could be performed with local anesthesia, Fedor Krause in Berlin and Otfrid Foerster in Breslau pioneered a procedure which was as much a therapeutic intervention as a neurophysiological experiment (Foerster, 1923; Krause, 1911). Later, the neurosurgeon Wilder Penfield perfected this technique at the Montreal Neurological Institute by morphing Ferrier's chart to the famous homunculus of the cortical representation of bodily functions (Penfield & Rasmussen, 1950). Immediately before moving to Canada, Penfield spent six months with Foerster in Breslau where he trained to explore by electrical stimulation the brain of the awake patient during surgery. The issue was to determine the functional specificity of particular regions of the brain in order to save them during surgery. Penfield would open a patient's skull, expose the surface of the brain, and map its functional topography by stepping down point by point with his stimulating electrode along the brain's gyri, hereby eliciting the respective physiological responses. For this, not only had the operating theater been converted into a special electrophysiological laboratory, but both the operating team and the patient embarked together on an expedition into new territory. The tip of the stimulating electrode acted as voyager into the lands usually hidden not only inside the skull but deep in the vaults of the psyche. In Penfield's hands came the surprise, when the electrode entered *Memoria*, the lands of yesterday; Penfield perfected neurostimulation to a form of time travel.

Besides eliciting various forms of motor response, such as the movement of a finger, arm, or toe that confirmed the (by then) well known topography of the motor cortex, Penfield's electrode triggered quite different reactions when less well characterized areas of the brain were investigated, in particular structures further down the temporal cortex. Here, Penfield's electrical explorations resulted in the patient undergoing experiences of a forced déjà vu; again and again, the stimulating electrode acted as a memory activator bringing back a distinct "single recollection," not "a mixture of memories or a generalization" (Penfield, 1952, p. 180). But in contrast to the sensory or the motor cortex, there was no apparent topographical correspondence between the position of the electrode and the experiential content or the emotional quality of a memory re-actualized and hence Penfield could not control the experience with his electrode. Instead, the recollections evoked by the electrical activation of the temporal cortex retained the rich details of the original experience—and often even more strongly compared to the habitual act of remembering. The memories forced into the patient's consciousness were experienced not only as present but often as "more real" than regular memories, and the patients retained some sense of a mixed reality, "somehow doubly conscious of two simultaneous situations" (Penfield, 1952, p. 184). Apart from these generalizable characteristics, it was entirely left to the positioning of the electrode whether the stimulation elicited the permanent boredom of an office life or a happy family reunion across oceans.

For Penfield, the accuracy and comprehensiveness of the memory apparatus he had accessed by means of the stimulating electrode were most remarkable. The individual's ability to actively recall particular episodes may have been observed to differ from patient to patient, but deep down in their brains every detail of their lives was being recorded with the precision of a machine. The automatic operations of this memory system could best be described as the action of a kind of tape recorder, continuously recording the events of a life, from birth to death. However, this was a very special type of recorder that did not register the events as they occurred in the outside world, but from the internal perspective of the subject—as Penfield aptly described by differentiating the metaphor in terms of its technological validity vis-à-vis the experiential quality:

> The subject feels again the emotion which the situation originally produced in him, and he is aware of the same interpretation, true or false, which he himself gave to the experience in the first place. Thus, evoked recollection is not the exact photographic or phonographic reproduction of past scenes and events.
>
> (Penfield, 1952, p. 183)

The experiments exemplified William James' notion of a continuous "stream of consciousness" (James, 1892, chapter IX), though this had also once just been a metaphor. Now, experimental research had proven that somewhere inside the head there was a material structure which acted like a storage system and automatically preserved impressions from every moment of a life; though not the impressions in terms of the way they reached the body and stimulated the sensory organs, but in their perceived, emotionally charged and evaluated form. Technological advances now permitted access to this memory processing unit as a biological system, because it operated, in all likelihood, as an electrophysiological process and hence proved accessible by electrical stimulation.

The phonograph and the photograph were not quite the right models here, but they worked so well because they captured precisely what memory was not: an objective representation of the outside world. The memory system, by contrast, was an automatic registration unit for the subjective interpretation of sensory information. According to Penfield, the memory process did not work like a photograph or a phonograph, but it resembled some kind of psychic tape recorder. Penfield could thus embed his experiments in a set of material and metaphorical connections where technological models mediated perfectly between the two—the metaphor of the stream of consciousness and the electrophysiological process, or "the word and the world" (Morgan & Morrison, 1999).

Late in his life, after switching his career from the operating to the lecture theater, Penfield made a famous sketch connecting his experimental clinical observations with his general philosophical ideas, again by means of metaphorical connections. In this sketch, a chain of metaphors bridged the anatomy of the brain and its interior parts to "—M I N D—", situated far away from the hippocampus and even outside the cortex (see Figure 5.1). The graphic symbol of the "key of access" connected the right and

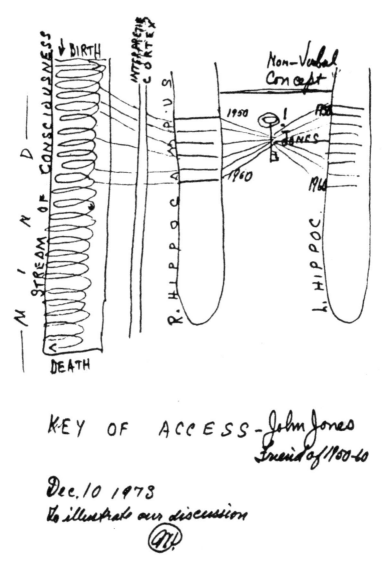

Figure 5.1 Sketch by Wilder Penfield of memory mechanism inside brain; Penfield Archives, Osler Library, McGill University, Montreal, Canada. Reproduced by permission of the Literary Executor of the estate of Wilder Penfield.

left halves of the brain, and lines formed structural as well as logical bridges from the key symbol to the different sections of the hippocampi and from there through—or under—the cortex to the "stream of consciousness" that was the mind. In this sketch, the theory and the tool blended in new ways. Here, the stream of consciousness spanning life and death had become a spiral capturing on its strip of film, like a psychic video machine, all experiences awaiting recollection in the form of active retrieval or forced recollection by means of electric stimulation.

When the brain or some of its functions are being compared to technological inventions such as the camera, the phonograph, or the tape recorder, these models stand in for specific functions attributed to the brain. Each of the models accentuates a particular, functional aspect of the brain. Yet the historical sequence of different technological models mobilized in order to grasp the brain's functional capabilities also characterizes the investment in brain research for addressing the more fundamental questions about the nature of being human. For more than 200 years now, the brain has increasingly been mobilized as the central organ for addressing the *condition humaine* (Hagner, 1997). Investigating the arguments and ways in which particular models were adopted to explain the brain illuminates the complex history of this branch of research and elucidates how machines served for explanatory purposes inside and beyond the respective fields of technical application. In addition, it amounts to a rich and changing history of how scientists and the general public looked to the brain to answer vital humanist questions. Models in brain research typically mediated between questions of meaning, function, and significance on the one hand and the world of organic structures and mechanical functions on the other. Seen in this way, technological models are media in the multiple sense of the word, in that they transform and transmit information according to their technological specifications, and mediate between the world of biological function and the meaningful realm of day-to-day experience. They open a channel to the operations of the brain, which is structured by their technical functionality as well as by their cultural significance (Keller, 2000).

Compared to other branches of medicine or the life sciences, brain research has engaged in a particularly active dynamic with regard to metaphors and models; the result is an ongoing process of technological metamorphosis that shows little sign of abating (Draaisma, 2000). In contrast, for example, to the metaphor of the pump that was consolidated long ago in the physiology of the heart, the brain appears to be an unstable organ that has been compared to a wide and variable range of objects. The argument that is typically brought forward for this heterogeneity is the intrinsic complexity of the brain as a natural object whose very nature is apparently so much more than a single, simple device. Such a naturalistic argument is, however, more difficult than it may appear at first glance and should be treated with some caution by the historian—as becomes apparent when the brain is compared with organs such as the liver. Viewed at the level of molecular and biochemical processes, for example, there are probably very few organs surpassing the complexity of the liver's synthesizing machinery. But in the Western tradition, the liver has lost much of its once valuable cultural significance; and hence its biological intricacy has been increased by biochemical investigations without much notice from circles outside the life sciences. Epistemologically speaking, the plain argument of anatomical design is not exclusively empirical and hence not trivial. The complexity so easily assumed to apply to the brain is very much an attributed quality, reflecting a cultural expectation and a desire to invest the brain with the elevated status of delivering compelling answers about human nature and intelligence.

Analyzing the contemporary media of the interwar period, Walter Benjamin famously spoke of the *optical unconscious* and declared film to be the central medium providing access to it "just like psychoanalysis did to the psychical" (Benjamin, 2008; Krauss, 1993). Benjamin's idea of a materially grounded unconscious invites a fresh conceptualization not just of film but of media in general and a reading of their currently electronic mode of operation, for example, as the cultural unconscious of modernity. Where has this process led to today and what will be the next step beyond the internet and augmented reality? Though Lenin did not yet know of the internet in his famous formula of communism as "Soviet power plus electrification," he quite rightly pointed to the manifold effects and consequences of electrification beyond the narrow limits of the technology of the 1920s. Since Galvani's discovery of the spark of life, models and metaphors have entered the human body in many ways; Mary Shelley's *Frankenstein; or, the Modern Prometheus* (Shelley, 1818/2008) provided a blueprint still valid today for the way in which social and organic lifeworlds are transformed by research in science and technology.

The Brain as Communication Technology: A Humanist's Utopia

The metaphorical appropriation of the body by electrical tools began with the simple analogy between cable and nerve fibre and provided the material basis for comparing the nervous system with telegraphy. In the later years of the nineteenth century, scientists had already begun to lament "this frequently used metaphor;" as, for example, psychologist Wilhelm Wundt wrote in his *Grundzüge der physiologischen Psychologie* (Wundt, 1874, p. 346). Speaking of the cable network as the "nervous system of the state"—or vice versa of the body's "telegraphy system"—scientists and their audiences used the analogy in both directions (Otis, 2001). However, these intriguing analogies amounted to more than just rhetorical figures. Given the particular concerns of the nineteenth century for debates on time and progress, it may come as no surprise that such analogies were integrated into an evolutionary account of the history of technology, which claimed that all technological inventions could be traced back to biological principles. For the German philosopher Ernst Kapp, the similarities between biological and engineering solutions simply proved that technology in general was nothing but an unconscious externalization of the body's intrinsic principles of operation, an "organic projection" as he called this mechanism of technogenesis (Kapp, 1877, p. 140). As a consequence of this process of externalization, the technological principles—and thus the biological operations of the body—become accessible to scientific exploration and intervention, setting in motion a process of open-ended perfection, a co-evolution of man and machine that has fascinated media theorists ever since. A century later and digesting the impact of the emergence of television, Marshall McLuhan reiterated Kapp with his famous statement:

> With the arrival of electric technology, man extended, or set outside himself, a life model of the central nervous system itself.
>
> (McLuhan, 1994, p. 43)

The idea of electric technology as the life model of the nervous system appears to be directly illustrated in a popular book dating from the mid-1920s. Here, the simple operation of a bell, for example, provided the perfect example of the reciprocal relation between biological model and its technical mirror image, because electrical impulses

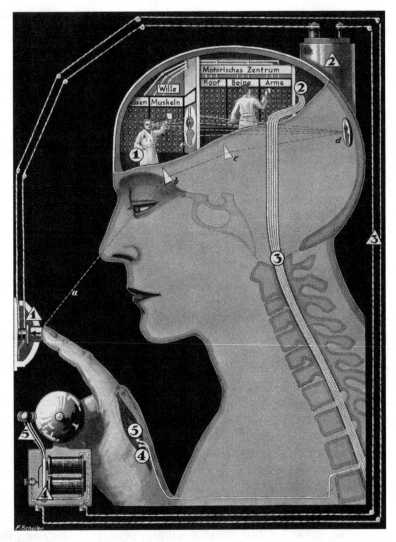

Figure 5.2 Comparison of electric and nervous circuitry (Kahn, 1924–31, vol. 2, table XVII), with kind permission from E. Kahn, (c) Deschitz (www.fritz.kahn.com).

travel through nerves and muscles the same way they proceed through the bell's circuits (see Figure 5.2). In so far as this image showed two electric circuits—one in the outside world driving the bell and another inside the body driving the hand pushing the knob—it simply aligned body parts and technical details in a graphic explanation of common analogies; but it did not mobilize the technology to directly replace the body function. The electric magnet pushing and pulling the lever was depicted right next to the muscle and nerve, but it was still the biological muscle and not its electrotechnical counterpart that moved the finger. This was decisively different, however, for the brain where, in the image, room-sized switchboards stood in for the respective centers of volition and execution. In retrospect, the replacement of cognitive control with something limited like a switchboard hardly seems ingenious. Looking

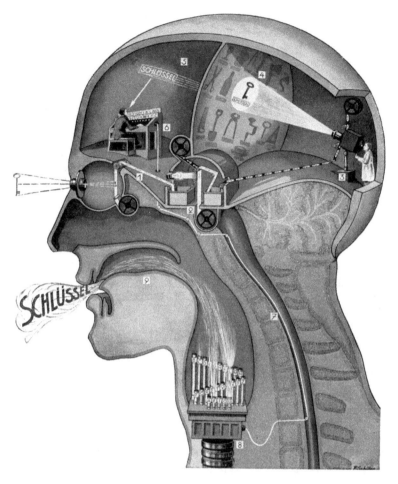

Figure 5.3 Mental cinema: the processing of visual information by the brain (Kahn, 1924–1931, vol. 4, table VIII), with kind permission from E. Kahn.

back at an image such as this from the distance of three quarters of a century, one may smile at the naivety and simplicity of its technological solution. The heavy modernism of such images underlines their datedness today, since no biological structure ages as quickly as obsolete technology. Indeed, the telegraphy office was soon to be replaced by the computer and later the internet.

In its time, however, the point was less to postulate the brain as a telephone exchange than to elucidate an important aspect of its functionality by means of this analogy. Even for the most general audiences of the 1920s, it was quite evident that the brain was capable of processing many more and varied sensations than could be explained by the single analogy of the telephone. During the 1920s, new communication technologies such as the radio and film added further options for spelling out the functional identity of sensory processing and media technology (see Figure 5.3). The physiological processes when sitting in front of the screen in a movie theater, for example, could be formulated as the eyes taking pictures like a camera that were sent to a processing station in the brainstem before reaching the visual center in the back of the head. These pictures were then projected on to higher visual centers further up front in the brain to be deciphered, before sending an impulse to the steering of the larynx. The question of what exactly was being shown pales here in view of the many amalgamations of body and technology that depict the human body as a wonderful machine which technology can scarcely touch.

The examples here are taken from Fritz Kahn's magnum opus, the popular textbook *Das Leben des Menschen*, which appeared between 1924 and 1931 in five lavishly illustrated volumes with more than 1500 images. The book sold widely and enabled Kahn, a physician by training, a second career as a popular science writer, first in Weimar Germany and later, after his emigration, in the USA. Kahn's recipe for success was the combination of a lucid style of writing with intriguing visuals portraying the body and its functions as machine ensembles. In this way, Kahn's world of images and Kapp's organ projection almost merged. Unlike Kapp however, Kahn reversed the explanatory strategy. While Kapp suggested that technical inventions were based on a preconscious familiarity with the functional principles of the human body, Kahn explained bodily structures and functions by comparing them with everyday technology—even if this reservoir of functional analogies itself required further explanation. Where Kapp—and later McLuhan—speculated on epistemic connections between technology and the body, Kahn made the body's mysterious, organic interior familiar by means of common gadgetry, as if a form of techno-literacy had the potential to reconnect with the body's machinery in new ways. The flood of images showed how the modern human would understand him or herself through the invention of technical devices. Very few may have known exactly how to operate a switchboard or any other of the machines depicted, but seeing them in operation secured the possibility of a perfect explanation. In the visual language of Kahn's images, human beings would ultimately reveal their identity in the construction of sophisticated technology.

With his popular images Kahn visualized a romantic utopia of industrialization on two different levels. Firstly, human ingenuity in instrument making and machine building had allegedly reached a stage where machines epitomized the complexity of the human body and, secondly, the process of technological civilization should ultimately arrive at an enculturation of nature into technology (Borck, 2007). In this way Kahn extended the classical Enlightenment programme of cognitive self-reflexivity and moral autonomy to the body; technological advances enabled a radically new form of "know thyself,"—the technological explanation of bodily processes. According to Kahn, this technological enlightenment did not undermine human freedom and liberty, but resulted in the utopia of a seamless functionality in truly perfected technology. Kahn's machines did not leak or produce waste while the workers and operators diligently pursued their jobs; this was the happy paradise of industrial production.

Form and Function Beyond Technology

In the idea of a perfect technology, Kahn's industrialization of mind and body met with another strand in the history of models that explained the brain's functions: the logical machine. But before exploring this further, it is important to widen the scope of analysis in order to escape the technological determinism implicit in the arrangement of the chapter so far. Since the invention of the first true automata, machines have been debated as models of body and brain fairly independently of the underlying philosophical framework (Canguilhem, 1992). This is most famously exemplified in the dualism of René Descartes' posthumous *De Homine* (Descartes, 1633/1972) in contrast to the radical materialism of Julien Offray de LaMettrie's scandalous essay *L'homme Machine* (Jauch, 1998; LaMettrie, 1748/1912). Technological models, however, did not have a monopoly in the discourse on the nature of the living and the human body; and the current interest in the blurred boundaries between man and machine must not obscure the historian's awareness of the fact that alternative arguments also attracted widespread attention. For many brain researchers of the early twentieth century, for example, it was out of the question that mind or brain should be meaningfully compared to a mechanical machine because of its categorically different biological nature (Harrington, 1996). This argument gained momentum after the publication of evolutionary theory and with discoveries in cellular biology. As a consequence, researchers revived older ideas of a unique and universal biological principle such as irritability or movement and created from them a remarkable psychobiology, linking highest mental functions to the most basic but specifically biological properties of primitive cellular organisms such as amoeba (Schloegel & Schmidgen, 2002).

The so-called protoplasmatic theory of life (Geison, 1969; Lidforss, 1915; Roux, 1915), for example, appealed to neuroscientists because such a monistic approach to the realm of living phenomena could equally be applied to the appropriation of substances by a microorganism as to the apprehension of sensations by neurons (Borck, 1999). In addition, this theory was very well suited to explaining learning and adaptation, phenomena chronically difficult to account for with the machine metaphor because of a lack of appropriate tools. In light of the protoplasmatic theory, in contrast, learning and memory resulted from the dynamic plasticity of the neurons in the central nervous systems which formed new connections in an endless process, while others were withdrawn—as in sleep or forgetting.

Among the proponents of such biological models of logical reasoning and cognitive action counted a number of prominent neuroscientists such as Santiago Ramón y Cajal (1895), Auguste Forel (1894), or Theodor Meynert. Meynert is particularly interesting here and deserves further exploration, because he began with an abstract and geometrical model for the brain as the basis for his theory of mind and only later combined it with the protoplasmatic theory of brain action. His case hereby underlines the manifold metaphorical resources upon which neuroscientists drew for their comparisons and it illustrates how different analogies could be combined in more complex models—in this case all non-mechanical.

Initially, Meynert argued for the perfection of the brain on the basis of its geometrical shape, by linking it to a sphere that Plato had already identified as the form of perfection. To this macroscopic model, he added microscopic biological details. While he maintained that the nerve cell bodies that were located in the surface of the spheres were the seat of consciousness, Meynert differentiated the three different types of fibre that made up the brain's fibrous interior: these were the sensory, receiving stimuli from the outside world, the motor initiation action, and the so-called association type, building an internal communication structure within the brain itself. By means of this tripartite fibre system and the nerve cells themselves, the brain formed a special instrument of apperception, association, and projection. The anatomical structure, Meynert concluded, provided the physical basis of the brain's cognizing operations (Meynert, 1865, pp. 48–55); this was Meynert's basic conceptualization of the brain as a psychological apparatus. Two decades later, Meynert superimposed as an active, functional principle the protoplasmatic theory onto this anatomico-physical model:

> Just as the medusae stretch out their feelers into the world and take possession of their prey through tentacles, so this composite protoplasmatic being, which is the cortex, possesses centripetally-conducting extensions, the sensory fibers of the nervous system, which we may consider its feelers, and motor fibers, which are its tentacles.
>
> (Meynert, 1884, p. 127)

The metaphor of a physical apparatus for projection and association translated anatomy into psychology whilst the protoplasmatic theory provided the basic, vital principle for describing such phenomena as the psychophysiologically active, neuroanatomical details.

For Meynert, who continuously refined the use of metaphors in his brain theory, the concept of ideal shape and of the protoplasmatic actions of nerve cells did not serve a merely rhetorical role, explaining an otherwise well defined physiological action (Black, 1962). Quite the contrary, the metaphoric models provided mere anatomical observations with their epistemic significance for brain theory (Hesse, 1966); thus, they were a crucial part in Meynert's work as a teacher, scientist, and writer (Meinel, 2004). In fact, the model sparked further anatomical investigations, the metaphoric language of projection and association serving Meynert as a starting point for developing a specific brain preparation technique (see Figure 5.4). With this technique, he was able to demonstrate the different fibre types making up the connectivity within the brain as asserted by his theory (Guenther, 2009). In Meynert, the metaphors mediated back and forth between anatomy and meaning, in a spiraling process of the practical and the conceptual (Klein, 2003).

To map out the extent of the impact that biological and political metaphors have in brain theory (Draaisma, 2000) would be beyond the scope of this chapter, whose focus is on tools. But Meynert provides a good example of how neuroscientists relied on metaphors for explaining aspects of brain function and mental activity outside the realm of technological models. While Meynert engaged geometrical and biological models, others used tools specifically in order to highlight their difference to brains. The explanatory strategy of metaphors and models could thus be extended beyond the limitations of a particular technological model, when the very limitations explained the specificity of the brain, typically in terms of its super-technical functionality—as Penfield did with his differentiation of the human memory system from a standard phonograph or camera. In this way, the model could be utilized as an analogy together with a *differentia specifica* in order to demonstrate the superiority of the brain. In a similar tradition, the brain has been compared with various instruments of wonder such as, for example, a musical instrument (Kassler, 1994). The most famous of these—though not a musical instrument—is obviously Charles Sherrington's

Figure 5.4 Theodor Meynert's special preparation technique for highlighting the interconnectivity within the brain (Meynert, 1884, p. 43).

"enchanted loom," which modeled the brain on what was then, technologically, a most complex machine, the Jacquard loom, "where millions of flashing shuttles weave a dissolving pattern, always a meaningful pattern though never an abiding one" (Sherrington, 1953, p. 178).

The master, however, of the strategy of explaining the psyche in technical terms beyond the physical space of a technological model, was Sigmund Freud. Arguing that the realm of psychic processes must not be conflated with the physical space of the brain's anatomy, Freud broke with both the reductionism of the machine theory and the monism of psychobiology thereby transforming the traditional ontology of Cartesian dualism to the epistemology of psychoanalytic theory. As a well versed writer and with a solid grounding in the nineteenth-century neuroanatomical tradition of his teacher Meynert, Freud used analogies and technological models for elucidating the flaws of any naturalistic theory of the brain (Borck, 1998). He mobilized models and metaphors precisely as imperfect tools, which can be studied in his many strategic comments on the limits of particular visual metaphors and functional analogies.

A famous example is Freud's comparison of the psyche with the city of Rome in *Civilization and its Discontent* (Freud, 1930/2001). The point of comparison is not the large number of ancient buildings, nor their existence to this day, but the very non-imaginability of a Rome that has been preserved in all buildings ever built there; such a city may be conceived of but can no longer be visualized. In order to gain clarity on its significance, the analogy has to be driven beyond the limits of visual representation. For Freud, the narrative quality and logical structure of language allowed access to the specific dimension of time—so central to psychoanalysis—in a linguistic representation which built on the physical model but left it, metaphorically, behind. The pictorial language of the illustrations inevitably fell behind the mature concept, but this very failure illustrates the complexity of figurative thought. Freud's concluding remark on the Rome comparison stressed his strategically inverted use of topographical metaphors (Freud, 1930/2001, p. 71): "Our attempt seems to be an idle game. It has only one justification. It shows us how far we are from mastering the characteristics of mental life by representing them in pictorial terms." Freud developed a visual argument that the psychical apparatus can only be described in language and not represented by anatomical visualizations, because in his theory the psyche followed the structural logic of a symbolic space and not the anatomy of a geometrical topography.

A particularly telling example for this explanatory strategy and hence for the usefulness of technical models in psychophysiological brain research is given in Freud's short *Note Upon the "Mystic Writing-Pad"* (Freud, 1925/1961) that inspired Jacques Derrida, in turn, to a long reflection about the primacy of writing (Derrida, 1961/1980). In his *Introductory Lectures on Psycho-Analysis* and elsewhere, Freud had compared the psychical apparatus with a microscope or telescope, and had already intentionally located the unconscious in virtual spaces such as the point of refraction (Freud, 1917/1971). In the *Note*, the psychical apparatus has finally become a little toy, a writing pad. Carefully, Freud studied the various details and layers of this child's toy, comparing each with a somewhat similar aspect of the psyche. But, in the final step, Freud transcended the realm of material technology and moved from the physical model to the linguistic by reflecting on the assumed functionality of a truly "mystic" writing pad—which obviously the psychic apparatus is. Whatever the brain does, the psychical apparatus is a peculiar system for receiving, storing, and reactivating various kinds of traces written on its surface. For Freud, the psyche is an inscription device.

The Brain as Writing Apparatus and Symbolic Machine

While Freud speculated about writing as the primary psychic operation, electroencephalography allowed the brain to literally inscribe its activity onto paper. It seemed as if the world of brain structures, nerve fibres, and electric potentials would blend into the world of meaning, life, and sense (Borck, 2008). Brain wave recording may not have come with its own technological model, but it offered as an analogy for the brain's workings a very powerful cultural technique. In retrospect, the absence of a technological model for the brain in electroencephalography proved particularly fertile ground, because it provided the necessary space for the computer, the most powerful brain model of the twentieth century, to later be inserted (Borck, 2005). Calculating machines had their very own history of metaphors with the "brains of brass" dating back to long before the availability of the first electric machines with thousands of vacuum tubes and "computing" as a professional field (Spufford & Uglow, 1997).

The next step in blending brains and computers followed on theoretical grounds when Alan Turing formulated a simple, yet universal, algorithm of problem solving

(Turing, 1936) and McCulloch and Pits described the circuitry of logical neuronal nets (Kay, 2001; McCulloch & Pitts, 1943). Later, machines became available that used electricity for logical operations and calculation, and the gap between the world of material processes and symbolic action appeared to have closed (Latil, 1956). From this moment on, brains were discussed as biological instantiations of such machines; brains had become computers.

In addition, the interdisciplinary evolution of cybernetics—the thinking in terms of "control and communication" across the mechanical–biological divide so fashionable shortly after the end of World War II—provided the perfect framework for a new wave of brain modeling in terms of control technology and steering devices, from W. Ross Ashby's *Design for a Brain* to John von Neumann's theory of automata (Ashby, 1954; von Neumann, 1966). Others like the British cybernetician William Grey Walter indulged in soldering and tinkering with simple electro-mechanical creatures that mimicked human behaviour (Hayward, 2001; Walter, 1953). As Walter demonstrated with his famous tortoises, intentionality and teleology could already be perfectly simulated by means of basic mechanical devices. There was no divide between physical, biological, and psychological processes.

Although computers were still immobile, garage-sized technological systems that did not physically resemble the brain, they quickly dominated brain theory. According to Norbert Wiener, brains resembled computers not only with regard to their calculation capabilities, but even in that they used a similar mechanism for data processing. Computers used electricity for their operations just as neurons communicate by electric impulses as physiology had revealed; similary there was a correspondence between the all-or-none principle and the digital code in the computer, as the inventor of the word "cybernetics" pointed out when he became interested in brain waves (Wiener, 1961). Whilst Wiener's idea did not really stand up to scrutiny, it was not the last time that brains would be mistaken for computers during the twentieth century (Churchland & Sejnowski, 1999). The computer dominated much of brain research and the public understanding of mind and brain throughout this period. Paradoxically, the celebrated victory of IBM's Deep Blue over Kasparov in 1997 made the hitherto deeply engrained analogy of brain and computer look superficial and falter, coinciding as it did with an increasing awareness of the machines' clumsiness in doing something beyond calculation, for example bodily movements. Today, only 10 years later, it is already hard to reconstruct how it was that the computer so easily assumed the role of central metaphor in brain research over such a long period of time. Maybe one day we will look back on the computer as the most convenient and common form of misunderstanding the brain in modern history.

The computer was not, however, just one more step in a long sequence of models (although it is this too). In a certain sense, the universal logical machine was the ultimate model that, up to now, has proven irreplaceable. What newer tool or gadget could possibly replace it? What could stand in as the next, central metaphor? The iPhone can be said to be much more versatile than the desktop computer, let alone the room-sized forerunner with which this analogy began; in addition it symbolizes a trend towards ubiquitous computing—human beings also have the option to use their brains in every situation, wherever they find themselves. Nevertheless, the iPhone does not suit as the up and coming model of the brain, because, in essence, it is just a handsome tool that ill fits the status of human nature's mirror image. Similarly, the worldwide web is sometimes said to be the brain's next top model (Mayer-Kress & Barczys, 1995), and yet with its physical dispersion, the internet is too intangible to serve as a model as the computer once did. If this trend holds, the entire strategy of analogizing brains with tools appears to be entering a new phase with the expiration of the computer metaphor. Ironically, the computer seems to have lost its significance as comparative tool in spite of its qualities of availability and versatility—as if its epistemic value as central icon waned in reverse relation to its omnipresence.

The Plastic Brain: What You See is What You Get

With the disappearance of the computer metaphor, the model changed sides and became a tool. For a long time the computer has proved indispensible for brain research, though no longer as cognitive resource but as technical instrument. This transition marks more than just an episode in the history of the brain; it is the end of the history of the brain machine (Jeannerod, 1985). The decline of the computer as central metaphor (which worked well as long as the emphasis was on functional similarity rather than physical resemblance) coincided with a second, potentially more significant cultural shift, which may have accelerated the first: the rise of a new visualization technique to the status of most prominent research methodology.

Functional imaging currently attracts enormous attention among the scientific community and the general public alike, because it allows the simultaneous visualization of structural as well as functional details in a single image, showing distinctively task-specific brain activation. Functional brain imaging has opened a new chapter in the history of brain research—the entry point of "brainhood," the positioning of the neurosciences as the universal frame of reference for addressing human nature (Vidal, 2009). The availability of this method for studying basic neurophysiology in relation to complex cognitive tasks and fundamental philosophical questions has certainly contributed to the shift in research towards higher mental functions and the emergence of such new fields as social neuroscience or neurophilosophy. In addition, the new imaging culture of isolating specific brain areas as the centers for particular psychic functions has also fostered a renaissance of localizationism—quickly denounced by some as "neophrenology" (Uttal, 2001). The precision of spatio-temporal information regarding brain activation facilitated an ontologization of different brain states by replacing the functionally abstract view of the brain-as-computer model with the concreteness of "the brain at work" (Hagner, 1996).

Another dynamic triggered by this technology is perhaps more problematic with regard to its socio-epistemic implications than the fragmentation of psychic processing into discrete units. The revived localizationism replaced the mediation of metaphors and models with the immediacy of an artificially real brain image, allegedly revealing the functional activity of the psyche within the brain. The debate about the short-comings of particular models for the brain, or the appropriateness of the machine metaphor in general, appears to be nostalgically futile and pointless now that the neurosciences can offer human societies brain images instead of machine models.

Models and metaphors may fail or betray, but they typically operate in the differentiality between the object and the concept, while images as objectifying representations always already tend to conflate the object with its representation. Here the future challenge for the neurosciences emerges. While psychoanalysis—as well as cybernetics— operated in an uncertain zone beyond the ontology of Cartesian dualism, the new imaging sciences engage the epistemological reductionism of materialist approaches but, at the same time, increase ontological complexity by constructing ever more subtle substrates of emotional, social, or cognitive states (Pickersgill, 2009). Brain research has thus moved into a space that Paul Valéry once characterized as "the interior of thinking" where there is "no thinking" (Valéry, 1973, p.124). Today, all kinds of fascinating brain images invite us to take false-colored pictures from the interior of thinking as an answer to the question of what thinking is. Ironically, the neurosciences transferred mental life onto the screen with the slogan "we are our brains;" the realism of the images testifies to the enormous effort to turn brains into media machines.

Once, metaphors and models participated actively in the neuroscientific research process; they inspired new, experimental approaches that resulted in technological tools or new models and they mediated the significance of such undertakings. Metaphors and models operated as multiple mediators in the zone where nature and culture articulate. In short, metaphors and models shaped the "neuroculture" of each period of brain research. Their multiple functions were essential in generating fresh perspectives in brain theory and in pointing to new directions for research. If today we "are" our brains, this mediation has been made redundant and the brain has become the medium and message. This points to a major cultural transformation compared to the long history of the machine paradigm that thrived on the very difference between proposed theory and the generally shared view—it was only a metaphor. From Descartes and LaMettrie to McCulloch and Walter, the specific potential of any metaphor or model—as well as its potentially scandalous nature—resulted from the shared assumption that being human, living as a person, meant something different than having a brain. This crucial difference between the model and that which it models is being eroded in the raison d'être of our present neuroculture. Ever more perfect visualizations and simulations characterize current practice in the neurosciences where the artificial has become indiscriminately real, animated, and alive.

The neurosciences, however, are a vast and dynamic field, which will continue in all likelihood to churn out surprising effects beyond today's imaginings. The sheer vastness and heterogeneity of the field seems likely to prevent the neurosciences from imminently uniting under a single paradigm and coherent brain theory. The very progress of the neurosciences undermines any stable sense of explaining mind, brain, and psyche. In a few decades, others will smile at today's naiveté. The real challenge in brain research is not to mistake today's solutions for the final answers. If the realism of today's world of imaging has replaced older modeling fantasies, then a new need for appropriate metaphors arises in order to maintain society's creative and humanistic potential against the perfected brain media of what-you-see-is-what-you-get. Models and metaphors obviously differ greatly in their liberating potential, their political overtones and individual or social grounding. The metaphor of the computer, for example, inspired the typically male fantasy of an intellectual life as pure information processing to be preserved electronically in the spaces of large technological systems.

Later the stipulated hard-wiring of the brain in conjunction with the rise of genetic determinism mirrored the rigidity of the Cold War era. More recently, the emphasis shifted towards neuroplasticity and neuroenhancement, now calling for new training strategies and smart drugs.

Where is the true place for imagination in the polarity between the potentialities of plasticity and the reductionistic realism of neuroimaging? The *Library of Babel*, Louis Borges' wonderfully claustrophobic novel, encapsulates a potential answer (Borges, 2000). In one reading, any book is just a predeterminate sequence of letters; every possible book has already been written and sits on a shelf in Borges' library. Another meaning of the same metaphor starts with the active process of reading rather than the ready product of the book. Since reading is a creative act that activates a text and constitutes a new meaning each time, no book is a mere representation of something already given but an opening (Haverkamp, 1996). Metaphors are more than just rhetoric; they are linguistic tools for finding orientation in complex worlds, as Freud and Sherrington masterfully demonstrated with their imaginative metaphors that push brain theory beyond representation. In today's neuroculture, the responsibility has shifted to the neuroscientists to keep alive the metaphors we live by.

References

Ashby, R. (1954). *Design for a brain*. New York: John Wiley & Sons.

Benjamin, W. (2008). The work of art in the age of its technological reproducibility. In W. Jennings, B. Michael, T. Doherty & J. Levin (Eds.), *Walter Benjamin: The work of art in the age of its technological reproducibility, and other writings on media* (pp. 19–55). Cambridge, MA: Belknap Press of Harvard University Press.

Black, M. (1962). *Models and metaphors: Studies in language and philosophy*. Ithaca: Cornell University Press.

Borck, C. (2008). Recording the brain at work: The visible, the readable, and the invisible in electroencephalography. *Journal of the History of the Neurosciences, 17*, 367–379.

Borck, C. (2007). Communicating the modern body: Fritz Kahn"s popular images of human physiology as an industrialized world. *Canadian Journal of Communication, 32*, 495–520.

Borck, C. (2005). *Hirnströme. Eine Kulturgeschichte der Elektroenzephalographie*. Göttingen: Wallstein Verlag.

Borck, C. (1999). Fühlfäden und Fangarme. Metaphern des Organischen als Dispositiv der Hirnforschung. In M. Hagner (Ed.), *Ecce Cortex. Beiträge zur Geschichte des modernen Gehirns* (pp. 144–176). Göttingen: Wallstein Verlag.

Borck, C. (1998). Visualizing nerve cells and psychic mechanisms: The rhetoric of Freud"s illustrations. In G. Guttmann & I. Scholz-Strasser (Eds.), *Freud and the neurosciences: From brain research to the unconscious* (pp. 75–86). Wien: Verlag der Österreichischen Akademie der Wissenschaften.

Borges, J. L. (2000). The Library of Babel. In *The total library: Non-fiction 1922–1986* (pp. 214–216). London: Penguin Press.

Brazier, M. (1961). *A history of the electrical activity of the brain: The first half-century*. London: Pitman.

Canguilhem, G. (1992). Machine and organism. In J. Crary & S. Kwinter (Eds.), *Incorporations* (pp. 45–69). New York: Urzone.

Churchland, P. S., & Sejnowski, T. J. (1999). *The computational brain*. Cambridge, MA: MIT Press.

Derrida, J. (1980). Freud and the scene of writing. In J. Derrida (Ed.), *Writing and difference*. Paris: Éditions du Seuil. (Original work published 1967).

Descartes, R. (1972). *Treatise of man*. (T. S. Hall, Trans.). Cambridge, MA: Harvard University Press. (Original work published 1633).

Draaisma, D. (2000). *Metaphors of memory: A history of ideas about the mind*. Cambridge: Cambridge University Press.

Ferrier, D. (1876). *The functions of the brain*. London: Smith.

Ferrier, D. (1873). Experimental researches in cerebral physiology and pathology. *West Riding Lunatic Asylum Medical Reports, 3*, 30–96.

Foerster, O. (1923). Die Topik der Hirnrinde in ihrer Bedeutung für die Motilität. *Deutsche Zeitschrift für Nervenheilkunde, 77*, 124–139.

Forel, A. (1894). *Gehirn und Seele*. Bonn: Strauss.

Freud, S. (2001). *The future of an illusion, Civilisation and its discontents and other works, (1927–1931)*. [Standard edition, vol. 21]. London: Vintage. (Original work published in German, 1930).

Freud, S. (1971). *Introductory lectures on psycho-analysis, (1915–17)*. [Standard edition, vols. 15 and 16]. London: Hogarth. (Original work published in German, 1917).

Freud, S. (1961). *The ego and the ID and other works, (1923–1925)* [Standard edition, vol. 29]. London: Hogarth. (Original work published in German, 1925).

Fritsch, G., & Hitzig, E. (1870). Über die elektrische Erregbarkeit des Grosshirns. *Archiv für Anatomie, Physiologie und wissenschaftliche Medizin, 37*, 300–332.

Geison, G. (1969). The protoplasmatic theory of life and the vitalist-mechanist debate. *Isis, 60*, 273–292.

Guenther, K. (2009). *A body made of nerves: Reflexes, body maps and the limits of the self in modern German medicine*. Dissertation, Harvard University, CA, Mass.

Hacking, I. (1983). *Representing and intervening: Introductory topics in the philosophy of natural science*. Cambridge: Cambridge University Press.

Hagner, M. (1993). Die elektrische Erregbarkeit des Gehirns. Zur Konjunktur eines Experiments. In H. Rheinberger & M. Hagner (Eds.), *Die Experimentalisierung des Lebens: Experimentalsysteme in den biologischen Wissenschaften 1850/1950* (pp. 97–115). Berlin: Akademie-Verlag.

Hagner, M. (1997). *Homo cerebralis: Der Wandel vom Seelenorgan zum Gehirn*. Berlin: Berlin Verlag.

Hagner, M. (1996). Der Geist bei der Arbeit. Überlegungen zur visuellen Repräsentation cerebraler Prozesse. In C. Borck (Ed.), *Anatomien medizinischen Wissens* (pp. 259–286). Frankfurt am Main: Fischer.

Harrington, A. (1996). *Reenchanted science–Holism in German culture from Wilhelm II to Hitler*. Princeton, NJ: Princeton University Press.

Haverkamp, A. (1996). *Theorie der Metapher*. Darmstadt: Wissenschaftliche Buchgesellschaft.

Hayward, R. (2001). The tortoise and the love-machine: Grey Walter and the politics of electroencephalography. *Science in Context, 14*, 615–641.

Hesse, M. (1966). *Models and analogies in science*. Notre Dame: University of Notre Dame Press.

James, W. (1892). *Psychology*. New York: Holt.

Jauch, U. (1998). *Jenseits der Maschine. Philosophie, Ironie und Ästhetik bei Julien Offray de La Mettrie (1709–1751)*. Munich: Carl Hanser-Verlag.

Jeannerod, M. (1985). *The brain machine: The development of neurophysiological thought*. Cambridge, MA: Harvard University Press.

Kahn, F. (1924–31). *Das Leben des Menschen. Eine volkstümliche Anatomie, Biologie und Entwicklungsgeschichte des Menschen.* Stuttgart: Kosmos.

Kapp, E. (1877). *Grundlinien einer Philosophie der Technik. Zur Entstehungsgeschichte der Kultur aus neuen Gesichtspunkten.* Braunschweig: Verlag Georg Westermann.

Kassler, J. C. (1994). Man–a musical instrument: Models of the brain and mental function before the computer. *History of Science, 22,* 59–92.

Kay, L. (2001). From logical neurons to poetic embodiments of mind: Warren S. McCulloch's project in neuroscience. *Science in Context, 14,* 594–614.

Keller, E. F. (2000). Models of and models for: Theory and practice in contemporary biology. *Philosophy of Science, 67,* 72–86.

Klein, U. (2003). *Experiments, models, paper tools: Cultures of organic chemistry in the nineteenth century.* Stanford, CA: Stanford University Press.

Krause, F. (1911). *Chirurgie des Gehirns und Rückenmarks nach eigenen Erfahrungen.* Berlin: Urban & Schwarzenberg.

Krauss, R. (1993). *The optical unconscious.* Cambridge, MA: MIT Press.

LaMettrie, J. O. de (1912). *Man a machine.* La Salle, IL: Open Court (Original work published 1748).

Latil, P. de (1956). *Thinking by machine: A study of cybernetics.* London: Sidgwick and Jackson.

Lidforss, B. (1915). Protoplasma. In C. Chun & W. Johannsen (Eds.), *Die Kultur der Gegenwart, Teil 3, Abt. 4, Bd. 1: Allgemeine Biologie* (pp. 218–264). Leipzig: Teubner.

Mayer-Kress, G., & Barczys, C. (1995). The global brain as an emergent structure from the worldwide computing network, and its implications for modelling. *Information Society, 11,* 1–27.

McCulloch, W., & Pitts, W. (1943). A logical calculus of the ideas immanent in nervous activity. *Bulletin of Mathematical Biophysics, 5,* 115–133.

McLuhan, M. (1994). *Understanding media: The extensions of man.* Cambridge, MA: MIT Press.

Meinel, C. (2004). Molecules and croquet Balls. In S. Chadarevian & N. Hopwood (Eds.), *Models: The third dimension of science* (pp. 242–275). Stanford, CA: Stanford University Press.

Meyer-Drawe, K. (1996). *Menschen im Spiegel ihrer Maschinen.* Munich: Fink.

Meynert, T. (1884). *Psychiatrie. Klinik der Erkrankungen des Vorderhirns begründet auf dessen Bau, Leistungen und Ernährung.* Wien: Braumüller.

Meynert, T. (1865). Anatomie der Hirnrinde als Träger des Vorstellungslebens. In M. Leidesdorf (Ed.), *Lehrbuch der psychischen Krankheiten* (pp. 45–73). Erlangen: Ferdinand Enke.

Morgan, M., & Morrison, M. (1999). Models as mediating instruments. In M. Morgan & M. Morrison (Eds.), *Models as mediators: Perspectives on natural and social sciences* (pp. 10–37). Cambridge: Cambridge University Press.

Nietzsche, F. (1872). *The birth of tragedy out of the spirit of music,* (I. Johnston, Trans. & modified from German by C. Borck). Retrieved from http://records.viu.ca/~Johnstoi/Nietzsche/tragedy_all.htm (viewed October 28, 2009).

Otis, L. (2001). *Networking: Communicating with bodies and machines in the nineteenth century.* Ann Arbor: University of Michigan Press.

Penfield, W., & Rasmussen, T. (1950). *The cerebral cortex of man: A clinical study of localization of function.* New York: Macmillan.

Penfield, W. (1952). Memory mechanisms. *Archives of Neurology and Psychiatry (AMA), 67,* 178–198.

Pickersgill, M. (2009). Between soma and society: Neuroscience and the ontology of psychopathy. *BioSocieties, 4*, 45–60.

Ramón y Cajal, S. (1895). Einige Hypothesen über den anatomischen Mechanismus der Ideenbildung, der Association und der Aufmerksamkeit. *Archiv für Anatomie und Physiologie, Anatomische Abteilung, 367*–378.

Roux, W. (1915). Das Wesen des Lebens. In C. Chun & W. Johannsen (Eds.), *Die Kultur der Gegenwart, Teil 3, Abt. 4, Bd. 1: Allgemeine Biologie* (pp. 175–187). Leipzig: Teubner.

Schloegel, J., & Schmidgen, H. (2002). General physiology, experimental psychology, and evolutionism: Unicellular organisms as objects of psychophysiological research, 1877–1918. *Isis, 93*, 614–645.

Shelley, M. (2008), *Frankenstein or the modern Prometheus*, the original two-volume novel of 1816–1817. Oxford: Bodleian Library.

Sherrington, C. (1953). *Man on his nature*. Cambridge: Cambridge University Press.

Spufford, F., & Uglow, J. (Eds.). (1997). *Cultural Babbage: Technology, time and invention*. London: Faber and Faber.

Uttal, W. R. (2001). *The new phrenology: The limits of localizing cognitive processes in the brain*. Cambridge, MA: MIT Press.

Turing, A. (1936). On computable numbers, with an application to the Entscheidungsproblem. *Proceedings of the London Mathematical Society, 42*, 230–265.

Valéry, P. (1973). *Cahiers*. Paris: Bibliothèque de la Pléiade.

Vidal, F. (2009). Brainhood, anthropological figure of modernity. *History of the Human Sciences, 22*, 5–36.

Viets, H. (1938). West Riding, 1871–1976. *Bulletin for the History of Medicine, 6*, 477–487.

von Neumann, J. (1966). *Theory of self-reproducing automata*. Urbana: University of Illinois Press.

Walter, W. (1953). *The Living Brain*. London: Duckworth.

Wiener, N. (1961). *Cybernetics: Or control and communication in the animal and the machine* (2nd ed.). Cambridge, MA: MIT Press.

Wundt, W. (1874). *Grundzüge der physiologischen Psychologie*. Leipzig: Wilhelm Engelmann.

Young, R. (1970). *Mind, brain and adaptation in the nineteenth century: Cerebral localization and its biological context from Gall to Ferrier*. Oxford: Clarendon Press.

Zago, S., Ferrucci, R., Fregni, F., & Priori, A. (2008). Bartholow, Sciamanna, Alberti: Pioneers in the electrical stimulation of the exposed human cerebral cortex. *Neuroscientist, 14*, 521–528.

6

The Neuromance of Cerebral History

Max Stadler

Once upon a time, when quizzed about the recent advances in the sciences of human brain, "Dr. Felix" was quick to reply that he rather had begun to "feel like Buck Rogers." And, he added, "we are just on the threshold," lest anyone doubt it: "where will we go—I don't know. But it is so far, so fast, that our wildest dreams are likely to be ultraconservative"(Coughlan, 1963a, p. 106).

The year is 1963, "Dr. Felix," Robert H. Felix, director of the National Institute for Mental Health, Bethesda, and the pages of *LIFE* magazine the outlet which broadcast, lavishly illustrated as always, Felix's wildest dreams (indeed, not only Felix's). "ESB" for instance, or Electrical Stimulation of the Brain, one read, was very high up on the list of those things likely to arrest even the ultraconservative imagination. This "electronic tool" promised to modulate the brain's electrical circuits—at will: "ways to "operate" directly on unhealthy emotions;" induce them—rage, fear, aggression, anxiety, happiness, a "well-oriented drive to attack and destroy;" heal sex criminals, compulsive overeaters, and those suffering from "shaky palsy" alike. Already, truculent monkeys were easily converted from "bad-tempered dictator to ... benign and tolerant philosopher" (Coughlan, 1963a, p. 100). If that hadn't been enough, readers were assured that more potent and dramatic even should prove the "chemical side of the matter," and "chemical mind-changers" in particular: the "startling" hallucinogens and, all the more familiar to readers of *Life*, all those "psychic energizers," "mood elevators," and "tranquillizers" (Coughlan, 1963b). In 1963, the day drew near when human personality would be "change[d] and maintain[ed] ... at any desired level," loneliness, depression, gloominess, and pessimism removed from society, and (a more ambiguous prospect) a "single pound" of LSD clandestinely making its way into "say, New York City's or Moscow's water supply" might easily "produce a temporary 'model psychosis' in the whole population" (Coughlan, 1963b).

Almost half a century later, the imaginary futures of neuroscience look altogether less Pynchonesque, but the neuroscientific Buck Rogers are still—or again—among

Critical Neuroscience: A Handbook of the Social and Cultural Contexts of Neuroscience, First Edition.
Edited by Suparna Choudhury and Jan Slaby.

us, even if, by virtue of sheer numbers they now would seem to resemble not the lone hero Buck but his foes, the innumerable Tiger Men from Mars (in the US, what began as an affair of less than 500, the Society for Neuroscience (SfN), now sports more than 40,000 members) (Doty, 1987, pp. 431–432; SfN, 2009). All this, no need to reiterate here, much to the excitement—or alarm—of no small number of critics and commentators; not to mention, the growing number of parasitic discourses and hyphenated disciplines grafting the "neuro" onto anything which might usefully profit from such a timely interdisciplinary alliance—anything, that is, from aesthetics to law.

Quite necessarily, a "critical" neuroscience would also have to operate within this heterogeneous web of discourses, actors, institutions, and emergent practices; and quite inevitably, this critical project would always seem to be at risk, despite all the good intentions, of thereby reinforcing the sense of exigency, or of merely reproducing the rhetoric, images, and futures advanced by neuroscientists, neuroenthusiasts, and neuroskeptics alike (a delict "neuroethics" indeed has been accused of being guilty) (De Vries, 2007; Hedgecoe, forthcoming). Any pretence of being critical would indeed seem to involve, at the very least, some reflected awareness of the shape of the discourse one is addressing. In this chapter, therefore, I shall be concerned, in broadly historical terms, with one of the elements traversing this complex discursive web: one perhaps all-too-obviously central as to be questioned, or sidelined—the brain. It is the brain, or rather the brain-centeredness of our accounts of what seems to be at stake, that I wish to confront with if not exactly a critical then certainly an unsuitable past. This chapter, in a way, is about the merits of being, from the vantage point of the history of science, not a historical program of how-to-be-critical. It offers an unromantic view from the history of science, not a historical program of how-to-be-critical.

It is certainly not immediately apparent why (or how) those pondering the "potential implications" of contemporary neuroscience should be particularly concerned with, say, the wild dreams induced in the 1960s by the electronic tools of brain science; nor, for that matter, what exactly the psychedelic threats to the geopolitically significant water supply systems were—all the less so, because the dramatic expansions of the hard-to-fathom complex of activities we refer to as the neurosciences are very recent history at best: "It is so far, so fast." Naturally, perhaps, much "neurotalk" tends to be future-oriented. The historical imagination, at any rate, has played no really explicit, or critical, role in what is, to be sure, a highly varied set of discourses. And yet, if the latter-day Buck Rogers naturally, as it were, keep their stern eyes on the future of the neurouniverse rather than on the subtleties of its past, it does not necessarily mean legitimately so; and history not being made explicit, does not necessarily mean that it is absent or that this is the case and, more specifically, that there is a tendency to endorse—somewhat uncritically—a romantic, brain-and-mind-centered vision of neuroscience's pasts, is the argument that I shall develop in this chapter.

The point, in brief, will be to deflate the notion that historically speaking neuroscience is best, and naturally, imagined as solely and essentially revolving around man's cerebral nature.[1] Itself of fairly recent vintage, this brain-centric vision has arguably structured (not least) much of the historical narratives of neuroscience which have become

---

[1] Of course, one could easily adduce a list of items where a little historical reflection might prove enlightening, quite irrespective of whether or not this historical imagination is indeed "brain-centered." In

available in recent decades. It is this vision, or historical figure, that the *neuromance* in the title gestures at: a cerebral romanticism inscribing the neurosciences, wittingly or not, into an age-old, anthropological quest of ultimate significance, the final capstone on the long-winded path to human nature exposed (the more obvious examples for this kind of sweeping narrative would include: Changeux, 1997; Clarke & Dewhurst, 1996; Corsi, 1991; Finger, 2001; Gross, 1998; Poynter, 1958).

As I shall argue, problematizing the neuroscientific past might mean thinking somewhat less romantically about the neurosciences instead; it might mean, that is, to disengage our historical imagination a little more from that very organ that has so profoundly come to define the image of neuroscience—the brain. Accordingly, I shall be less concerned in the following with an object-lesson in the illuminating (or exposing) deployments of neurohistory; or, for that matter, with chemical mind-changers and electrical stimulation in the post-World War II period per se; rather, more historiographically, and more inclusively, with the kinds of pasts conjured up in the first place. The difficulties of turning such historical niceties into "critical" ones (lest we celebrate too early) are, however, compounded by a host of issues which speak not merely to the tendency of romanticizing the brain; not least, they point to a tendency built into science studies, a field always prone to elevating its object—Science—into perhaps too central a force in matters of societal change; indeed, as I shall conclude, it may inadvertently run counter to the object of "critique:" feed rather than deflate the neuroscientific exigency—becoming Buck Rogers.

Neuromance

Though Dr Felix won't concern us here much further, let us briefly return to the scene painted at the outset. As a historical picture of brain science in the early 1960s, the above is, of course, little more than a caricature. One should certainly not imagine the 1960s brain as especially comic; neither, perhaps, as simply superseded in its at times bizarre enthusiasm; nor, however, was this little vignette meant to intimidate a deeper resonance with contemporary neuroscience. What the phantasm of unlimited brain-control unfolding in the above was meant to invoke was not a historical situation so much as a historical gesture: a quite typical maneuver on the part of the historians of neuroscience. In fact it is a—or perhaps the—primary mode of understanding and constructing this history: that the brain was not in the news for the first time either then or today will hardly be news for readers following the—more professional—historical literature. More properly, it is the project of showing—the novelty rhetoric of much of the contemporary "neurotalk" notwithstanding—that the brain and its sciences were always fundamentally cultural objects and, as such, have histories long pre-dating the much more recent advent of the neurosciences.

It is not too difficult to see how the above might fit into such picture, say, of the Cold War American brain: one showing the brain, its sciences, and the ways they mattered deeply entangled with the cultural, economic, and political fault-lines of the

fact, surprisingly little of this type of analysis has actually happened. Not least here, however, disinvesting in the brain/mind drama could be crucial, certainly for analytical purposes.

times. The specter of being brain-washed, or the antics of CIA-funded neuropsychiatrists might be familiar (see Alder, 2007; Littlefield, 2009; McCoy, 2006), as might the sky-rocketing use of amphetamines, tranquillizers, and anti-depressants in the 1950s and early 1960s, busily cultivated by a burgeoning pharmaceutical industry (Herzberg, 2008; Rasmussen, 2008; Tone, 2008).

There is indeed a very good case to be made that it was then, in the middle decades of the twentieth century, that the *central* nervous system began definitely to shape not merely the discourses, but also the practices, surrounding the nervous: lobotomy, the EEG, ESB, "electronic brains," a fast-growing range of wholly new substances promising a cure for the mentally ill and relief for the melancholy masses—Benzedrine, Miltown, LSD, chlorpromazine; the steady, well-engineered growth of psychosomatic medicine since the 1930s; chemical warfare worth its name (from the scientific, neurophysiological point-of-view, that is); "death" on the verge of being rethought as "brain death." All this would have contributed to the rising scientific and public salience of the "living brain" in the 1950s and 1960s (Belkin, 2003; Borck, 2005; Braslow, 1997; Crowther-Heyck, 1999; Pressman, 1998; Schmaltz, 2006).

More significant, however, for my purposes than the possible feel of déjà-vu is that, in fact, we lack anything in the way of a comprehensive picture of the developments at issue: the neurosciences in the second half of the twentieth century. And if, as seems plausible enough, there was indeed a significant shift around 1950 in matters of the brain—both culturally as well as an object of experimental science—the point of the following is not to improve on such a picture, or to belabor a necessarily somewhat arbitrary point of origin. Rather, by looking more closely at the formative decades just prior to the institutional, post-1970s crystallizations of "neuroscience," it is my aim to explore the limits of the historical maneuver above; of imagining, that is, the history of neuroscience in overly cerebral, brain-centric terms.

But first, it will be appropriate to dwell a little longer on these latter terms. The express concern with the cultural dimensions of the brain is indeed, and hardly surprisingly, what is most salient about the recent accumulation of historical literature on the neurosciences—as a historical occupation (and label) itself nearly contemporaneous with the run-up to the Decade of the Brain. "Surely the rising star of body parts in the 1980s" must have been the brain, as feminist historian Elaine Showalter noted in 1987 (Showalter, 1987, p. 39). The ensuing decade saw the creation, notably, of a *Journal of the History of the Neurosciences* and an *International Society for the History of Neuroscience*. From the mid-1990s, the US Society for Neuroscience launched a series on the *History of Neuroscience in Autobiography*, now grown to six volumes. And quite apart from such concerted efforts, it has become easier than ever to turn up remains relating to your favorite branch of neuroscientific prehistory in the vast, digital seas of the internet. While one would be hard pressed to detect the traces of an over-arching master-plan in these quite diverse activities in tradition-building, overall, the framing is perceptively different than the kind of history writing still prevalent well into the 1980s.[2] The history of this young science has become grafted

---

[2] Take the case of phrenology: though its ghost is still, or again, haunting the makers of fMRI images today, such polemic instrumentalizations are very unlike the deep interest historians have shown for the

onto a history of the brain, as much, perhaps, as the latter has been re-imagined through the lens of "modern neuroscience." The casual collapse of the one (neuroscience) into the other (brains) begins, but hardly ends, with the *Wikipedia* entry on "Neuroscience" (at the very beginning of knowledge). Its history section will carry you, and almost perfectly reproduces, another entry: "History of the Brain". Politically-correct-enough, it informs at length about contributions from "non-Western" science, but otherwise tells a familiar and edifying, if not particularly subtle, plot: Ancient Egyptian surgeons, Aristotle, Galen, Descartes, and on to the "modern period" which sets in with a number of great, nineteenth-century figures (the twentieth is generously skipped over): Golgi, Ramón y Cajal, Du Bois Reymond, and Helmholtz at the cellular end; Broca, Jackson, and Brodmann at the cortical one ("History of the Brain," 2010).

But Wikipedia is only one such symptomatic case, and probably not the most authoritative one. The tendency is widespread, and although I will focus here on the more academic kind of histories, I do not mean to single out, or prioritize, the latter when referring to the "historical imagination." Biographies of distinguished members of this or that medical specialty—for instance neurologists, psychiatrists, neurophysiologists, psychologists, and so on—are more likely to be subsumed now under the label "neuroscientist" (Söderqvist, 2002). Meanwhile, the label "behavioral sciences," once providing a similarly salient, omnivorous but disparate umbrella, that—spilled well over into the social sciences, has lost much of its former relevance in structuring narratives (see esp. R. Young, 1966). In other cases (and here we are coming closer to the kind of memory work at issue here), personae and events, should they fit less obviously into the (self)images of the neurosciences, recede practically into obscurity or remain at a safe distance, remembered as exponents of other, less obviously brain-and-mind-centered disciplines—say, molecular biology or biochemistry.

Since the 1980s, not only have the sciences of the brain been refashioned as neuroscience (or rather neuroscience has been fashioned as the new, and true, brain science). Importantly, the registers employed and theoretical tools mobilized by academic historians engaging with the brain have also mutated, alongside significant ideological re-constellations and a new sophistication in the profession, generally. Some of these new horizons will be familiar—the turn to the "local" and the much celebrated attention that was now being paid to the (equally local) practices of science, for example. Particularly important here is another, related theme which featured prominently in these historiographical departures since the 1980s: the increasingly culturalist orientation and habits of mind that historians of science brought to their subject matter. Somewhat ironically, it is this culturalism which, despite its utter productiveness in re-envisioning science's pasts, has been—inadvertently—complicit in what I called the romantic tendencies in the neurohistorical

---

matter as late as the 1960s and 1970s (Cantor, 1975; Cooter, 1985; Shapin, 1975, 1979; Wyhe, 2004); then, phrenology's rehabilitations as something quite other than "pseudo-science" functioned in a very different socio-political climate, more likely to be directed, by the waves of Marx-reading scholars, against the illegitimate powers and pretensions of the behavioral sciences and psychiatry (rather than, say, the much more nebulous prospects of a neuroenhanced, posthuman future).

imagination; inadvertently because, in terms of history, much of what has been written about the brain in the past few decades, had been driven by decidedly skeptical attitudes towards the new and growing visibilities of something called "neuroscience." The way to proceed has been to write histories of the brain and its sciences in the idiom of "culture;" and noteworthy too, it has been to study not the recent genesis of neuroscience, but periods prior to World War II.

Works such as Anne Harrington's *Medicine, Mind and the Double Brain* (1987) were among the first to react to "the burden of a wide variety of social, moral and philosophical concerns" which, Harrington wrote, the brain was made to carry in the late twentieth century, "when the explanatory possibilities of the brain sciences [were] widely perceived as almost limitless"—again (Harrington, 1987, p. 5, p. 285). As Harrington then set out to show (along with a growing number of fellow historians), such perceptions were not so fundamentally new. Moreover, it was shown that notions of the brain's structure and functioning could not be dissociated from the moral and social norms and discourses structuring, say, nineteenth-century industrial society, and the conditions that reproduced class, gender and racial divisions: and even though historians took pains not to draw explicit parallels concerning the contemporary "relationship between ideas of brain functioning and the social and political order," little doubt was left that the late twentieth-century resurgence of brain-talk should not be exempt to similar relativizing scrutiny. Indeed many of the same concerns—the naturalization-through-brains, in brief, of the social order or the "cultivation" of the cortex—were beginning to drive similarly ambitious projects (Borck, 2005; Hagner, 1992; R. Smith, 1992; Weidmann, 1999).

Whether Victorian mad doctors or interwar brain eugenicists, we are beginning to possess an increasingly fine-grained picture of just how consistently the brain has served within the last two or three centuries as the projection site of social, moral, and political spaces. These histories, sophisticated and scholarly, were for good reasons, explicitly not meant to be understood as histories of "neuroscience." Rather, they were advanced as—to be sure, timely—cultural histories of the brain. This was a quite different endeavor in so far as the goal here was rarely to recover origins, precursors, or to simply chart the evolution of neuroscience's embryonic ideas and concepts. It was to expose, if you will, the historicity and historical specificity of discourses that locate human nature in the brain; and it was to expose the complex ways in which such knowledge claims were culturally mediated, a maneuver which was not meant to yield straightforward continuities with contemporary neuroscience (and its quite distinct cultural contexts). Whether such subtle points are always registered as such may be an entirely different matter, of course, particularly once we factor in how such cultural history may function within the broader force-fields that define neuroscience's past (a connection which is always made, after all, and one which it has become difficult not to make).

To be sure, processes of naturalization, or representations of the brain are by no means the sole preoccupation of historians of neuroscience. But even when the historical object was ostensibly not the brain but less dramatic entities—the story of chemical nerve transmission, for instance (a relatively well-charted episode)—existing narratives have been remarkably resilient in omitting those agents that consistently propelled such knowledge (Dupont, 1999; Valenstein, 2005). These agents—insecticides, chemical warfare, the pharmaceutical industry, psychiatry—would indeed not seem to sit easily with the image of the fundamental-science-of-the-mind that neuroscience has accrued (Russell, 2001; Schmaltz, 2006). Others have been more impressed by neuroscience as an instance of modern biopolitics, or by the persistent recurrence of the past and the social in the concepts, practices, and rhetoric of the latest, current installment of neurofurore. But in these cases too, it was the brain/mind which figured as the—unquestioned—vanishing point (Abi-Rached & Rose, 2010; Dumit, 2003; Littlefield, 2009; Maasen & Sutter, 2007; Vidal, 2009). When, for example, Fernando Vidal argues, convincingly, that the "ideology of brainhood"—the modern notion that human beings, or persons, essentially are their brains—was intellectually prepared in the early modern period (far from being something "caused" by recent advances in neuroscience), intellectual history and history of brain science may enter an antithetical and asymmetrical relationship; the assumption still is that it is "human nature" that must be at stake, reproducing rather than challenging the inflated rhetoric of much neurotalk (Vidal, 2009).

I am schematizing terribly, of course, when collapsing a range of very different positions, approaches, and agendas into a single line of unearthing neuroscience's past. Still, it is worth pondering what arguably unites this historical discourse, and what arguably unites it too with the much broader realm of memory-work centering on contemporary neuroscience (which would span early and influential interventions such as Gardner (1985) to the more recent additions to the corpus by Craver (2007), Gross (2009), or Kandel (2006)—the former a case of philosophical rather than historical under-laboring, attesting to the confusing "mosaic" that is neuroscience, an epistemic coherence, slipping in the brain as the virtual entity holding it all together). What is common to all of these is the focus on the brain/mind—as a cultural construct; in terms of a history of ideas; a series of progressive, scientific departures; as part of a philosophical (mind/body) epic. Or here, in this affirmation of neuroscience's phantasmic, discursive glue is where one needs to locate the limits—and for critical purposes, short-comings—of what gets floated under the label "history of neuroscience." Even the more skeptical, cultural-historical maneuvering is all too easily turned on its—neuroromantic—head.

By culturalism, then, I do not mean here the theoretical commitments of a very peculiar historical approach but rather, the consensual way of doing history of science today. It might mean (a culturally-informed) "intellectual," "discourse," or "conceptual" history, though more often it would now also imply attention being paid to the "local:" the situatedness of knowledge claims, laboratory cultures, visual and representational technologies, and so on. Characterized negatively, "cultural" here means not least paying attention to the cultural and local, largely at the expense of political, social, economic, and, in this specific case at least, recent history. Let us, then, not take too seriously the allusion in the above to social and political conditions. Like the majority of contributions to the *Journal of the History of Neurosciences*, like the bulk of practitioners' histories and like Wikipedia, what has emerged over the last few decades as the cultural history of the neurosciences, despite the obvious differences between and within these genres, has a common, and overly romantic, point-of-reference: the brain. It is a perfectly legitimate project, but one that has curiously distracted from, even obscured, the less romantic dimensions of this past; and even more so, the conditions that have turned the neurosciences into a major scientific industry within the past few decades.

To see where this might be heading, consider briefly just how distant the spaces of contemporary neuroscience have become from however we end up imagining its pre-history. Impressive indeed is the very recency of neuroscience's emergence and its institutional expansions. According to one recent survey, no less than 81% of American neuroscience programs now in existence were founded only after 1975; 72% of the undergraduate programs even only after 1989. All the while, the neuroscience community has expanded dramatically, doubling since 1991; the number of (US) PhD degrees awarded rose steadily from 404 in 1996 to 584 in 2004 and 689 in 2005, and was estimated to be well over a 1000, or one in eight biomedical PhDs, by 2008—figures matched only by biochemistry (Association of Neuroscience Departments & Programs (ANDP), 2007; National Science Foundation (NSF), 2009). NERV, the Nasdaq/Neuroinsights Neurotech Index, to cite another sign of the times, claims that since 1999, "venture investment" in neurotechnology has nearly tripled, constituting now a $145 billion "global industry" (of which, hardly surprising, some 85% are made up of—entirely unromantic—"neuropharmaceutical revenues") (NeuroInsights, 2010).

Such numbers should be treated with tremendous care, as Paul Nightingale and others have argued, showing just how empirically unfounded all the talk of a biotech "revolution," or for that matter, a "neuro-revolution," in fact is (Hopkins, Martin, Nightingale, Kraft, & Mahdi, 2007). Here they may serve to bring home the point pressed above, illustrating how deeply at odds we should imagine this industrious "thing" called neuroscience to be with those historical narratives centering on the brain (or on human nature, or those which casually conflate the intellectual history of the mind/body problem with that of brain science). In this version, the coherence provided by the brain/mind as the entity structuring our narratives would quickly seem to lose its informativeness.

This other, untold story would be impressed instead by the conditions that have sustained neuroscience's growth, attempting to see it as a symptom rather than a cause. It would quite probably come to resemble, and blur with, those other, less stimulating histories which are now being treated under separate headings such as that of the pharmaceutical industry, psychiatry, mental health, or health care; more generally, it would blur with a domain which has become increasingly well-charted by STS scholars— biotechnology/biomedicine—including the concomitant transformations of the academic research sector since the 1980s (its neoliberalization/commercialization in particular) (for instance, Jasanoff, 2007; Mirowski & Sent, 2005; Shapin, 2008). Along different lines, such de-centered histories would also blur the disciplinary vision that follows the brain (or memory, or language) too closely into the laboratories. Instead, they would embed the neurosciences within much broader transformations in the (recent) history of science, while at the same time make neuroscientific knowledge production imaginable as a historically specific form of knowing—as an effect which coalesced at the intersections of various techno-scientific departures (none of which, sig-nificantly, would seem to be very romantic): molecular biology, of which brain-minded scientists have begun to dream by the 1950s, is an obvious case in point; computer science, physics, and engineering another (think of MRI and data analysis).

Construing the neurosciences as the grand, or detestable, finale to a history of the brain is a historical construction that on the whole has worked against, rather than for, arriving at a historically and empirically informed picture of contemporary neuroscience—whether or not it was advanced in terms of a heroic, eminent lineage; as the foil onto which to project progress; or, as a repressed past whose exposure would undermine neuroscience's claims to revolutionary novelty. I am not suggesting that this must necessarily be so, or that the one could, or should, replace the other. The strategic point, after all, of writing histories of the brain—or better yet, and more inclusively, of the nervous system—is ideally about establishing some de-familiarizing distance between the two. What I am suggesting, however, is that we should be more careful in crafting our stories, and that "the brain" effectively serves to conceal, rather than reveal, the mundane determinants of neuroscience's genesis. Especially the discourse of the new, human nature and society-transforming science of the brain/mind is something that the construction of which is itself in need of analysis, not a concept that should guide our analyses.

Cyber Romance

To illustrate, let us return, once more, to the middle of the last century. One of the more potent origin myths of, if not exactly neuroscience, then all the more emphatically, of the dawn of a new era of scientific engagement with the brain/mind is located at this juncture: World War II. And like many a myth, this one is not entirely without its plausibilities. For there can indeed be little doubt that in the wake of World War II, both human behavior and mental health—injured by a very recent, violent past and endangered by a fully-automatized future—had turned into fundamental problems of planetary dimensions. It would provide new opportunities, not least for those biomedical scientists skilled enough to profit from the rampant post war ideology of "basic science," "team work," and "interdisciplinarity."

Most famously, it was Vannevar Bush who then spelt it out for everyone: "basic research" now would be key (Bush, 1945). One of the more visible expressions in the world of biomedicine of this new optimism were the several institutions launched in response to the American National Mental Health Act of 1946 (Farreras, Hannaway, & Harden, 2004); elsewhere too, all the signs seemed to point towards progress, expansion, and fundamental science. In England, the Mental Health Research Fund was founded in 1946, its activities heavily slanted towards the newly coalescing field of "neurochemistry" (Bachelard, 1988; McIlwain, 1985); neurophysiology, a rather more academic specialty, also thoroughly emancipated from its institutional entanglements with medicine, spurned in its rigors by the wonders of electronics the recent war had thrown up (Chadarevian, 2002; Schoenfeld, 2006); more self-confident than ever, neuropsychiatrists and neurologists turned experimental and interventionist, zeroing in on the brain rather than its bodily, outwards manifestations (Braslow, 1997; Pressman, 1998). None of these tendencies, to be sure, was new and without precedent. But there was now a "climate favorable to all fields of research in any degree involving the brain, whether they begin or end with it," as the French electro-physiologist

Alfred Fessard would diagnose the situation in 1952, pondering the case for a projected International Brain Institute under the umbrella of UNESCO. Fessard did not even bother "to repeat the usual generalities about the importance to mankind of the intensive study of the brain" (Fessard, 1952).

"Neuroscience" as we know it barely existed. What existed, and what we may in retrospect identify as so many departures, elements, and events presaging its eventual coalescence were a myriad of scattered traditions, specialties, initiatives, institutions-in-the-making and new alliances not always, not yet, and not primarily structuring their practices around this common object, the central nervous system—the "most complex organ in the universe," as it turned proverbial at the time (see, for instance, Pfeiffer, 1955; Walter, 1953). Rather—and this is the thrust of the present section— to the extent that they did enroll the central nervous system, we had better look twice so as not to conflate heroic discourse and the (by and large) banality of scientific realities. Indeed, a great many instrumental figures once-to-be fashioned as makers of this nascent, neuroscientific future were well established by the time. Significantly, many of them, like Francis Schmitt, Ralph Gerard, Bernard Katz, or Stephen Kuffler, were raised in the prewar period in quite different circumstances, and were engaged with quite different, unspectacular objects: the central nervous system was not the horizon of their scientific doings and self-perceptions (Hodgkin, 1992; Katz, 1998; Libet, 1974; McMahan, 1990; Schmitt, 1990). Instead, they focused on bioelectricity and muscular metabolism: frogs' legs, the dog's heart, sea-urchins, nerve fibers obtained from the squid and other such lowly and peripheral things—the things that defined the realm of "excitable tissues" through which this dizzyingly heterogeneous lot of experimental physiologists had once roamed freely (Stadler, 2009).

It was not least from these circles that those would be recruited who would impose, beginning in the 1960s, an early identity onto a new label: neuroscience. By 1973 J. Z. Young, re-discoverer of the squid giant axon and another one of those instrumental figures, could ponder, when looking back on those amorphous days, that at last "we know our identity … we are [all] Neuroscientists" (Worden, Swazey, & Adelman, 1975, p. 40). But this, alas, is not the story we know; and certainly, it is not the one that has gripped the historical imagination. Neither is it any one of those other plotlines which we might want to bring to bear in this connection: the story of neurochemistry (which would be a far less academia-centered one than that of the self-appointed neuroscientists above); or that of neurology; cell-biology; or molecular biology. The grand narrative that exists is a different one. It is the story of cybernetics.

This one, to be sure, is not a story explicitly, or merely, about "neuroscience," even though the brain looms large in the great mass of literature that has accumulated on this major intellectual event (see Abraham, 2003; Dupuy, 2000; Edwards, 1997; Galison, 1994; Gerovitch, 2004; Hayles, 1999; Hayward, 2001; Heims, 1991; Husbands & Holland, 2008; Kay, 2001; Pias, 2004; Pickering, 2002a). "We selected from prompt action" (as the exemplary object of cybernetic theorizing), as John von Neumann—physicist, weapons-science spin-doctor, and game-theorist—pointed out to his cybernetic friend Norbert Wiener, in a letter in 1946, "the most complicated object under the sun—literally" ("Neumann to Wiener," 1946).

Famous enough their mission has become nevertheless; and because of, rather than despite, this forbiddingly complicated organ. Wiener and von Neumann thus belong to the core set of actors appearing again and again in our accounts of the brain/mind around mid-century: and there is no doubt that cybernetics' central protagonists were always quick to announce a new age of brain science; this, these maverick scientists believed to be unlocking with the aid of models, interdisciplinary inquiry, colorful teams composed of vagabond engineers, physiologists, and mathematicians, a new language (wrestled from the communication engineers) and novel, modern instrumentation. As one of them, Grey Walter, saw it: previous generations did not—and perhaps could not—dare to "accept the brain as a proper study for the physiologist;" instead, they had chosen simplicity: muscle, nerves, and sense organs, "often carried to the extreme … so as to eliminate all but a single functional unit" (Walter, 1953, pp. 27–28).

What Grey Walter preferred not to mention was just how powerfully this physiological extremism held sway, side-lining the brain and mind underneath and beyond the speculative loops spun by his cybernetic comrades. Worse, people "speculating along such lines," opined, for one, the then secretary of the British Medical Research Council, wrapping up the empiricist temper that fortunately curtailed the scientific mainstream—rarely produced "adequate data either to check or support their speculations" ("Mellanby (MRC) to Randall," 1949). The story of these speculations has been told often enough, at any rate, so that we can confine ourselves to some pertinent complications. Let us remind ourselves, briefly, of what is said to be at stake in the story of cybernetics' unfolding.

As the received stories have it, ontological certainties that were previously in place were effaced in the process: man/machine/animal; model and reality; natural and artificial were categories no longer commanding assent when, from the early 1940s, the cyberneticians inaugurated a new vision of the human aided by information theory, circuit diagrams and flow charts. Most relevant to my argument, it was here that not only was a new vision of the brain/mind in the making—the brain-as-computer, a model-making and information processing thing—but also that the "living" brain/mind was introduced as an object of experimental and quantitative study in the first place—perhaps, after half a century of "eclipse," as one historian put it (Kay, 2001); after a dark age of behavioristic superficialities, timid physiologists of the peripheral nervous system, and primitive research technologies, as cybernetics aficionados themselves liked to style it. A great many commentators—media theorists, literature critics, and historians—have explored cybernetics as this epistemic event, the latter diagnosed early on as no less than the "fourth discontinuity" (the fourth that is after the Copernican, Darwinian, and Freudian ones) (Mazlish, 1967). Tracing the reverberations of cybernetics into fields as far apart (or close) as literature, pedagogy, city planning, art, and ecology, their interests were not always, or primarily, confined to the brain, let alone brain science. In fact, few would claim that cybernetics exhausted, or even fundamentally shaped, the history of the brain/mind around mid-century. That cybernetics had basically "evaporated" by 1960, or that it made little contact with the "wet" biology of the brain and the scientific mainstream, is common knowledge.

Yet, more fundamentally than anything else, it was cybernetics that has served to frame historical narratives of brains and minds in the twentieth century (among others see Baars, 1986; Boden, 2006; Gardner, 1985). Certainly it is not too difficult to see why cybernetics would have assumed such a prominent position in the historical imagination, even if the computational brain, at least in its mid-century variant, would seem to have long lost its appeal (Borck, this volume). The ontological confusions routinely set into operation by cyberneticians—the figure of cyborg, especially— surely enough resonate with twenty-first century, technoscientific conditions of living as much as the renegade image of vanguard interdisciplinarity that cybernetics came to exemplify must appeal to anyone who discerns its vindication in the amorphous disciplinarity of contemporary science; here were computers and modeling practices elevated for the first time to the center of scientific activity; here was a technoscientific war, whose manifold implications in the origins of cybernetics continues to excite; here was a materialist, seductively (or seemingly) anti-humanist discourse of the mind and human nature attractive to both explorers of the brain as well as *Geisteswissenschaftler* suspicious of the traditions of meaning-and-subject-centered analyzing. Here, not least, was something recognizably "cultural" in the science: unlike those hordes of fairly monosyllabic, uncultured, and altogether unexciting specialists that had begun to inhabit the laboratories of biomedicine by 1950, the missionaries of cybernetics were not nearly so inhibited; on the contrary, they were vocal, imaginative, and out-reaching.

Whether they deflated head-on, "the Descartian split between mind and body" ("McCulloch to Gerty," 1943), brought inspiration to art and music, or mounted robotic spectacles at the Festival of Britain, the spectacular—the popular, philosophical, and techno-futuristic—was thus never far off in this cybernetic delirium of a universal science of control and communication (Dunbar-Hester, 2010; Pickering, 2002b); and never far afield either was the brain— albeit, on the whole, a somewhat virtual one: a brain modeled, theorized and imagined rather than a brain dissected and measured. The vision of the missionaries of cybernetics was a naturalistic one that could cause ideological alarm (as when communism or technological progress threatened to reduce men to mere automata) as much as the much-needed hope: for many, here was in the offer a model of postwar living: a world where life, peace, and truth would be matters of "communication" (Young, 1951); where scientists (and artists too) operated not unlike these model-generating brain-machines; and where the common people would ideally operate like scientists. For others this vision constituted a dangerously "new type of metaphysics;" "No Christian," as Wiener's old friend J.B.S. Haldane commented sardonically on the new "cerebralism," "after reading the first verse of St John's Gospel, can object to the emphasis laid on communication" (Haldane, 1952).

It would no doubt be difficult to imagine a cultural history of the nervous system in the period without cybernetics; it would, however, be equally mistaken to take the "cerebralism" and its cultural/intellectual effects (which it evidently had) *for* this history. It is the near inevitability with which this problematically cultural (and intellectual) vision figures in the stories we actually tell that is flawed—a function, more than anything, of cybernetics' public visibility. The romanticism, if you will, consists in the ways these stories tend to reproduce, rather than question, the dramatic categories prescribed by the cybernetic discourse itself—revolutionary departure, brain/mind/body, human nature, and so on. But, just as talk of an "information society" and its celebrated weightlessness (another feat routinely traced to the vicinity of cybernetics) consistently obscures the energies and materialities at work beneath the glitter of digital futures, so the rupture story of the cybernetic brain obscures the inconsistencies of the record; and similarly serves to locate the onset of a heroic endeavor—an interdisciplinary, materialistic science of the mind/brain—in the past: a comprehensible future that has already begun.

It is not, in fact, too difficult to imagine a somewhat different picture. Thus, whether historians have re-located the sources of what may be called cybernetic regimes of knowing (as opposed to the outpourings of the small coterie of self-professed cyberneticians) in interwar telephone engineering or the machinery of state bureaucracy (Agar, 2003; Beniger, 1986; Hagemeyer, 1979; Mindell, 2002; Noble, 1986); whether they have historicized the mid-century moment of model-mindedness and interdisciplinarity as a symptom of complex ideological circumstances (less so, the inevitable progress of knowledge spearheaded by maverick scientists) (Cohen-Cole, 2003, 2009; Crowther-Heyck, 2005); or whether they have shown even the "cyborgs" to be suspiciously absent from the annals of cybernetics (Kline, 2009), the result is less recognizably the plotline of a singular incision. All this invites reading the cybernetic discourse, in historical terms, as a symptom of much vaster (and mundane) sea changes—and, for our purposes, in ways that bring to the fore the multiple and non-convergent forces that shaped the sciences of the nervous system during this period.

In belittling discourses surrounding the brain, my aim is not to pit a dull history of "real" science against a cultural history of the nervous system in the period. The case that is being made is about taking more seriously the many and less obviously neuroscientific factors besides the brain that shaped the history of the nervous system; and, it is about being more scrutinizing in our attempts to locate "culture" (and the significance we want to bestow on it). We might then quickly arrive at a dramatically deflated and thoroughly cultural picture of cybernetics' significance, while at the same time come to better appreciate just how tangential her cerebral discourse may have been to whatever happened in the laboratories, or in most of them.

The case of Norbert Wiener, whose immense public presence as the Cassandra of the dawning age of automation has been thoroughly documented, is itself instructive in this connection (Hayles, 1999; Heims, 1980; Siegelman & Conway, 2004). And present Wiener was: by 1949, Wiener's notoriously difficult, formula-laden *Cybernetics* (1948) had sold a spectacular 13,931 copies and a more accessible version was already in commission. "PANDORA" or "CASSANDRA," Wiener's own preferred titles being "absolutely out of the question" ("from the publishing point of view"), Wiener's grim vision of man's technological future hit the shelves in 1950 as *The Human Use of Human Beings* ("Technology press to Wiener," 1949). Indeed, just how actively Wiener and allies were courted by journalists and the extent to which these medializations may have shaped the message and nature of the cybernetic project itself, is a dimension yet to be fully explored. Not least the discourse of "models" for which cybernetics rightly acquired fame, might then, on closer inspection, turn out to be less of the epistemic rupture that opened up fundamentally new spaces of scientific complexity (such as the brain). Rather, cybernetics might emerge as an effect, or condensation of the media-technological infrastructure with which it came interlaced. Its significance would reside in the light it casts on the mediations of postwar intellectual life; far less so, in what it tells us about the evolutions of brain (or neuro-) science.

Cybernetics thus would, as one such helpful scribe advised Wiener, "make the foremost story of the 20th century"—but only, that was, "if the essential element of CYBERNETICS could be reduced to simple symbols — blocks of wood, even" or, even better, "photographs:" "Channel[s]" that "would make the implications of CYBERNETICS amenable to presentation in dramatic and concrete terms with meaning for the average man" ("Jones to Wiener,"1948). Models, metaphors, visual aids, charts, analogies, and diagrams—the insignia of the cyberneticians—served purposes beyond the emphatically epistemic, as not least the then thoroughly professionalizing community of science-journalists would have come to appreciate. More than ever before, these devices were beginning to live precarious double lives as tools of communication, a problem felt in particular when they seemingly were needed most—when scientists ventured beyond their own disciplinary terrains, or, as happened with similarly increasing frequency, beyond their laboratories (Bowler, 2009). Such transgressions were programmatic to what cybernetics was and, as Bowker (1993) has shown, much of the cyberneticians' success was dependent on strategically exploiting an idiom of "universalism;" it would smooth the implantation of the cybernetic discourse in potentially any science. A more historical, and less sociological, approach would highlight instead how profoundly such "cybernetic strategies" were themselves parasitic on the verbal and visual technologies that were then being floated. Models and related verbal and visual technologies of communication were not the exclusive domain of the cyber scientist. Advertisers, journalists, and educators in particular had by then generated an impressive armature of models, visual aids, and other technologies of persuasion (Buxton, 1999; Lagemann, 2000; Seattler, 1990). "To tell the truth [was] not enough" as Patrick Meredith, science teacher turned director of the Visual Education Centre, Exeter, explained in 1948, "it must be communicated" (Meredith, 1948).

By no means were model strategies the proprietary format of the cyberneticians, even though they may have been particularly adept at the task. Cybernetic missionary J. Z. Young was "highly stimulating … [and] quick, vigourous, imaginative," unlike the "usual scientist," as one BBC employee judged ("Notes on J. Z. Young," 1948). It was such "really first-rate science popularizer[s]" who excelled—much to the pleasure of the BBC—at bringing closer to the postwar public the most recent conflations of minds, brains, and machines. It is unsurprising, then, that the likes of Young or Grey Walter were routinely given the opportunity to weigh in on the general "spate of brain talks" which were hitting the airwaves at the time. Here the man of "average, not exceptional intelligence" was offered "synoptic glimpses" of difficult subject matter—not least, the many models, analogies, and other "illustrations" of "the way information is conveyed from one creature to another" ("Draft outline," 1949).

It is, in part, the fact that such symbiotic relations as the one between Wiener and the press, or Young and the BBC, were by no means exceptional which renders the cybernetic discourse highly problematic as a historical account of brain science (or of scientific modeling, or of technological evolution). Just as cybernetics amalgamated rather than originated vast amounts of (futuristic) knowledge, so the format of its presentation is best construed as parasitic on a set of fairly banal practices and developments. In fact, even this would be saying too much, in as much as the postwar publicity in matters of the cerebrum was vastly more encompassing than the cybernetic story would seem to suggest—an ideological playground and confrontation space for all manner of learned neurologists, philosophers, anthropologists, and laboratory scientists.

The story of the cybernetic discovery of the brain/mind is nothing, in other words, that could simply serve to contextualize the stories we tell, let alone a story that penetrates deeply beyond the surfaces of postwar cerebral culture. Neither is the business of models the only such fairly unromantic dimension. Much the same could be said, for instance, about "interdisciplinarity" (Cohen-Cole, 2009; Stadler, 2009); their mediations were not merely a matter of journalistic pasttimes either. When Norbert Wiener—appalled by the rumors of hordes of war-traumatized Americans and, yet more disconcerting, of housewives now " 'practicing' 'dianetic therapy' upon each other"—pondered filing an infringement lawsuit against the "dianetics boys" in the early 1950s, it may have been a signal of just how deeply cybernetics expressed, rather than informed, the cultural climate of the times. (This confusion was in fact only "understandable, since both sets of postulates," as Ron Hubbard helpfully explained it to Wiener, "do both stem from electronic engineering" ("Hubbard to Wiener," 1950; "Wiener to Schuman," 1950).

Even allowing for the complications introduced into the picture by what was generously glossed over here, namely, the more seriously incommunicable strategies to which cyberneticians availed themselves—statistics, mathematical models, and information theory—the story of cybernetics begins to look significantly different when re-embedded in its historical conditions of possibility. By the same token, the standard cybernetic story is not very illuminating as a guide to the mundane and less stimulating world of the average neurophysiological laboratory (or asylum, or neurological clinic). This world has been largely obscured from our view, and among the reasons, as we have seen, are the complex entanglements of the cybernetic vision with its own popularity. The general picture we have of postwar developments as viewed from within the various, traditional disciplines cyberneticians attempted to colonize, and from which they themselves operated (most of the time), is thus blurry at best. However, and as if to return to the stories which have been less successful in shaping our historical imagination, it may have been precisely these less spectacular departures which then aided the inauguration, in less visible and spectacular fashion, of this new identity whose history I have attempted to disentangle from the adventures of the brain and mind. One influential lineage at least of this new species managed, let us note here, to elevate in the process its profoundly and instructively non-cerebral doings towards new and cerebral horizons. It is the story of the so-called neuroscientists of the first hour.

For the likes of them, a trajectory such as that of Ralph Gerard—sometimes credited for having coined, towards the late 1950s, the term "neuroscience"—would have been far from atypical: by the end of the war, and already internationally famed, Gerard had turned Professor of Physiology at the University of Chicago, and chairman of the Physiology panel of the Office of Naval Research. He soon re-emerged as Director of Laboratories at the University of Illinois Neuropsychiatric Institute and in 1955 went on to become a member of the Mental Health Research Institute in Ann Arbor (Libet, 1974). Always of a somewhat holistic bent, Gerard was also a "core group" member of the Macy Conferences on Cybernetics; yet even for all his intellectual vitality, Gerard's immense scientific reputation was built on different, and definitely less metaphysical, grounds: notably forays into the heat production of muscle and nerve, and later, into the electrophysiology of resting potentials in single muscle cells. These were hardly the raw materials for an epic of the brain and mind. Similarly, take the case of Francis

Schmitt, whose central place in the annals of neuroscience as the man behind the so-called Neuroscience Study Program had been secured early on (see Swazey, 1975).

There was "urgency in effectuating [a] quantum step in an understanding of the mind," as Schmitt announced by 1963. The required "entirely new type of science," on Schmitt's mind, would in turn better be fundamental—a "biophysics of the mind" (cited in Swazey, 1975, p. 529, p. 532). Though the label was soon eschewed (evidently), there is indeed little in Schmitt's utterly unromantic oeuvre that would seem to predispose him to having paved the path towards neuroscience—as long as we construe them that is, in overly brain-centric terms: like Gerard and a great many other instrumental figures, Schmitt was brought up between the wars on the biophysics of nerve and muscle—frogs, squids, and other such lowly materials. When the entrepreneurial Schmitt arrived at MIT in 1941, his ambitions began even more definitely to concentrate on the mushrooming (and bewildering, many-faceted) research-field which then went under the name of "biophysics." In no time, as Nicolas Rasmussen has shown (1997a, 1997b), Schmitt turned his MIT facilities into a world center of electron microscopy. His wartime projects—supported in part by the rubber and leather industries—on wound healing and the structure of collagen and rubber, set the pace for Schmitt's more recognizably biophysical future. This future, significantly, converged less on the mind than on the biophysics of muscle and nerve; and it converged, secondly, on Schmitt's passionate engagement with this new science called "biophysics" (a mission which notably resulted in a grandiose, month-long international conference in Boulder, Colorado in 1956).

Importantly, historians of this curiously amorphous science have shown just how indistinguishable and undifferentiated in its biophysical hey-day, the future transdisciplinary ventures of molecular biology, bioengineering and neuroscience were (Chadarevian, 2002; Gaudillière, 2002; Rasmussen, 1997a). Indeed, it would be difficult to image a terrain more distant from the epic of the cultivation of the brain than the mix of collagen, keratin-fibres, polymers, leather, muscles, and squid-nerve which Schmitt, for one, assembled together—with ease. "We encounter little difficulty in securing grants-in-aid ... for fundamental biol[ogical] research related to medicine," as Schmitt had approvingly noted (Schmitt, 1954a). Indeed it was here that the new alliances were being forged between people, as Schmitt said, "working ... on the molecular level" ("Minutes, NIH," 1956),—electrophysiologists, molecular biophysicsts, physical chemists,—and, after all, the brain.

Indeed laboratory scientists of the fundamental kind now encountered few difficulties when tapping into the social and cultural concerns haunting the postwar world. Prominent among them was, as Schmitt, a skilful propagandist and money-raiser, put it one more than one occasion, "the almost staggering problem of mental health (said to compromise more than half of all the health problems of the nation)"(Schmitt, 1954b). And it was to much more palpable (if less publicly visible) effect than the cyberneticians that the likes of Schmitt translated such compromising facts into concrete realities. A considerable chunk of Schmitt's sprawling biophysics program at MIT was thus paid for by the Commonwealth Fund which, as Schmitt quickly discerned, was one of the many agencies then developing a "considerable interest in psychiatry, particularly as it bears on social problems" ("Schmitt to Dean

G. Harrison," 1950). The MIT-Commonwealth program would quickly "stabilize" at 20–25 postdoctoral fellows a year, Schmitt's "young turks" soon circulating by the dozen (Sizer, 1956). By 1954 some sixty "medical men" alone had gone through the process, serving the "far flung attack" on the problems of biomedicine.

Schmitt was remarkably (and exceptionally, it must be said) successful in inserting his stronghold of fundamental biology as a central node into the local network of Boston hospitals and research institutions. It served, not least, the need to thoroughly instill into biomedical minds the "quantitative methods of biophysics and biochemistry." And, bizarre though it may seem, it did not so appear to contemporary eyes and mindsets deeply, even naively optimistic about the powers of science and technology. "The spirit of the times," said Schmitt; "such [was] the nature of pure research that one cannot predict the particulars," reported the *Rhode Islander* in summer 1952, the "MIT squid project" consuming its entire, over-sized cover page: "Important clues to the functioning of the human nervous system may be uncovered. ... In any case, the frontiers of knowledge, as we consider them, will be pushed back a little further" (H. Smith, 1952).

Conclusion

The story, or stories, of pushing back these frontiers, and of these nascent neuroscientific identities as well still needs to be told. It is unlikely that it would turn out to be an epic revolving around the brain and mind, let alone around human nature. This chapter has dealt with only one such lineage and the aim, to be sure, was not to advance yet another myth of origins. Rather, it was my intention to sketch a space of inquiry into the nervous system that is all-too-easily glossed over in these necessarily manifold origins of neuroscience, devoid as it was for the most part, of the brain, of "culture" (certainly in the emphatic sense), and of the intellectual excitement surrounding cybernetics and the puzzles of the mind–body problem. Instead of a grand narrative of human nature transformed—or reduced—to the brain, this story would indeed seem to lack such a center; or if there was one, it more likely would revolve around squid, muscle potentials, molecules, sea snails, and other such uncerebral entities— and of how it came about that they became so closely allied to human memory, mind, or language.

Once we rid ourselves of the idea that the neurosciences fundamentally and always revolve around the essentials of human nature, our questioning might, in turn, take on a less dramatic but more constructive tone. Complicating the conceptions—and this would prominently include, the empirical picture—of neuroscience's past and present conditions of operation might help us move beyond, for instance, the "linear" and quasi-deterministic models of technoscientific change that implicitly inform much of the hype (or scares) surrounding the ascent of the neurosciences and our imagined, neurocultural futures. Similarly, one might then come to question more soberly whether the quite typical, apologetic, and polarizing constructions of ground-breaking but innocent neuroscientists on the one hand, and a merely sensation-hungry media landscape on the other, adequately reflect the political economy of science in the twenty first century (or indeed, the complex entanglements of any kind of knowledge production). Then again one might wonder, as we primarily wondered here, to what degree the departure-rhetoric and brain-centric images of contemporary neuroscience might be complicit in the debatable constructions of—and assumptions about— neuroscience's pasts.

Being "critical" would begin rather than end here. The romance of the brain that was at issue here is, needless to say, at best one such matter at stake. As the case may illustrate, however, it is not necessarily "neuroscience"—a problematically vague construct enough—that is to be singled out for critique, let alone neuroscientific research. Brain-centric discourses abound. Thus, when today's neuroskeptics feel their intuitions about human nature being offended, or neuroethicists feel obliged to sound out her imminent devastations, such interventions all too often operate in seeming ignorance of the historical malleability of this very nature, and on an impoverished view of the putative transformative agency—science—that is being accused, or celebrated. Likewise, it is often difficult to resist the impression of historical naiveté when observing the neuroenthusiast proliferation of "interdisciplinary" ventures, as if science and the humanities had ever suffered from cross-pollutions (albeit pollutions relegated, more often than not, to the trash bins of intellectual history). Perhaps, in the current times of neoliberal academia-government, when the less natural sciences perceive growing difficulties in justifying their existence, it might be advisable not to overzealously accelerate the leveling of voices by casting one's lot with the mirror neurons; if "human nature" is under siege these days, it may after all have to do less with the "potential implications" of neuroscience, than the diminishing space and prestige that other, less neuroscientific voices will be given in the twenty first century.

Yet, to end on a more self-critical note, there are perhaps few reasons to be overly expectant about what history and science studies can achieve in the critical direction; at least in so far as one demands, intellectually or otherwise, to go beyond slogans such as that science is somehow cultural and social, and beyond the rehearsal of positions and gestures of exposure that have long become history themselves—think, for instance, of Foucault-inspired analyses (half a century old by now), and the kind of reflex-like manner in which genealogical modes of analysis tend to be invoked. The "culturalism" at issue in this chapter is only one element in the way that the postmodern, left-leaning canon that has shaped so much of intellectual life in the latter half of the twentieth century has not only been thoroughly established, but also, has come to lose the subversiveness it may once have had (Anderson, 2009; Cooter, 2007; Latour, 2004).

Perhaps academia and the reality of critique has always been more impotent and conservative than one would like to think, but the considerable *Umwertung der Werte* at stake here certainly makes it no easier to envision what a critical neuroscience might profitably draw from the fields of science studies or history of science (and what not). The perceived inability of telling or arriving at big pictures of developments is one example; the absence of being able to communicate with either scientists or a broader public another. More disturbing perhaps is the sense that the very conceptual ammunition of science studies has somehow lost its critical impetus; or, at any rate, that it has come to curiously resemble the complexity-increasing (rather than merely reductive) vocabulary of the technosciences. Perhaps, as Bruno Latour proclaimed not long ago, the once seemingly stable dichotomies of nature/culture, fact/artifact, or knowledge/power have indeed become so very unstable and so universally appropriated as non-dichotomies, that "explanations resorting automatically to power, society, discourse have outlived their usefulness": outlived, that is, their critical force (Latour, 2004).

If so, surely this is one further reason to find limitations in the overly culturalist mode of imagining neuroscience's pasts—it also makes it all the more difficult to envision something of a positive program. There is no doubt that there is something deeply disturbing about the new cerebro-biologism that creeps into all manner of social domains, and not least into the humanities themselves, be it under the guise of neurointerdisciplinarity or the questionable promise of a "third" culture. Equally there is no doubt then that it is not something called "neuroscience" that deserves to be singled out for critique, let alone daemonized, but the conditions that serve to render it a perhaps overly self-confident and increasingly hegemonic discourse about human affairs. Things surely are not all bleak in matters of being critical. A quite minimal list of items would probably include: let's not follow too closely on the heels of those promulgating overly simplistic and futuristic assumptions about (neuro)science—as when the latter is equated, for all practical purposes, with those things that happen in an academic laboratory, or things that are novel and innovative, obscuring the established and workable. Furthermore, and more curiously, let's not follow too closely on the heels of the observers of science themselves; they all too easily fall, after all, into the habit of seeing only science—theirs is a tendency to overestimate the very relevance of science (or technology) in processes of societal change at the expense of other factors—as if it was science which had the power to actually "define" and "make-up" things, persons, and beings; as if it was truly "world making." Like the romance of the brain, after all, such elevations may bestow upon science, or neuroscience, a significance that may be undeserved; and hence be counter-productive, in terms of being critical.

References

Abi-Rached, J., & Rose, N. (2010). The birth of the neuromolecular gaze. *History of the Human Sciences, 23*(1).

Abraham, T. (2003). From theory to data: Representing neurons in the 1940s. *Biology and Philosophy, 18*(3), 415–426.

Agar, J. (2003). *The government machine: A revolutionary history of the computer.* Cambridge, MA: MIT Press.

Alder, K. (2007). *The lie detectors: The history of an American obsession.* New York: Free Press.

Anderson, W. (2009). From subjugated knowledge to conjugated subjects: Science and globalisation, or postcolonial studies of science? *Postcolonial Studies, 12*(4), 389–400.

ANDP (2007). 2007 ANDP Survey. Retrieved from: www.andp.org/newsite/surveys/surveys.htm.

Baars, B. (1986). *The cognitive revolution in psychology.* New York: The Guilford Press.

Bachelard, H. (1988). A brief history of neurochemistry in Britain and of the Neurochemical Group of the British Biochemical Society. *Journal of Neurochemistry, 50*(3), 992–995.

BBC, (1949). Draft outline of six suggested talks. File "William Grey Walter, 1948–1962". BBC Archives, Reading, UK.

BBC, (1948). Notes on J. Z. Young. (undated, c.1948) in File "Prof. J. Z. Young, Talks 1946–1959". BBC Archives, Reading.

Belkin, G. (2003). Brain death and the historical understanding of bioethics. *Journal of the History of Medicine and Allied Sciences, 58*(3), 325–361.

Beniger, J. R. (1986). *The control revolution: Technological and economic origins of the information society*. Cambridge, MA: Harvard University Press.

Boden, M. (2006). *Mind as machine : A history of cognitive science*. Cambridge, MA: MIT Press.

Borck, C. (2010). Toys are us; Models and metaphors in brain research. In S. Choudhury and J. Slaby (Eds.), *Critical neuroscience: Between lifeworld and laboratory*. Oxford: Wiley-Blackwell.

Borck, C. (2005). *Hirnstroeme: Eine Kulturgeschichte der Elektroenzephalographie*. Goettingen: Wallstein.

Bowker, G. (1993). How to be universal: Some cybernetic strategies, 1943–1970. *Social Studies of Science, 23*, 107–127.

Bowler, P. (2009). *Science for all: The popularisation of science in early twentieth-century Britain*. Chicago: University of Chicago Press.

Braslow, J. (1997). *Mental ills and bodily cures. Psychiatric treatment in the first half of the twentieth century*. Berkeley: University of California Press.

Bush, V. (1945). *Science: The endless frontier*. Washington, D. C.: US Government Printing Office.

Buxton, W. (1999). Reaching human minds: Rockefeller philanthropy and communications, 1935–1939. In T. Richardson & D. Fisher (Eds.), *The development of the social sciences in the United States and Canada: The role of philanthropy*. Stanford, CT: Ablex.

Cantor, G. (1975). A critique of Shapin's social interpretation of the Edinburgh phrenology debate. *Annals of Science, 32*(2), 245–256.

Chadarevian, S. D. (2002). *Designs for life: Molecular biology after World War II*. Cambridge: Cambridge University Press.

Changeux, J. (1997). *Neuronal man: The biology of mind*. Princeton: Princeton University Press.

Clarke, E., & Dewhurst, K. (1996). *An illustrated history of brain function: Imaging the brain from antiquity to the present* (2nd ed.). San Francisco: Norman.

Cohen-Cole, J. (2009). Cold War salons, social science, and the cure for modern society. *Isis, 100*(2), 219–262.

Cohen-Cole, J. (2003). *Thinking about thinking in Cold War America* (Doctoral dissertation, University of Princeton, 2003).

Cooter, R. (2007). After death/after-"life": The social history of medicine in post-postmodernity. *Social History of Medicine, 20*(3), 441–464.

Cooter, R. (1985). *The cultural meaning of popular science: Phrenology and the organization of consent in nineteenth-century Britain*. Cambridge: Cambridge University Press.

Corsi, P. (1991). *The enchanted loom: Chapters in the history of neuroscience*. Oxford: Oxford University Press.

Coughlan, R. (1963a, March 8). Control of the brain, part I. Behavior by electronics. *Life*, 90–106.

Coughlan, R. (1963b, March 15). Control of the brain, part II. The chemical mind changers. *Life*, 81–94.

Craver, C. (2007). *Explaining the brain: Mechanisms and the mosaic unity of neuroscience*. Oxford: Oxford University Press.

Crowther-Heyck, H. (2005). *Herbert A. Simon: The bounds of reason in modern America*. Baltimore: Johns Hopkins Press.

Crowther-Heyck, H. (1999). George A. Miller, language, and the computer metaphor of mind. *History of Psychology, 2*(1), 37–64.

De Vries, R. (2007). Who will guard the guardians of neuroscience? Firing the neuroethical imagination. *EMBO reports, 8*, S65–S69.

Doty, R. (1987). Neuroscience. In J. Brobeck, O. Reynolds & T. Appel (Eds.), *The history of the APS: The first century, 1887–1987* (pp. 427–434). Washington, D. C.: APS.

Dumit, J. (2003). *Picturing personhood: Brain scans and biomedical identity.* Princeton: Princeton University Press.

Dunbar-Hester, H. (2010). Listening to cybernetics: Music, machines and nervous systems, 1950–1980. *Science, Technology & Human Values, 35*(1), 113–139.

Dupont, J. (1999). *Histoire de la neurotransmission.* Paris: Presses Universitaires de France.

Dupuy, J. (2000). *The mechanization of the mind: On the origins of cognitive science.* Princeton: Princeton University Press.

Edwards, P. N. (1997). *The closed world: Computers and the politics of discourse in Cold War America.* Cambridge, MA: MIT Press.

Farreras, I., Hannaway, C., & Harden, V. (Eds.). (2004). *Mind, brain, body and behavior: Foundations of neuroscience and behavioral research at the National Institute of Health. Biomedical and health research.* Amsterdam: IOS Press.

Fessard, A. (1952). Plan for the establishment of an international brain institute, Unesco Document Code: NS/BR/3. Retrieved from http://unesdoc.unesco.org/ulis/cgi-bin/ulis.pl?catno=149004.

Finger, S. (2001). *Origins of neuroscience: A history of explorations into brain function.* Oxford: Oxford University Press.

Galison, P. (1994). The ontology of the enemy: Norbert Wiener and the cybernetic vision. *Critical Inquiry, 21*(1), 228–266.

Gardner, H. (1985). *The mind's new science: A history of the cognitive revolution.* New York: Basic Books.

Gaudillière, J. (2002). *Inventer la biomédecine: La France, l' "Amérique et. la production des savoirs du vivant (1945–1965).* Paris: Editions La Découverte.

Gerovitch, S. (2004). *From newspeak to cyberspeak: A history of Soviet cybernetics.* Cambridge, MA: MIT Press.

Gross, C. (2009). *A hole in the head: More tales in the history of neuroscience.* Cambridge, MA: MIT Press.

Gross, C. (1998). *Brain, vision, memory: Tales in the history of neuroscience.* Cambridge, MA: MIT Press.

Hagemeyer, F. (1979). *Die Entstehung von Informationskonzepten in der Nachrichtentechnik. Eine Fallstudie zur Theoriebildung in der Industrie- und Kriegsforschung* (Doctoral dissertation, Freie Universität, Berlin, 1979).

Hagner, M. (1992). The soul and the brain between anatomy and Naturphilosophie in the early nineteenth century. *Medical History, 36*(1), 1–33.

Haldane, J. (1952). Reviews. *British Journal for the Philosophy of Science, 3*(9), 103–105.

Harrington, A. (1987). *Medicine, mind, and the double brain: A study in nineteenth-century thought.* Princeton: Princeton University Press.

Hayles, N. (1999). *How we became posthuman: Virtual bodies in cybernetics, literature and informatics.* Chicago: University of Chicago Press.

Hayward, R. (2001). The tortoise and the love-machine: Grey Walter and the politics of electroencephalography. *Science in Context, 14*(4), 615–641.

Hedgecoe, A. (forthcoming) Bioethics and the reinforcement of socio-technical expectations. *Social Studies of Science.*

Heims, S. (1980). *John von Neumann and Norbert Wiener: From mathematics to the technologies of life and death.* Cambridge, MA: MIT Press.

Heims, S. (1991). *Constructing a social science for postwar America: The cybernetics group, 1946–1953.* Cambridge, MA: MIT Press.

Herzberg, D. (2008). *Happy pills in America: From Miltown to Prozac.* Baltimore: Johns Hopkins Press.

History of the Brain. (2010). In *Wikipedia.* Retrieved from http://en.wikipedia.org/wiki/History_of_the_brain.

Hodgkin, A. (1992). *Chance and design: Reminiscences of science in peace and war.* Cambridge: Cambridge University Press.

Hopkins, M., Martin, P., Nightingale, P., Kraft, A., & Mahdi, S. (2007). The myth of the biotech revolution: An assessment of technological, clinical and organisational change. *Research Policy, 36*(4), 566–589.

Hubbard to Wiener. (1950, August 21). Norbert Wiener papers (MC22), Box 8, Folder 123. MIT Archives, Cambridge, MA.

Husbands, P., & Holland, O. (2008). The Ratio Club: A hub of British cybernetics. In P. Husbands, M. Wheeler & O. Holland (Eds.), *The mechanical mind in history* (pp. 91–148). Cambridge, MA: MIT Press.

Jasanoff, S. (2007). *Designs on nature: Science and democracy in Europe and the United States.* Princeton: Princeton University Press.

Jones to Wiener. (1948, November 17). Norbert Wiener papers (MC22), Box 6, Folder 86. MIT Archives, Cambridge, MA.

Kandel, E. (2006). *In search of memory: The emergence of a new science of mind.* New York: W. W. Norton.

Katz, B. (1998). Sir Bernard Katz. In L. R. Squire (Ed.), *The history of neuroscience in autobiography* (Vol. 1). Washington, D.C.: Society for Neuroscience.

Kay, L. (2001). From logical neurons to poetic embodiments of mind: Warren S. McCulloch's project in neuroscience. *Science in Context, 14*(4), 591–614.

Kline, R. (2009). Where are the cyborgs in cybernetics? *Social Studies of Science, 39*(3), 331–362.

Lagemann, E. (2000). *An elusive science: The troubling history of education research.* Chicago: University of Chicago Press.

Latour, B. (2004). Why has critique run out of steam? From matters of fact to matters of concern. *Critical Inquiry, 30*(2), 225–248.

Libet, B. (1974). R. W. Gerard Born October 7, 1900–Died February 17, 1974. *Journal of Neurophysiology, 37*, 828–829.

Littlefield, M. (2009). Constructing the organ of deceit: The rhetoric of fMRI and brain fingerprinting in post-9/11 America. *Science, Technology & Human Values, 34*(3), 365–392.

Maasen, S., & Sutter, B. (Eds.). (2007). *On willing selves: Neoliberal politics and the challenge of neuroscience.* London: Palgrave.

Mazlish, B. (1967). The fourth discontinuity. *Technology and Culture, 8*(1), 1–15.

McCoy, A. (2006). *A question of torture: CIA interrogation from the Cold War to the War on Terror.* New York: Henry Holt and Co.

McCulloch to Gerty. (1943, November 3). McCulloch Papers, Folder "Gerty". APS Library, Philadelphia.

McIlwain, H. (1985). In the beginning: To celebrate 20 years of the International Society for Neurochemistry (ISN). *Journal of Neurochemistry, 45*(1), 1–10.

McMahan, U. (1990). *Steve: Remembrances of Stephen W. Kuffler.* Sunderland, MA: Sinauer Associates.

Mellanby (MRC) to Randall. (1949, March 29). Randall papers, RNDL 2/2/1. Churchill Archives Centre, Cambridge.

Meredith, G. (1948). Visual aids–Context–Its significance and its visual generalization. *ELT Journal, 2*(4), 97–101.

Mindell, D. A. (2002). *Between human and machine: Feedback, control, and computing before cybernetics.* Baltimore: Johns Hopkins Press.

Mirowski, P., & Sent, E. (2005). The commercialization of science and the response of STS. In E. Hackett, O. Amsterdamska, M. Lynch & J. Wajcman (Eds.), *Handbook of science and technology studies.* Cambridge, MA: MIT Press.

Neumann to Wiener. (1946, November 29). Norbert Wiener papers (MC22), Folder 72, Box 5. MIT Archives, Cambridge, MA.

NeuroInsights. (2010). *Neurotech clusters 2010: Leading regional leaders in the global neurotechnology industry 2010–2020.* Retrieved from: http://www.neuroinsights.com/neurotechclusters2010.html.

Noble, D. (1986). *Forces of production: A social history of industrial automation.* New York: Oxford University Press.

NSF. (2009). *Survey of earned doctorates.* Retrieved from: http://www.nsf.gov/statistics/survey.cfm.

Pfeiffer, J. (1955). *The human brain.* London: Victor Gollanz.

Pias, K. (Ed.). (2004). *Cybernetics–Kybernetik. The Macy-Conferences 1946–1953. Volume II / Band II: Essays and documents.* Berlin: Diaphanes.

Pickering, A. (2002a). Cybernetics and the mangle: Ashby, Beer and Pask. *Social Studies of Science, 32*(3), 413–437.

Pickering, A. (2002b). The tortoise against modernity: Grey Walter, the brain, engineering and entertainment. In *Experimental cultures: Configurations between science, art, and technology, 1830–1950* (pp. 109–122). Berlin: Max Planck Institute for the History of Science, preprint 213.

Poynter, F. (1958). *The history and philosophy of knowledge of the brain and its functions: An Anglo-American symposium, London, July 15–17, 1957.* Oxford: Blackwell.

Pressman, J. (1998). *Last resort: Psychosurgery and the limits of medicine.* Cambridge: Cambridge University Press.

Rasmussen, N. (2008). *On speed: The many lives of amphetamine.* New York: New York University Press.

Rasmussen, N. (1997a). *Picture control: The electron microscope and the transformation of biology in America, 1940–1960.* Stanford: Stanford University Press.

Rasmussen, N. (1997b). The mid-century biophysics bubble: Hiroshima and the biological revolution in America, revisited. *History of Science, 35*(109), 245–293.

Russell, E. (2001). *War and nature: Fighting humans and insects with chemicals from World War I to Silent Spring.* New York: Cambridge University Press.

Schmaltz, F. (2006). Neurosciences and research on chemical weapons of mass destruction in Nazi Germany. *Journal of the History of the Neurosciences, 15*(3), 186–209.

Schmitt to Dean G. Harrison. (1950, November 15). Schmitt papers (MC154), Box 1, Folder 49. MIT Archives, Cambridge, MA.

Schmitt, F. (1990). *Never-ceasing search.* Philadelphia: American Philosophical Society.

Schmitt, F. (1956). *Minutes January 14–15 1956 (Meetings of the biophysics and biophysical chemistry study section. NIH).* Schmitt papers (MC154), Box 21, Folder 1. MIT Archives, Cambridge/MA.

Schmitt, F. (1954a). *Perspectives in life sciences (Talk, 1954).* Schmitt papers (MC154), Box 11, Folder 15. MIT Archives, Cambridge/MA.

Schmitt, F. (1954b). *Responsibilities and opportunities in experimental biology and medicine (Lecture MS).* Schmitt papers (MC154), Box 11, Folder 15. MIT Archives, Cambridge/MA.

Schoenfeld, R. L. (2006). *Exploring the nervous system: With electronic tools, an institutional base, a network of scientists.* Boca Raton, FL: CRC Taylor & Francis.

Seattler, P. (1990). *The evolution of American educational technology.* Englewood, CO: Libraries Unlimited.

SfN (Society for Neuroscience). (2009). About membership. Retrieved March 6, 2010, from http://sfn.org/index.aspx?pagename=membership_AboutMembership.

Shapin, S. (2008). *The scientific life: A moral history of a late modern vocation.* Chicago: University of Chicago Press.

Shapin, S. (1979). The politics of observation: Cerebral anatomy and social interests in the Edinburgh phrenology disputes. In R. Wallis (Ed.), *On the margins of science*, Sociological Review Monograph, 27 (p. 1979).

Shapin, S. (1975). Phrenological knowledge and the social structure of early nineteenth-century Edinburgh. *Annals of Science, 32*(3), 219–243.

Showalter, E. (1987, October 11). Review of the book *Medicine, mind, and the double brain* by Anne Harrington. *New York Review of Books, 39.*

Siegelman, J., & Conway, F. (2004). *Dark hero of the information age: In search Of Norbert Wiener–Father of cybernetics.* New York: Basic Books.

Sizer. (1956). A review of the program for training in Medical Research. Schmitt papers (MC 154), Box 1, Folder 51. MIT Archives, Cambridge, MA.

Smith, H. (1952, August 3). It's a case of pure nerve. *The Rhode Islander,* 3–5.

Smith, R. (1992). *Inhibition: History and meaning in the sciences of mind and brain.* Berkeley: University of California Press.

Söderqvist, T. (2002). Neurobiographies: Writing lives in the history of neurology and the neurosciences. *Journal of the History of the Neurosciences, 11*(1), 38–48.

Stadler, M. (2009). *Assembling life. Models, the cell, and the reformations of biological science, 1920–1960* (Doctoral dissertation, Imperial College, London, 2009).

Swazey, J. (1975). Forging a neuroscience community: A brief history of the neurosciences research program. In F. Worden, J. Swazey & G. Adelman (Eds.), *The neurosciences: Paths of discovery.* Cambridge, MA: MIT Press.

Technology press to Wiener. (1949, October 26). Norbert Wiener papers (MC22), Box 6, Folder 104. MIT Archives, Cambridge, MA.

Tone, A. (2008). *The age of anxiety: A history of America's turbulent affair with tranquilizers.* New York: Basic Books.

Valenstein, E. (2005). *The war of the soups and the sparks: The discovery of neurotransmitters and the dispute over how nerves communicate.* New York: Columbia University Press.

Vidal, F. (2009). Brainhood, anthropological figure of modernity. *History of the Human Sciences, 22,* 5–36.

Walter, G. W. (1953). *The living brain.* New York: Norton.

Weidmann, N. (1999). *Constructing scientific psychology: Karl Lashley's mind-brain debate.* Cambridge: Cambridge University Press.

Wiener, N. (1948). *Cybernetics. Or control and communication in the animal and the machine.* New York: The Technology Press.

Wiener to Schuman. (1950, August 14). Norbert Wiener papers (MC22), Box 8, Folder 122. MIT Archives, Cambridge, MA.

Worden, F., Swazey, J., & Adelman, G. (Eds.). (1975). *The neurosciences: Paths of discovery.* Cambridge, MA: MIT Press.

Wyhe, J. V. (2004). Was phrenology a reform science? Towards a new generalization for phrenology. *History of Science, 42*(3), 313–331.

Young, J. (1951). *Doubt and certainty in science.* Oxford: Clarendon Press.

Young, R. (1966). Scholarship and the history of the behavioral sciences. *History of Science, 5*(1), 1–51.

7

Empathic Cruelty and the Origins of the Social Brain

Allan Young

Versions of Human Nature

The topic this afternoon is … human nature in the age of biotechnology. The subject crops up … in our conversations and is very often just below the surface … [and] it's worth our while to … think about how to think about human nature in an age of genomics, in an age of neuroscience and what might be possible in the way of altering it and ultimately what those alterations might mean and whether they would be a good thing.

Leon Kass, chairman of the session on Human Nature and Its Future,
convened on 6 March 2003 by the (USA)
President's Council on Bioethics

The term "human nature" commonly refers to a bundle of innate and universal human faculties and dispositions that distinguish humans from other animals and normal people from various kinds of abnormal people. Within Western societies, opinions about what comprises human nature are not uniform: they have changed over time and they vary among groups within populations. Until recently, it was permissible to speak of a canonical version of human nature, represented in the operations of key social institutions and sectors of knowledge production, notably the social and behavioral sciences, biomedicine, psychiatry, and the law.

The canonical version originated in Enlightenment debates about the nature of the mind and its relation to earlier conceptions such as the soul (Ryle, 1949). For convenience I will refer to this version as "Human Nature 1.0." In its most basic form, Human Nature 1.0 is associated with four features:

1. Mind is the body's command center, theater of self-awareness, and agency of self-identity and continuity.
2. Normal people are rational.

Critical Neuroscience: A Handbook of the Social and Cultural Contexts of Neuroscience, First Edition.
Edited by Suparna Choudhury and Jan Slaby.
© 2012 John Wiley & Sons, Ltd. Published 2016 by John Wiley & Sons, Ltd.

3. Normal people are self-interested: seeking pleasure and gratification and avoiding pain and distress.
4. Minds are self-contained: "mental happenings occur in insulated fields, known as 'minds', and there is, apart from telepathy, no direct causal connection between what happens in one mind and what happens in another. Only through the medium of the public, physical world can the mind of one person make a difference to the mind of another. The mind is its own place and in his inner life each of us lives the life of a ghostly Robinson Crusoe. People can see, hear, and jolt one another's bodies, but they are irremediably blind and deaf to the workings of one another's minds and inoperative upon them." (Ryle, 1949, p. 13)

Recent developments in cognitive and social neuroscience research have encouraged scientists and their audiences to re-contextualize these features. The new version, still emerging, can be called "Human Nature 2.0" and can be summed up as follows:

1. Mind becomes a visible epiphenomenon of the social brain. It is, however, too early to talk about the relationship between mind and brain in strictly deterministic terms.
2. Humans are rational agents, as assumed in Human Nature 1.0. At the same time, minds and brains (the seat of human agency) are products of a higher and more stringent kind of rationality: natural selection. A person can be considered rational to the extent that, on a given occasion, his or her intentions, purposive behavior, and the material results of his or her goal-directed action are consonant ("sensible") and proportionate according to the standards of his or her community. Natural selection is rational in that it is determined by a ruthless cost–benefit calculus (reproductive success).
3. The "hedonistic calculus" of Human Nature 1.0 is unaffected.
4. The most striking difference between the two versions concerns the mechanisms through which minds/brains communicate. In version 1.0, minds know other minds only indirectly, through signs and symbols, encoded in language, gestures, and purposive behavior. In version 2.0, there is an additional mechanism: minds are routinely in direct contact, via neural resonance, mirroring, and empathy.

The Prehistory of Empathy

The idea that brains and minds might interpenetrate reprises nineteenth-century medical discourse and debate on suggestion, hypnosis, and mental contagion. Jean-Martin Charcot claimed that his clinical studies of mesmerism, hysteria, and psychogenic trauma had led him to believe that the hypnotic state is evidence of a biological diathesis. His claim was contested by Hippolyte Bernheim (1891), who believed that hypnosis is a form of suggestion, and that suggestibility is both universal and normal, notwithstanding the observation that highly suggestible people are more credulous than others. Bernheim defined "suggestion" very broadly, as an "act by which an idea is introduced into the brain and accepted by it." It occurs in two forms: in *hetero-suggestion*, ideas pass from one mind to another; in *auto-suggestion*, ideas emerge spontaneously within the mind, where they become associated with particular sensations, emotions, and images. Auto-suggestion might be a source of distress and even psychosomatic disorders, but it is not intrinsically pathological (Bernheim, 1980 [1891] p. 18).

Bernheim's conception of hetero-suggestion was the basis for Gustave Le Bon's influential monograph, *La psychologie des foules* (1895). Le Bon believed that French society was undergoing a massive and unfortunate transformation, that could be traced to the accession of the masses (classes populaires) to political power. The masses want to "utterly destroy society as it now exists, with a view to making it hark back to that primitive communism which was the normal condition of all human groups before the dawn of civilization" (Le Bon, 1895/2002, pp. ix–xi; Nye, 1975; van Ginneken, 1992). Rationality is a trait of the civilized, autonomous individual. The masses are not individuals in this sense, but rather creatures of a formation called the "crowd" (la foule). Once an individual is part of a crowd, that person "acquires, solely from numerical considerations, a sentiment of invincible power which allows him to yield to instincts which, had he been alone, he would perforce have kept under restraint." That person's mind and brain become permeable to other minds and brains, and his or her conscious personality is lost. That individual now descends the evolutionary ladder. "Isolated, he may be a cultivated individual; in a crowd he is a barbarian ... He possesses the spontaneity, the violence, the ferocity ... of primitive beings [and can] be induced to commit acts contrary to his most obvious interests" (Le Bon, 1895/2002 p. 6).

Le Bon accepts Bernheim's thesis that, to varying degrees, everyone is suggestible. However, he has no interest in the expression of suggestibility in unexceptional circumstances. And this makes him different from Bernheim. Indeed Boris Sidis, a Harvard psychiatrist and authority on suggestion, criticized Bernheim for defining the trait so broadly as to include most mental activities (Sidis, 1898). In practice, Bernheim did very little to challenge the idea of the autonomous, self-contained individual. When he discusses suggestibility, he mentions contagious yawning, the psychosomatic symptoms that can be induced by auto-suggestion, and his efforts to reverse these symptoms through clinical hetero-suggestion. Human Nature 1.0 remains unaffected.

During the same period, an analogous notion emerged in Germany. Theodor Lipps identified a psychophysical process (Einfühlung), superficially similar to Bernheim's notion of "indirect" suggestion, a spontaneous response to sensory stimuli producing an "inner imitation." This is Lipps' description: I observe someone's facial expression of affect and "there exists within me a tendency to experience in myself the affect that naturally arises from that gesture." When there is no obstacle, the tendency is realized and the subjective meaning of the affect becomes my experience of the affect. "Einfühlung" is "positive" when it does not conflict with my own character and "negative" when there is conflict. Even when there is conflict my tendency to experience his affective state remains. Thus, when a person stares at me in an arrogant way "I experience within myself the arrogance contained in that look. ... My inner being objects; I feel in the arrogant look ... a denial of my personality." Within myself,

I resist the negative "Einfühlung" and it is this effort that contributes (developmentally) to the ontogenesis of the self. It enables subjectivity to separate from the selves that it observes and, so, empathically experiences. (Jahoda, 2005; Lipps, 1903, p. 193, Pigman's translation; Pigman, 1995).

Lipps' conception of positive "Einfühlung" is similar to the idea of "sympathy" described by earlier writers, notably David Hume (whose work Lipps translated for publication in Germany) and Adam Smith (Penelhum, 1993). In 1909, Edward Titchner introduced Lipps' notion to Anglophone readers as "empathy," but with a significant alteration. Unlike Lipps, he makes an explicit distinction between empathy (the capacity to fully comprehend the situation of the observed individual) and sympathy (the capacity to share the feeling of the observed individual). By the 1930s, Titchner's distinction had entered psychological discourse and, soon afterward, was absorbed into the everyday language of educated people. The distinction is both analytical and moral. Empathy is a morally neutral state—I comprehend Zande witchcraft beliefs without wishing to promote them. Sympathy readily blends into compassion and perhaps an impulse to improve the situation of the observed individual. Thus, the credibility of the self-contained mind is unaffected.

It is a mistake to suppose that ideas about human nature might have evolved differently had it not been for Titchner's interference. Edmund Husserl adopted Lipps' notion of "Einfühlung" for his phenomenology. From the beginning of life, he wrote, human subjectivity comprises intersubjectivity: a relation between self and other in which the other is apprehended by means of a primitive holistic process of "pairing" occurring at the level of the body. But Husserl retains the "primordial ego" as the foundation for this process; he writes about intersubjectivity without interpenetration (Moyn, 2005, pp. 58–62). Freud mentions Lipps and "Einfühlung" in *Jokes and their Relation to the Unconscious* (1905). By equating "Einfühlung" to the observer's *cognitive* identification with the other's perceptions and intentions, Freud similarly tailors it for a Cartesian ego (Pigman, 1995).

To summarize: nineteenth-century and early twentieth-century investigations of suggestibility, hypnosis, "Einfühlung," and empathy did not undermine confidence in Human Nature 1.0 or its representative, the autonomous, self-contained individual. The more serious challenge dates to the 1980s, when it becomes possible, for the first time, to see the mind at work inside the brain.

Empathy and Mirror Neurons

Interest in empathy and embodiment has revived as a consequence, in part, of the discovery of the so-called "mirror neurons." The initial mirror neuron research was conducted on rhesus monkeys and utilized an invasive technology permitting scientists to detect and trace the activation of single neurons in the brain's motor cortex. Subsequent research on humans employed non-invasive technologies—most often fMRI—that image the activation of populations of neurons rather than individual cells. In these experiments, the subject observes goal-directed behavior being performed by someone else. The sensory input activates a "neural matching system" in the observer's motor cortex. His activation pattern mirrors the pattern in the performer's brain, and it matches the pattern in his own brain whenever he performs this action. Subjects were asked to passively read action words such as "lick," "pick," and "kick," and fMRI showed mirroring in cortical regions that are activated when tongues, fingers, and feet produce these actions. Similar effects were produced when subjects were asked to imagine themselves or other people performing designated behavior, including expressed emotion. Thus "mirror neurons can be thought of as a sensory-motor gateway for forming *an internal representation of the observed person's state and intents* based on their body language, facial expressions, actions, and so on" (Dinstein, Thomas, Behrmann, & Heeger, 2008, my italics; see Grush, 2004).

Mirror neurons operate in tandem with brain regions and networks responsible for:

1. selecting the movements that will be mirrored on a given occasion, and
2. inhibiting the performance of the mirrored movements.

These two operations are invisible in most laboratory experiments, since they are designed to focus the subject's attention on a single, unambiguous behavior. But life outside the laboratory is more complicated. Multiple actors and actions may simultaneously enter the observer's sensory field. Elements in the field may stimulate imagined events and recall episodic memories each of which can, in turn, become a target for mirroring. Further, many actions remain ambiguous until cognitive processing puts them into context and, only then, makes it possible to infer a goal.

The human neural matching system supports four phenomenological states:

1. The observer experiences mirror neuron activation passively in a state called "resonance."
2. Neural activation engenders a spontaneous and involuntary re-enactment of observed behavior and emotions. This state includes emotional contagion, contagious yawning, and the so-called "chameleon effect."
3. The observer uncouples his mirrored neural representation and projects it onto its source, that is, as a cognitive, conative, or emotional state of the individual being observed. The ability to objectify uncoupled representations is called "perspective-taking."
4. The uncoupled representation is objectified (made explicit) and is accessible to the observer as a resource for "true imitation."

The states are likewise evolutionary and developmental stages. The ability to uncouple mirrored representations (stages 3 and 4) requires the development of structures and networks outside the mirror neuron system. Non-human primates and other mammals get to the second stage, but no further. Normal children are capable of perspective taking and true imitation by the age of four. Perspective taking is a precondition for "mind reading." This seems to be a distinctively human capacity that enables us to interpret other people's intentions, predict their behavior, and attempt to manipulate them. (While other mammals lack this ability, there is compelling empirical evidence that some bird species—notably corvids—are adept mind readers and agents of deception).

Perspective taking is the basis for self-conscious empathy. For many writers, mirroring is an intrinsically empathic event, and this view helps to explain the recent explosion of interest in empathy in cognitive and social neuroscience, neuropsychiatry, developmental and evolutionary psychology, anthropology, moral philosophy, evolutionary biology, neuroeconomics, neuroethics, neuroaesthetics, and popular science journalism. Here is an excerpt from an article by Daniel Goleman, writing in 2006 (10 October) in the *New York Times*:

> The fledgling field of social neuroscience is [now] figuring out the brain mechanics [of] the circuitry that underlies the urge to help others in distress. ... Mirror neurons operate like a neural WiFi, activating in our own brains the same areas for emotions, movements and intentions as those of the person we are with. This allows us to feel the other person's distress or pain as our own [and we are] moved to help relieve it. Those who feel another's distress most strongly are most likely to help; those less moved can more easily ignore someone else's distress.

Goleman's excerpt reports the consensus view in social neuroscience and is consistent with Lipps' original notion (Einfühlung) that the observer can be said to embody the target of his gaze (Lipps, 1903; Carr, Iacoboni, Dubeau, Mazziotta, & Lenzi, 2003; de Vignemont & Singer, 2006; Fogassi, Ferrari, Gesierich, Rozzi, Chersi, & Rizzolatti, 2005; Hein & Singer, 2008). However, the phenomenon goes beyond Lipps' vision. The target's sensory-motor representations have penetrated the observer's brain: the correspondence between brains is identity and not analogy. There is another significant difference with the past. Titchner and later social psychologists made a distinction between empathy and sympathy (compassion). Goleman, however, presumes that empathy is not just pro-social but is also morally positive (disposing people to benevolence), a view held by the majority in social neuroscience at the time (2006).

The Social Brain

The term "social brain" recurs throughout cognitive and social neuroscience literature (Adolphs, 2003; Brüne, Ribbert, & Schiefenhövel, 2003; Johnson et al., 2005). The brain is doubly social: it enables and inclines humans to engage in complex forms of social interaction, and it is the product of our ancestors' five million year adaptation to social life. The two meanings of social are bridged by the brain's capacity for empathy and mind reading and the biological hardware (notably the mirror neuron system) that serves these functions. The social brain also comprises three evolutionary narratives:

- the narrative of the Jacksonian brain;
- the narrative of other minds; and
- the narrative of the one and the many.

The narratives, whose beginnings date back to the seventeenth century, are explorations of the brain's biological and sociological origins, its architecture, its interface with the mind, and the ways in which researchers might penetrate its recesses.

They are neither "mere stories" nor the "historical background" to the real business of neuroscience. Because they are integral to the business, I will describe them one by one, with an occasional detour.

The Narrative of the Jacksonian Brain

In the Croonian Lectures on the Evolution and Dissolution of the Nervous System (1884), the neurologist John Hughlings Jackson described the nervous system as comprising a hierarchy of sensory-motor "centers" acquired incrementally as evolutionary adaptations. At the bottom of the hierarchy are the oldest centers— spontaneous, inflexible, reflex-like. The older centers are inhibited and controlled by centers acquired later. When a control center is disabled (by disease or alcohol, for example), previously inhibited centers are released to perform their evolved functions, and the effect is expressed in symptoms, syndromes, and mental states. These released functions are called "positive" symptoms; a "negative" symptom, such as paralysis, results from the loss of a function. This process, which retraces the nervous system's evolutionary path in reverse order, is called "dissolution." A patient with delirium tremens who sees non-existent rats and mice is exhibiting a positive symptom consequent to shallow dissolution, leaving several evolutionary layers unaffected. On the other hand, a case of epileptic mania, characterized by the explosive discharge of energy and the so-called "dreamy state" that follows grand mal seizures are products of deep dissolution reaching lower evolutionary layers.

Thus the selection of appropriate neuropsychiatric disorders and positive symptoms allows researchers to explore the brain's evolutionary architecture. Hughlings Jackson's clinical interest focused on epilepsy and aphasia, and his most extended observations concern these disorders. Following his death in 1911, interest in the Jacksonian brain declined, the exceptions being W.H.R. Rivers in Britain, Paul McLean in the United States (his "triune brain" reiterates the Jacksonian scheme), Henri Ey in France, and arguably Sigmund Freud in *The Interpretation of Dreams*. Interest in the evolutionary meaning of mental disorders reemerged in the 1960s (Price, 1967), stimulated by developments in sociobiology and (later) evolutionary psychology. These writers were generally more interested in the architecture of the mind rather than the brain, and their work spanned many conditions, including depression, postpartum depression, antisocial personality disorder, generalized anxiety, schizophrenia, agoraphobia, and animal phobias. In these accounts, each disorder reveals its distinctive evolutionary origin. There is no grand narrative, rather the mind comes together as a mosaic of evolutionary events and dispositions. The Jacksonian brain is different in this regard. It reemerges (anonymously) in the 1990s, concurrent with the availability of functional neuroimaging technology, the consequent discovery of the human mirror neuron system, and the widespread conviction that empathy and mind reading are core features of human nature and its evolutionary history. To investigate empathy and mind reading, however, one requires an appropriate assortment of normal and abnormal brains. Three disorders are especially suited to the job: schizophrenia, autism spectrum disorders, and psychopathy.

In one respect, the social brain and Jacksonian brain are quite different. Hughlings Jackson believed that every mental state has a correlative nervous state: the highest link of the purely physical chain of sensory-motor structures. The two states occur in parallel; philosophers of mind call this "property dualism." He explicitly rejected Descartes' doctrine of dual substances and likewise "materialists" who claimed that every mental state can be reduced to a discrete neural state. Hughlings Jackson called his position the doctrine of concomitance. The term is rarely used today, but the problematic—the brain–mind nexus—continues to attract the attention of philosophers, including John Searle, Jerry Fodor, and Daniel Dennett.

Into the 1990s, reductionists lacked an effective technology and research program to bridge mind and brain. This makes the social brain special: it provides a bridge based on three kinds of empathy, namely motor empathy, emotional empathy, and cognitive empathy. Mirror neurons are a subpopulation of motor neurons that extend to brain regions associated with emotional and cognitive empathy. Thus, social brain research has the possibility of delineating a "purely physical chain of sensory-motor structures" extending to the conscious mind, leaping over the doctrine of concomitance. Further evidence is provided by continuing experiments in which participants' brains are imaged while they complete carefully designed cognitive tasks or, alternatively, while they observe emotionally evocative stimuli.

The Narrative of Other Minds

The size of the human brain is an evolutionary puzzle. Our ancestors split from the great apes six million years ago. During this period, the ancestral human brain quadrupled in volume. The metabolic costs of the human brain are enormous: it constitutes 2 % of total body weight and consumes 15 % of cardiac output and 20 % of body oxygen. These demands are ceaseless and inflexible. A brief shortfall results in neuronal death, resulting in a debilitating and permanent loss in functioning. It can be assumed that the evolutionary growth of the brain reflects an adaptive advantage: the benefits were consistently greater than the metabolic costs. During the initial stage, benefits were caloric and a product of improved adaptation between the organism and the physical environment. Efforts to model the evolution of hominid brains indicate that increasing costs would eventually exceed environmental benefits. How then did the expanding brain pay for itself?

The early history of the hominid brain is about adaptation between organism and physical environment; the subsequent history is about brains adapting to other brains. The process is described as a *cognitive arms race* (Barton & Dunbar, 1997; Byrne & Whiten, 1988; Dunbar, 2003). It began with the emergence of a unique hominid mind-reading capacity: the ability to detect the intentions and predict the behavior of other members of one's group. The next stage was the emergence of so-called "cheaters" who used mind reading to manipulate other members. Cheaters would have had an adaptive advantage and therefore multiplied. In time the proportion of cheaters would increase to the point that social life would become unpredictable and regress to the previous, more primitive stage. This did not happen because the brain evolved a "cheater detector" capacity. This, in turn, could only be a transient solution, since a new generation of opportunistic individuals would exploit this capacity to cheat a new generation of victims. Once again cheaters thrive, social life grows unpredictable, and so forth. Devolution is avoided with the emergence of cheater-detector 2.0. And so on, over millions of years, until arriving at the current version of the human brain.

The cognitive arms race is dependent on the ability of individuals to detect the intentions of others and predict their behavior—in other words, "mind-reading." The responsible mechanism is the human mirror neuron system, which has its own evolutionary history (see Fadiga, Fogassi, Pavesi, & Rizzolatti, 1995, for the discovery of mirror neurons; Fogassi, Ferrari, Gesierich, Rozzi, Chersi, & Rizzolatti, 2005; Fogassi, Gallese, & Rizzollati, 2002; Gallese, 2001, 2003, 2006; Gallese & Goldman, 1998; Iacoboni, Molnar-Szakacs, Gallese, Buccino, Mazziotta, & Rizzolatti, 2005; see Jacob, 2008; Jacob & Jeannerod, 2005; Kohler, Keysers, Umlitá, Fogassi, & Rizzolatti, 2002; Rizzolatti & Arbib, 1998; Rizzolatti & Craighero 2004; Tettamanti, Buccino, Succaman, Gallese, Danna, Scifo, & Perani, 2005; Singer, 2006):

Stage one: The observer's mirror neurons *resonate* with the neurons of the agent performing a goal-directed action. A transient "primary representation" of the neural activation pattern is produced in the observer's brain. The brains of non-human primates did not evolve beyond this stage. Emotional contagion is possible, but not emotional empathy.

Stage two: The primary representation can be uncoupled from the transient experience and copied inside the brain. This is the neural basis for perspective taking. Cognitive and emotional empathy are now possible.

Stage three: Copies are archived and provide the brain and mind with a library of action patterns. True imitation becomes possible.

The phylogenetic series is replicated in the cognitive development of normal children.

The Problem of the One and Many

Human Nature 1.0 poses an evolutionary puzzle. How did aggregates of autonomous, self-interested individuals—our remote ancestors—coalesce into stable, self-reproducing societies? And once formed, how did the earliest groups evolve into complex social formations?

Thomas Hobbes' thesis was that our ancestors were guided by reason and driven by fear to surrender their private right to use force to a sovereign power that would exercise its strength in the interest of collective peace and defense (Sahlins, 2008 p. 13). Freud's solution in *Totem and Taboo* (1913) is a two-tier hierarchy maintained by the violence and authority of a consummately selfish and insatiable patriarch. A parallel solution has been observed among baboons: the hierarchy is stable, the alpha male is similarly violent and insatiable, but the position of individuals within the hierarchy is fluid. John Price, a founding father of Evolutionary Psychology (EP), believes that their situation is very close to the condition of the earliest humans. He sees the legacy of this Paleolithic adaptation in the epidemiology and symptomatology of major depression (Price, 1967). Adam Smith offered a third solution. In *The Theory of Moral Sentiments* (1759/2002) he writes that man is doubtlessly selfish, but his self-love is tempered by an imaginative capacity to place himself in the situation of others and by his innate concern for their happiness and misery. This explains the naturalness of pity and compassion. In *The Wealth of Nations* (1789/1937), he responds to the further question of how the earliest groups might have evolved into more complex formations. It is through a human propensity to exchange one thing for another: goods, gifts, and assistance.

The solution given in the evolutionary narrative of the social brain comes close to Adam Smith's account. It emphasizes similar propensities: empathic mind reading and exchange. As you will see, it has problems staying on course.

The Narrative of the One and Many

This narrative begins with the riddle of altruism. Population biologists define altruism as behavior in which individuals sacrifice or reduce their own reproductive chances in favor of other members of their group. If this behavior is genetically determined, then altruistic individuals should eventually disappear. This does not occur. The riddle is solved by kin selection theory, which says that altruism is adaptive if the frame of reference is the survival and reproduction of genes rather than individuals. If this is so, then altruism is limited to the altruist's relatives, who share some of his genes (Axelrod & Hamilton, 1981).

The great leap forward in social evolution is the emergence of reciprocity, a behavior that incorporates non-kin in networks of mutually advantageous exchanges. Like mind reading, social life evolved dialectically (Bernhard, Fischbacker, & Fehr, 2006; Boyd, Gintis, Bowles, & Richerson, 2003; Nowak & Sigmund, 2005; Rosas, 2008; Simpson & Beckes, 2006). Reciprocity creates the possibility of "free-riders," individuals who take but do not reciprocate; they enjoy benefits without costs. The situation recalls the story about deceivers. Non-reciprocators have a reproductive advantage (they get calories without expending energy) and eventually replace reciprocators; social life regresses. This did not happen, however, because of another evolutionary development: the emergence of punishment in the form of retribution or ostracism. Non-reciprocation becomes expensive. Punishment is also expensive for enforcers, who may themselves become targets for retaliation and the disaffection of their own kin and neighbors. Since enforcers jeopardize their own reproductive success, punishment is properly called "altruistic punishment."

Punishment solves a riddle but is also the source of a riddle. Why would a rational individual—someone innately self-interested and capable of calculating cost-benefits—become an enforcer whose material benefits are hypothetical? This individual's potential payoff may be in the distant future, and the future costs of his or her actions are unpredictable. Even if they eventually gets his or her fair share, that person cannot know whether this would have happened without his or her intervention. Therefore, the enforcer's expectation of material rewards can provide only a weak motive for practicing altruistic punishment.

Neuroeconomics—a hybrid of experimental economics and social neuroscience—opened the way to a solution with a landmark experiment, "The neural basis of altruistic punishment," published in the journal *Science* (Camerer & Fehr, 2006; de Quervain et al., 2004; Fehr & Camerer 2007; Fliessbach, et al., 2007; Knoch, Pascual-Leone, Meyer, Treyer, & Fehr, 2006; Lanzetta & Englis, 1989; Singer et al., 2006). The experiment was organized around "the dictator game." One participant is given a sum and told to divide it among other players as he wishes. In subsequent rounds, similar sums are given to the other players. Some players violate cultural standards of fairness and keep an excessive portion for themselves. Participants can punish these so-called "defectors" by withholding payment when the opportunity arises. However, the enforcer must reduce the amount that he or she pays him- or herself. Thus the enforcer's behavior is altruistic and pro-social: it contributes to the stability of the network.

Neuroimaging technology (positron emission tomography) was used during this experiment to observe the enforcers' brains in action. Images showed activation of the caudate nucleus of the dorsal striatum, a "reward center" (pleasure) associated with dopamine excretion. Activation was correlated with the enforcer's anticipation of punishing the defector; intensity of activation correlated positively with severity of the punishment. In other words, the enforcer's brain empathically mirrors the imagined (anticipated) distress of the target and, at the same time, delivers pleasure. (The capacity of the brain to mirror imagined distress has been demonstrated in participants who were asked to imagine someone else in physical pain (Jackson, Brunet, Meltzoff, & Decety, 2006; see also Lamm, Nusbaum, Meltzoff, & Decety 2007; Ochsner et al., 2008; Singer, 2006). Parallels with Bernheim's speculations on autosuggestion should be obvious.

Schadenfreude

The part played by the imagination in the operation of empathic cruelty can be seen directly in a recent study by Takahashi and collaborators (2009). The study concerns the emotions of envy and "Schadenfreude," conceived as two sides of one coin. Envy is described as a painful emotion, characterized by feelings of inferiority and resentment, and produced by the individual's awareness of another person's superior quality, achievement, or possessions. Schadenfreude is characterized as a pleasurable emotion, produced by awareness that a misfortune has fallen to a person who is envied or otherwise resented.

The Takahashi group recruited 19 male and female students for their research. Prior to fMRI scans, the participants were asked to read descriptions of three fictive students. (Participants and fictive students were matched for gender.) The first student (A) is the "protagonist:" participants are expected to view students B and C from A's perspective. The protagonist is depicted as someone with only average abilities, social endowments, personal achievements, possessions, and prospects. Student B is depicted as someone who is superior and successful in these respects and in the life domains that are important to the protagonist (and participant). Student C is depicted as superior and successful, but in domains that are not important to the protagonist.

During fMRI scans, participants silently read scripts pertaining to A, B, and C. The phase one scripts described the successes and advantages enjoyed by B and C. Participants rated the sentences according to how envious the events made them feel (1 = no envy, 6 = extreme envy). Phase two scripts described various misfortunes that spoiled events and prospects for the fictive students. Participants were asked to report the intensity of their pleasure—or Schadenfreude—regarding each outcome. Thus, they provided two responses: subjective appraisals of their emotions, and images of neural activation.

An earlier neuroimaging study (Eisenberger, Lieberman, & Williams, 2003) showed that physical pain and "social pain" (in the experiment, self-reported distress caused by social exclusion) are associated with the same region of the brain, the anterior cingulate cortex (see also Lieberman & Eisensberger, 2009). The Takahashi research shows that intense envy (focused on student B) produces a similar activation. On the other side of the coin, intense Schadenfreude (likewise focused on student B) is associated with activation of the ventral striatum, described as "a central node of reward processing." Thus, the Schadenfreude effect imaged in this research replicates the events inside the enforcer's brain in the de Quervain study.

Empathic Cruelty and Human Nature

Altruistic punishment persisted throughout the long period following the emergence of social networks based on reciprocity. And it can be assumed that motivation for altruistic punishment was transmitted across generations as a heritable disposition. The rise of state societies, markets, and institutions for regulating exchange reduced the importance of reciprocity and the role of altruistic punishment. But these developments were too recent to affect the disposition to punish, and it can be considered an aspect of human nature. In *The Concise Oxford Dictionary*, "cruelty" is defined as "having pleasure in another's suffering." If so, the disposition can be called "empathic cruelty."

Recall Daniel Goleman's account of mirror neurons; where he represents empathy as *intrinsically pro-social and morally positive*. This view pervades social neuroscience. "Empathy allows us to understand the intentions of others, predict their behavior, and experience an emotion triggered by their emotion. In short empathy allows us to interact effectively in the social world. It is also the "glue" of the social world, drawing us to help others and stopping us from hurting others" (Baron-Cohen, Knickmeyer, & Belmonte, 2005; see also Iacoboni & Dapretto, 2006; Lawson, Wheelright, & Baron-Cohen, 2004, p.163; Wheelright, 2006; Williams, 2001)

Simon Baron-Cohen, an authority on autism spectrum disorders, writes that human evolution has produced polar types of brains: a female brain with highly developed empathic capacities, and a male brain adapted to manipulating objects and creating systems. Empathy originated as a pro-social adaptation allowing Paleolithic females to detect the wants of pre-verbal children and the moods of the potentially dangerous males with whom they lived. On the other hand, autistic individuals are characteristically poor empathizers. The epidemiology of the disorder is biased towards males: the ratio is 5 to 1, and 10 to 1 with high functioning autistic disorder. We should think of autism as a disorder of the extreme male brain.

According to Baron-Cohen, people respond to suffering in these three ways:

1. The observer's response mirrors the sufferer's distress.
2. The observer's response is culturally appropriate but does not mirror the suffering, for example the observer responds with sadness to the sufferer's pain.
3. The observer takes pleasure in the sufferer's condition.

Baron-Cohen equates "empathy"—the glue of the social world—with the first two responses. He explicitly excludes the third. He does not consider a fourth possibility, where the observer mirrors the sufferer's distress while taking pleasure in the sufferer's condition. Why? Is "empathic cruelty" a contradiction in terms? De Quervain's research suggests otherwise.

Empathic Psychopaths

> Psychopathy can be considered one of the prototypical disorders associated with empathic dysfunction. Reference to empathic dysfunction is part of the diagnosistic criteria of psychopathy. The very ability to inflict serious harm to others repeatedly can be, and is, an indicator of a profound disturbance in an appropriate "empathic" response to the suffering of another.
>
> (Blair, 2005, pp. 707–708)

Recent research by Jean Decety and collaborators (2008) utilized eight adolescents diagnosed with "aggressive conduct disorder" (CD) and eight matched controls. The classification "conduct disorder" is limited to young people, generally males. Aggressive CD people have a record of inflicting pain on others. Participants' brains were scanned with an fMRI apparatus while they watched videos of people experiencing pain resulting from an accident or someone else's intentional action. Brain images showed that the pain matrix in the CD brains is activated to a significantly greater extent than in the normal brains. They also showed greater activation in the striatum— "part of the system implicated in reward and pleasure." Regions associated with the regulation of emotion were activated to a lesser extent than in the normal brains, and it is assumed that similar activation patterns occur when CD adolescents actually inflict pain on others.

The brain images show that "highly aggressive antisocial youth enjoy seeing their victims in pain and ... may not effectively regulate positively reinforced aggressive behaviour" (Decety, Michalska, Akitsuki, & Lahey, 2008)—in other words, behavior providing them with "enjoyment" or "excitement." CD brains and normal brains share an innate capacity for empathic cruelty. The difference between them is that CD brains are more empathic than normal brains, but also less capable of regulating the consequent emotion. (This is the favored hypothesis. An alternative hypothesis is that CD youths have a lower threshold for responding to situations of negative effect, including viewing pain in others, and are less able to regulate negative emotion. Distress induces renewed aggression, aggression inflicts more pain, the empathic experience of the pain heightens distress in the CD brain, and so on.)

In the same year, a research team (Fecteau, Pascual-Leone, & Théoret, 2008) investigated empathy and psychopathic tendencies in a non-psychiatric population. Male college students were asked to watch four videos: a human hand at rest; a Q-tip touching the hand at point X (over the first dorsal interosseus muscle); a needle inserted at point X, and a needle penetrating an apple. During viewing, motor cortex excitation was monitored by transcranial magnetic stimulation (TMS). This technology is able to localize and measure neural responses to pain within the sensory-motor system: muscles at point X mapped onto corresponding regions of the brain. Responses to the static hand video provided a base line. The Q-tip and needle videos elicited reduced motor cortex excitation; the effect was greatest in response to the needle video. The response is characterized as "empathic." Participants were also asked to complete a questionnaire, the Psychopathic Personality Inventory (PPI). High scores on "coldheartedness" (callousness, lack of guilt and lack of sentimentality) correlated with a greater reduction in cortical excitation.

Fecteau and collaborators cite research by Avenanti and collaborators (2005), researchers who followed a similar procedure except that participants were asked to complete a questionnaire measuring "empathic concern" and personal distress. The results are that "massive inhibition of corticospinal excitability affecting upper limb muscles" correlates with high empathy scores: greater reduction equals greater empathy. In other words, while everyone responds empathically to the needle video, the empathic response is more intense in participants with psychopathic tendencies. Thus, Fecteau's team and Decety's team reach a similar conclusion. (Note that this is the explanation for reduced neural excitation on these occasions: the response may be part of an evolutionary adaptation that helps the observer's corticospinal system "implement escape or freezing reactions" (Avenanti, Bueta, Galaty, & Aglioti, 2005, p. 958).

In common with Baron-Cohen, the Fecteau team seems reluctant to get to the bottom of the empathic cruelty business. The team visualizes empathy inside the psychopath's brain, and then asks how one should understand this finding given that "the psychopathic construct … is usually defined by a lack of empathy." Their solution is to conceive empathy as a two-step process. Step one produces an embodied (mirrored) simulation at a sensory level, facilitates mind-reading, and provides the psychopath with a "substantial advantage for manipulation or harm." According to DSM-IV, while "deceit and manipulation are general features" of the condition, these individuals "frequently lack empathy"—that is to say, emotional empathy and a benevolent attitude (American Psychiatric Association, 1994, pp. 645–657). According to the Fecteau team, these features, characteristic of true empathy, are produced during step two, when the simulation information needed for mind reading is made available for an emotional/affective response (pity, sorrow, remorse, outrage, for example). Thus, the exaggerated empathic response that fMRI unexpectedly visualized in the "coldhearted" participants is explained as the consequence of a defect in step-two processing that "might be maladaptive in psychopaths" (Fecteau, Pascual-Leone, & Théoret, 2008, p. 142).

Conclusion

Empathic cruelty is biologically indistinguishable from other empathic emotions, yet it was excluded from the discourses of social neuroscience until recently. Its discovery was adventitious and Human Nature 2.0, as revealed in the work of social neuroscience, seems to be moving in an unanticipated and possibly unwelcome direction. In *The Eighteenth Brumaire of Louis Bonaparte*, there is this familiar line: "Men make their own history, but they do not make it as they please." Karl Marx's point about making history might be applied equally to efforts aimed at reconstructing human nature.

References

Adolphs, R. (2003). Cognitive neuroscience of human social behaviour. *Nature Reviews Neuroscience, 4*, 165–178.

American Psychiatric Association (1994). *Diagnostic and statistical manual of mental disorders of the American Psychiatric Association* (4th ed.). Washington, D. C.: American Psychiatric Association.

Avenanti, A., Bueta, D., Galaty, G., & Aglioti, S. M. (2005). Transcranial magnetic stimulation highlights the sensorimotor side of empathy for pain. *Nature Neuroscience, 8*, 955–960.

Axelrod, R., & Hamilton, W. D. (1981). The evolution of cooperation. *Science, 211*, 1390–1396.

Baron-Cohen, S., Knickmeyer, R. C., & Belmonte, M. K. (2005). Sex differences in the brain: Implications for explaining autism. *Science, 310*, 819–823.

Barton, R. A., & Dunbar, R. I. M. (1997). Evolution of the social brain. In A. Whiten & R. W. Byrne (Eds.), *Machiavellian intelligence II* (2nd ed., pp. 240–263). Cambridge: Cambridge University Press.

Bernhard, H., Fischbacker, U., & Fehr, E. (2006). Parochial altruism in humans. *Nature, 442*, 912–915.

Bernheim, H. (1980). *Hypnotisme, suggestion, psychothérapie: Etudes nouvelles.* Paris: O. Doin. (Original work published 1891.)

Blair, R. J. (2005). Responding to the emotions of others: Dissociating forms of empathy through the study of typical and psychiatric populations. *Consciousness and Cognition, 14*, 698–718.

Boyd, R., Gintis, H., Bowles, S., & Richerson, P. J. (2003). The evolution of altruistic punishment. *Proceedings of the National Academy of Science, 100*, 3531–3535.

Brüne, M., Ribbert, H., & Schiefenhövel, W. (Eds.). (2003). *The social brain: Evolution and pathology.* London: John Wiley & Sons.

Byrne, R. W., & Whiten, A. (Eds.). (1988). *Machiavellian intelligence.* Oxford: Oxford University Press.

Camerer, C. K., & Fehr, E. (2006). When does "economic man" dominate social behaviour? *Science, 311*, 47–52.

Carr, L., Iacoboni, M., Dubeau, M. C., Mazziotta, J. C., & Lenzi, G. L. (2003). Neural mechanisms of empathy in humans: A relay from neural systems for imitation to limbic areas. *Proceedings of the National Academy of Science, 100*, 5497–5502.

Decety, J., Michalska, K. J., Akitsuki, Y., & Lahey, B. B. (2008). Atypical empathic responses in adolescents with aggressive conduct disorder: A functional MRI investigation. *Biological Psychology, 80*, 203–211.

de Quervain, D. J.-F., Fischbacher, U., Treyer, V., Schellhammer, M., Schnyder, U., Buck, A., & Fehr, E. (2004). The neural basis of altruistic punishment. *Science, 305*, 1254–1258.

de Vignemont, F., & Singer, T. (2006). The empathic brain: How, when and why? *Trends in Cognitive Sciences, 10*, 435–441.

Dinstein, I., Thomas, T., Behrmann, M., & Heeger. D. J. (2008). A mirror up to nature. *Current Biology, 8*, R13–R18.

Dunbar, R. I. M. (2003). The social brain: Mind, language and society in evolutionary perspective. *Annual Review of Anthropology, 32*, 163–181.

Eisenberger, N. I., Lieberman, M. D., & Williams, K. D. (2003). Does rejection hurt? An fMRI study of social exclusion. *Science, 302*, 290–292.

Fadiga, L., Fogassi, L., Pavesi, G., & Rizzolatti, G. (1995). Motor facilitation during observation: A magnetic stimulation study. *Journal of Neurophysiology, 73*, 2608–2611.

Fecteau. S., Pascual-Leone, A., & Théoret, H. (2008). Psychopathy and the mirror neuron system: Preliminary findings from a non-psychiatric sample. *Psychiatry Research, 160*, 137–144.

Fehr, E., & Camerer, C. F. (2007). Social neuroeconomics: The neural circuitry of social preferences. *Trends in Cognitive Science, 11*, 419–427.

Fliessbach, K., Weber, B., Trautner, P., Dohmen, T., Sunde, U., Elger, C. E., & Falk, A. (2007). Social comparison affects reward-related brain activity in the human ventral striatum. *Science, 318*, 1305–1308.

Fogassi, L., Ferrari, P. F., Gesierich, B., Rozzi, S., Chersi, F., & Rizzolatti, G. (2005). Parietal lobe: From action organization to intention understanding. *Science, 308*, 662–667.

Fogassi, L., Gallese, V., & Rizzolatti. G. (2002). Hearing sounds, understanding actions: Action representation in mirror neurons. *Science, 297*, 846–848.

Freud, S. (1946). *Totem and taboo: Resemblances between the psychic lives of savages and neurotics.* New York: Vintage. (Original work published 1913.)

Freud, S. (1960). *Jokes and their relation to the unconscious.* London : Routledge & Kegan Paul. (Original work published 1905.)

Gallese, V. (2006). Intentional attunement: A neurophysiological perspective on social cognition and its disruption in autism. *Brain Research, 1079*, 15–24.

Gallese, V. (2003). A neuroscientific grasp of concepts: From control to representation. *Philosophical Transactions of the Royal Society of London, B 358*, 1231–1240.

Gallese, V. (2001). The "shared manifold" hypothesis: From mirror neurons to empathy. *Journal of Consciousness Studies, 8*, 33–50.

Gallese, V., & Goldman, A. (1998). Mirror neurons and the simulation theory of mind reading. *Trends in Cognitive Science, 2*, 493–501.

Goleman, D. (2006). Friends for life: An emerging biology of emotional healing. *New York Times*, October 10.

Grush, R. (2004). The emulation theory of representation: Motor control, imagery, and perception. *Behavioural and Brain Sciences, 27*, 377–398, 425–442.

Hein, G., & Singer, T. (2008). I feel how you feel but not always: The empathic brain and its modulation. *Current Opinion in Neurobiology, 18*, 1–6.

Iacoboni, M., & Dapretto, M. (2006). The mirror neuron system and the consequences of its dysfunction. *Nature Reviews Neuroscience, 7*, 942–951.

Iacoboni, M., Molnar-Szakacs, I., Gallese, V., Buccino, G., Mazziotta, J. C., & Rizzolatti, G. (2005). Grasping the intentions of others with one's own mirror neuron system. *PLoS [Public Library of Science] Biology, 3*, e79.

Jacob, P. (2008). What do mirror neurons contribute to human social cognition? *Mind and Language, 23*, 190–223.

Jacob, P., & Jeannerod, M. (2005). The motor theory of social cognition: A critique. *Trends in Cognitive Sciences, 9*, 21–25.

Jackson, P. L., Brunet, E., Meltzoff, A. N., & Decety, J. (2006). Empathy examined through the neural mechanisms involved in imagining how I feel versus how you feel pain. *Neuropsychologia, 44*, 752–761.

Jahoda, G. (2005). Theodor Lipps and the shift from "sympathy" to "empathy." *Journal of the History of the Behavioural Sciences, 4*(2), 151–163.

Johnson, M., Griffin, R., Csibra, G., Halit, H., Farroni, T., de Haan, M., & Richards, J. (2005). The emergence of the social brain network: Evidence from typical and atypical development. *Development and Psychopathology, 17*, 599–619.

Kass, L. (2003). Session on human nature and its future. President's Council on Bioethics. Retrieved April 16, 2009 from http://www.bioethics.gov/transcripts/march03/session3.html.

Knoch, D., Pascual-Leone, A., Meyer, K., Treyer, V., & Fehr, E. (2006). Diminishing reciprocal fairness by disrupting the right prefrontal cortex. *Science, 314*, 829–832.

Kohler, E., Keysers, C., Umlitá, M. A., Fogassi, L., Gallese, V., & Rizzolatti. G. (2002). Hearing sounds, understanding actions: Action representation in mirror neurons. *Science, 297*, 846–848.

Lamm, C., Nusbaum, H., Meltzoff, A. N., & Decety, J. (2007). What are you feeling? Using functional responses magnetic resonance imaging to assess the modulation of sensory and affective during empathy for pain. *PLoS ONE, 2*(12), e1292. doi:10.1371/journal.pone.0001292.

Lanzetta, J. T., & Englis, B. G. (1989). Expectations of cooperation and competition and their effects on observers' vicarious emotional responses. *Interpersonal Relations and Group Processes, 56*, 543–554.

Lawson, J., Baron-Cohen, S., & Wheelwright, S. (2004). Empathising and systemising in adults with and without Asperger Syndrome. *Journal of Autism and Developmental Disorders, 34*, 301–310.

Le Bon, G. (2002). *The crowd: A study of the popular mind.* (Trans. *La psychologie des foules*). Mineola, NY: Dover. (Original work published 1895.)

Lieberman, M. D., & Eisenberger, N. I. (2009). Pains and pleasures of social life. *Science, 323*, 890–891.

Lipps, T. (1903). Einfühlung, innere Nachahmung, und Organempfindungen. *Archiv für die gesamte Psychologie, 1*, 185–204.

Moyn, D. (2005). *Origins of the other: Emmanuel Levinas between revelation and ethics.* Ithaca: Cornell University Press.

Nowak, M. A., & Sigmund, K. (2005). Evolution of indirect reciprocity. *Nature Reviews Neuroscience, 437*, 1291–1298.

Nye, R. A. (1975). *The origins of crowd psychology: Gustave Le Bon and the crisis of mass democracy in the third republic.* Bervely Hills, CA: Sage.

Ochsner, K. N., Jamil, Z., Hamelin, J., Ludlow, D. H., Knierirm, K., Ramachandran, T., & Mackey, S. G. (2008). Your pain or mine? Common and distinct neural systems supporting the perception of pain in self and other. *Social Cognitive and Affective Neuroscience, 2*, 144–160.

Penelhum, T. (1993). Hume's moral psychology. In D. F. Norton (Ed.), *The Cambridge Companion to Hume* (pp. 117–147). Cambridge: Cambridge University Press.

Pigman, G. W. (1995). Freud and the history of empathy. *International Journal of Psycho-Analysis, 76*, 237–256.

Price, J. P. (1967). The dominance hierarchy and the evolution of mental illness. *Lancet, 2*, 243–246.

Rizzolatti, G., & Arbib, M. A. (1998). Language within our grasp. *Trends in Neuroscience, 21,* 188–194.

Rizzolatti, G., & Craighero. L. (2004). The mirror-neuron system. *Annual Review of Neuroscience, 27,* 169–192.

Rosas, A. (2008). The return of reciprocity: A psychological approach to the evolution of cooperation. *Biology and Philosophy, 23,* 555–566.

Ryle, G. (1949). *The concept of mind.* London: Hutchinson.

Sahlins, M. (2008). *The Western illusion of human nature.* Chicago: Prickly Pear Press.

Sidis, B. (1898). *Psychology of suggestion: Research into the subconscious nature of man and society.* New York: Appleton.

Simpson, J. A., & Beckes, L. (2006). Reflections on the nature (and nurture) of cultures. *Biology and Philosophy, 23,* 257–268.

Singer, T. (2006). The neuronal basis and ontogeny of empathy and mind reading: Review of literature and implications for future research. *Neuroscience and Biobehavioural Reviews, 30,* 855–863.

Singer, T., & Frith, C. (2005). The painful side of empathy. *Nature Neuroscience, 8,* 845–846.

Singer, T., Seymour, B., O'Dougherty, J. P., Stephan, K. E., Dolan, D. J., & Frith, C. D. (2006). Empathic neural responses are modulated by the perceived fairness of others. *Nature, 439,* 466–469.

Smith, A. (2002) *The theory of moral sentiments* (1st ed.), K. Haakonssen, (Ed.). Cambridge: Cambridge University Press. (Original work published 1759).

Smith, A. (1937). *An inquiry into the nature and causes of the wealth of nations* (3rd ed.), E. Cannan, (Ed.). New York: Modern Library. (Original work published 1776).

Takahashi, H., Kato, M., Matsuura, M., Mobbs, D., Suhara, T., & Okubo, Y. (2009). When your gain is my pain and your pain is my gain: Neural correlates of envy and Schadenfreude. *Science, 323,* 937–939. Retrieved from www.sciencemag.org/cgi/content/full/323/5916/937/DC1.

Tettamanti, M., Buccino, G., Succaman, M. C., Gallese, V., Danna, M., Scifo, P., & Perani, D. (2005). Listening to action-related sentences activates fronto-parietal motor circuits. *Journal of Cognitive Neuroscience, 17,* 273–2781.

van Ginneken, J. (1992). *Crowds, psychology, and politics 1871–1899.* Cambridge: Cambridge University Press.

Wheelwright, S., Baron-Cohen, S., Goldenfeld, N., Delaney, J., Fine, D., Smith, R., & Wakabayashi, A. (2006). Predicting autism spectrum quotient (AQ) from the systemizing quotient-revised (SQ-R) and empathy quotient (EQ). *Brain Research, 1079,* 47–56.

Williams, J. H. G., Whitten, A., Suddendorf, T., & Perrett. D. I. (2001). Imitation, mirror neurons and autism. *Neuroscience and Biobehavioural Review, 25,* 287–295.

Part III

Neuroscience in Context

From Laboratory to Lifeworld

8

Disrupting Images

Neuroscientific Representations in the Lives of Psychiatric Patients

Simon Cohn

Frank, a 47 year old man, has been diagnosed with schizophrenia for almost 15 years. He's had spells in hospital, was even sectioned once, but under medication has been able to cope independently for much of the time. He was in hospital again—a severe episode—when he was approached by the neuroscience research department to volunteer to have a scan. He was told categorically that it would not affect his own treatment, and that it was simply to help in a long-term study, but in truth Frank was not convinced. He'd told me, before having it, that he secretly hoped it might show them what was really going on. He wanted to leave hospital, and felt that the scan might show not only what was wrong, but also what could be made right.

After the scan, about three months later when he was at home, I talked again to Frank. Unlike some of the other volunteers at various institutions around London, he had not been given a copy of the image, but he didn't seem to mind at all:

> I know that they couldn't give me the scan, because it must have showed them that I don't really have schizophrenia. This brain scanning is showing them things they don't want to know, or can't understand. But I know … it's showing them that we've all got schizophrenia—everyone.

I want to use Frank's comments to think about the neuroscientists and what they are doing. My research began in a number of brain imaging sites in London and a diversity of experiments conducted there over five years ago, which ranged from studying the "normal" brain to investigating the functional as well as structural abnormalities that might be associated with a wide range of mental illness. I therefore also managed to shadow psychiatric clinics at two London hospitals where imaging studies were conducted, but where the research hasn't yet actually had an impact on medical practice beyond screening for structural abnormalities. Increasingly, my focus became how many of these provisional investigations are taken up and interpreted by the

Critical Neuroscience: A Handbook of the Social and Cultural Contexts of Neuroscience, First Edition.
Edited by Suparna Choudhury and Jan Slaby.
© 2012 John Wiley & Sons, Ltd. Published 2016 by John Wiley & Sons, Ltd.

volunteer patients themselves, irrespective of any of the scientific claims. The brain image itself serves as an emblem of this—not merely because of the technical awe it invariably elicits, but because it is also a symbol of the even broader potential the new neuroscientific representations are rapidly being invested in.

Though the temptation might be to dismiss what Frank says—because of its naive and unreservedly conspiratorial tone—I want to argue that there is some truth to what he proclaims—not merely that the researchers don't exactly know what they are doing in some of their exploratory studies, but as he also implies, that unlike the rapid adoption of its claims by the patients themselves, perhaps there are intrinsic limitations preventing the new science from radically revising what he, and others, can be said to suffer from. It is this central point that offers a contribution to the more general project of *Critical Neuroscience*. The issue is not that neuroscience does not, or should not, play a significant role in psychiatry, but that at present at least its commitment to a simple reductionist paradigm is also affording the researchers a degree of naivety and lack of social awareness that is of concern. The effect is that, unlike traditional psychiatric encounters which, despite issues of power and inequality, are nevertheless inherently social interactions, the emerging role of neuroscience in psychiatry suggests the role of individual experts and doctors might be deferred by the apparently objective, and self-determining technology. Given the rapid rise of contemporary biological psychiatry, the issue therefore is not simply how diagnosis and treatment might alter in the near future—for example in revisions of the DSM and other formal protocols—but the extent to which new technical and scientific practices might alter the nature of the encounter between doctor and patient, and even more importantly, the patients' overall experience of illness itself.

As a way of addressing this, I will go on to describe how the promissory nature of contemporary neuroscience is already having an impact—if not yet on current psychiatric practice, then at least for the patients' own understandings of illness—some of whom participate in the experiments and have talked to me outside the laboratory setting. Centrally, the idea of a radically new way to conceive of their illness based almost wholly on a biological paradigm of legions and pathology potentially promises to disrupt the feelings and beliefs they have about their condition, and offers a way of redefining their sense of self and relationships with others. The point is, however, that this is done independently of the scientists and clinicians, who, through the highly technical nature of their work and the scientific paradigm they are committed to, retreat from view and have a far less explicit role in the construction of illness than psychiatry of the past. It is this ostensibly disembedded nature of neuroscientific facts that represents its unique potential in psychiatry.

A number of commentators have recently noted that the actual practice of producing any type of brain scan—whether PET (positron emission tomography), MRI (magnetic resonance imaging), or SPECT (single photon emission computed tomography)—involves a great deal of carefully orchestrated design and manipulation by a number of different scientists, physicists and statisticians and the aligned participation of the cohort of volunteers who are scanned (Jack & Roepstorff, 2003). Raw data not only need to be made digital and then cleansed of noise and artifacts, but then also have to be statistically mapped onto an existing three dimensional representation of the brain to produce an image of the various structures and regions for analysis (Beaulieu,

2002). Even saying that the final images are "constructed" largely fails to address just how much work is done in order to achieve the final images (see Joyce, 2008, for an excellent account of the various stages necessary to produce an MRI, for example). Yet the underlying point of all these observations is the fact that the final images appear, to the uninitiated, as straightforward representations much like other kinds of medical images.

More generally, the rapid rise of modern biology as a singular way by which people conceive of themselves has been noted by many contemporary commentators. For example, the notion of the "biological citizen" conveys the way people increasingly come to regard themselves, their actions, their choices, and responsibilities in terms of individualized scientific descriptions, particularly from the realm of genetics (Gibbons & Novas, 2007). But it is also suggestive of the ways in which society and the state might be reconfiguring the roles and responsibilities of individuals, especially in terms of health and risk of illness (Carter, 1975). Recently, it has been suggested that the influence of neuroscience and related pharmaceutical treatments is generating a specifically "neurochemical" notion of self (Rose, 2007). Consequently, some critics are alarmed at the extent to which this rapidly expanding technical knowledge of the human body suggests there is an escalating abstract value attached to people's biological constitution and tissue that might already be redefining what it means, from social and ethical perspectives, to be human (Waldby & Mitchell, 2006). But, apparently contrary to these concerns, what is intriguing is the degree to which such explanations appear to be being adopted so readily by people themselves: the drift towards the corporal as a sole explanation for all things seems to be a general trend, even if it is one that does not favor everyone.

The increasingly detailed understanding of brain structure and function is being used to suggest that a wide range of conditions, including bipolar and schizophrenia, that up until now had largely escaped definitive biological description, will soon be understood like any other condition—effectively eliminating the old opposition between mental and physical illness (see, for example, the overview by Rombouts, Barkhof, & Scheltens 2007). Currently, beyond correlating a few structural abnormalities, and suggesting a range of functional differences in different systems and brain pathways, little has been sufficiently stabilized to constitute a medical fact.

As Fleck originally notes, facts can be understood to be claims of truth that have a high correspondence with the reality under scrutiny, but that also are assertions that within a particular scientific thought community are able to resist any alternative explanations (Fleck, 1981). Thus, for a period of time at least, they establish some degree of certainty both through intrinsic and extrinsic properties (Löwy, 2008). I consequently want to suggest that the emergence of new knowledge of the brain is generating new ideas of illness and suffering that are already being taken up by some patients, amidst the hype and constant articles in newspapers and on TV that could be said to be generating a new kind of "brain identity" (Vidal, 2009). Yet, rather than adopting some radical sense of a "neurochemical self" or "biological citizen," the patients I came into contact with are trying to use the new representations in their current experience of illness. The result is that the new brain sciences are having even more complex effects than simply shifting clinical definitions of psychiatric conditions; they are radically redefining illness in the everyday lives of people who experience them.

Perhaps more than many other categories of distress, the biomedical classification of mental illnesses has often been hesitant (Shorter, 1997). Although medical science has pursued establishing a physical basis to such things as schizophrenia for over 150 years, insufficient knowledge and understanding of the nervous system has prevented few causal links from being established between pathology and disease. Given this, the desire from some quarters to augment or even replace the identification and taxonomy of conditions subjectively diagnosed largely by behavioral symptoms with more definitive, stable, and objective underlying pathology has historically always generated an anxious tension. But the rapid advances of neuroscience are being hailed by many as finally offering a biological paradigm that will not merely augment, but potentially replace, traditional psychiatry. As one of the scientists proclaimed to me, with a strong hint of conviction in addition to provocation, "the talking therapies of last century have had their day!"

Traditional Psychiatric Practice

There is a misleading, but nevertheless important, portrayal of the way a psychiatrist traditionally makes a definitive diagnosis or treatment decision: that at a specific moment a discrete conclusion is arrived at and the trajectory of the patient suddenly alters. In practice, unless there are some very unusual circumstances that demand a rapid judgment to be made for the safety of the sufferer or others, decisions are based on the accumulation of information over some time and from a wide variety of sources. Consultations with a sufferer frequently reveal previous diagnoses, other unexplained or unrecorded illness episodes, and disclose the broad entanglement of germane signs and those rejected as inconsequential from the narratives elicited from the sufferer. Relatives, sometimes explicitly, sometimes inadvertently, add to this accumulation of what a doctor frequently terms "background" information. At informal meetings, usually as they accompany the sufferer, they are frequently invited to comment about instances of unusual behavior, inconsistent actions, dramatic personality changes, mood swings, and so forth, to help build up a more "holistic" understanding of the particular circumstances. And, in addition, a psychiatrist will frequently see a person more than once: a tentative diagnosis being further confirmed by subsequent conversations and evaluations.

> For me, it's important not to jump to conclusions. To be honest, my initial instincts are nearly always correct—it's not a matter of making the wrong diagnosis. But there's something important in feeling my way—so that the patient is acclimatized to the whole thing, and so that they feel they have been part of the entire process.

This sense that diagnosis is actually a progressive procedure is further supported by the judgments of other medical professionals. Corridor conversations with other psychiatrists, scribbles in the ward book kept by nurses and nursing assistance, all help to "build up a complete picture." The metaphor of a picture is in fact one of also delineating a boundary frame; diagnosis is a process in which the person gets gradually converted into a unified "case" consisting only of coherent, relevant, and pertinent elements.

However, most doctors are as willing to refer to their own "experience" and judgment, as this more rigid checklist. As one put it, referring to a hospital-based diagnostic protocol:

> The manual is useful, because it provides a general checklist. But we've all been doing this for so many years, and things like the DSM has so many clauses and contrary indications, that for me it really is there to provide a report on a decision I have probably already made.

In other words, this psychiatrist openly uses the checklist post-facto, as a way of regimenting and making concrete more implicit evaluations made according to his own internalized skill as a psychiatrist over the years.

Beyond the reality that such things like making a diagnosis rarely occurs definitively but is a compound operation, I want to suggest what is important in this conventional practice is that the decision-making appears embodied in a single person and embedded in a range of social encounters—both with the patient and others. From the patient's perspective the psychiatrist serves as the nexus of medical reasoning— the diverse evaluations, the many other contributions, past components, are all brought together into a single narrative of the presenting condition. This personification is meaningful; not only does it confirm and reassure a patient that the psychiatrist is the focus of expert knowledge, but also that they can reflect how diagnosis has been based on evaluations of the concrete interactions with themselves. In other words, the very ambiguity of diagnosing something so various and indefinite as schizophrenia is actually contained and secured by the idea that it is, in the end, based on a particular person's judgment. The diagnosis thereby gains some degree of stability—precisely because it is characterized as a subjective evaluation of potentially indeterminate criteria.

> I just have to trust him, you know, my doc? I mean he's the expert, the one who knows these things. I know I'm not well—but he's the one who can say what I really have— what I am suffering from.

Of course, it is also this sense of the subjective nature of psychiatry that continues to be the basis for much of its criticism. It is not surprising that patients and their advocates have tended to concentrate on the dynamics of the consultation—the imbalance of power, features of secrecy and unspoken judgment, the application of norms of gender, class, and ethnicity—in order to argue that by being based on an individual assessment the social nature of illness and abnormality is articulated through the guise of medical science. I am in no way disputing any of these; nor am I claiming that an individual psychiatrist does not express particular bias and prejudice arising from their own cultural and historical background, and further, the high degree of socialization that medical training enforces. But perhaps what nevertheless has sustained psychiatry is the very fact that the multitude of factors and reasons, possible causes and signs, are all assembled in the undertakings of social interaction, since it is precisely the acceptance of subjectivity that endorses the actions as having "reason" at all. As one consultant summarized his work:

What I do is distil information and stories ... I'm always hesitant about diagnosing someone because we are all too aware of the stigma that it can bring, and the distress that even a label can have on a person or their family ... But diagnosis is about activating the most appropriate treatment options, and therefore it is actually inherently a process about ensuring that we are deciding the correct path a patient takes rather than anything as crude as "what they have."

New Knowledge and Old Psychiatry

For the remainder of this chapter, I want to examine the extent to which current neuroscientific clinical research might be radically altering how such illness is understood and conceptualized in the near future. Central to this is the extent to which the embodied and embedded[1] nature of what I have crudely typified as traditional psychiatric practice might be being replaced by a set of practices, technologies, and materials that do not merely claim some kind of "objectivity", but are able to do this precisely because they appear not to rely on the messy realities of social encounters. The current expectation that neuroscience will find the biological basis of a wide range of contemporary mental illnesses must be looked at not only in terms of the cultural context in which such an imperative has such significance, but the ways in which the reality that the scans reveal serves itself to endorse and confirm such objectives. In other words, the extent to which the highly compelling nature of the final images, that belie the many different processes, individuals, and decisions necessary in their construction, also implicitly determine what is actually considered to be medically relevant or not.

At the heart of this is the degree to which the notion of physicality can serve to encapsulate complex conditions that are currently described via a range of qualitatively different criteria, and whether an emerging biological paradigm divorced from subjective judgments and varied social encounters can provide anything meaningful for patients at all. It is now frequently said that through such techniques as MRI, the biology behind or beneath the symptoms will be revealed. Further, some hope that the somewhat untidy classification system of mental illness itself will be refined and made unambiguous as the physical basis is understood in greater and greater detail (Phillips, First, & Pincus, 2003). The slicing up of matter into smaller and smaller sections, and the increasing technological power to identify density or types of neurones, neurochemicals, and statistical averages of mass all serve not merely to isolate specific components that might robustly be associated with illness, but now to conceive of elements of illness as actually physically present. In so doing, it avoids the clutter of traditional psychiatry based on evaluating the person; much of this research is not even conducted by psychiatrists, but neuroscientists and experimental psychologists.

[1] The pairing of the term "embodied" and "embedded" are now commonplace in critical discussions relating to philosophy of mind. I want to adopt these terms, however, not to critique the assumption that the brain and the mind are synonymous, but that the neuropsychiatry can somehow escape the subjective interactions between clinicians and patients, and be based solely on a matching between validated scientific knowledge and an individual subject's brain.

However, what is immediately apparent in the neuroscientific research done at present is that although it frequently positions itself in opposition to existing talking-therapy psychiatry it is entirely dependent on it: virtually all the current neuroscience research for possible clinical application relies on individuals already diagnosed by standard techniques. Thus, though not explicit, the final object of study is the result of very traditional embedded interactions and judgments, which through brain imaging are made distant and rendered virtually invisible. Raising this issue with the researchers I met—that although they saw their research as radically shifting ideas of mental health, it nevertheless relied on traditional diagnostic techniques to identify volunteers, presumably pre-empting the disease-objects they were hoping to physically identify—they tended to reply as this professor did:

> You have to realize that this is currently the best method psychiatry has to diagnose … and in general it's proved pretty reliable, even though it's very inefficient and relies too much on the subjective expertise of one or two individuals. What we hope to do is address all this—make the whole thing more objective, more scientific and hopefully available for more people to use.

Although the reliability and validity of their own research is in effect dependent on the more general assumption that current psychiatric diagnosis offers an approximation of specific conditions that unquestionably exist beneath, it is believed that the new technology will be able to distil this to establish the definitive biological causation that traditional psychiatry could only ever approximate to. One neuroscientist offered this as the most direct response to my questioning:

> For me, it's all a question of how we can move "upstream"—so that instead of just finding correlates with conditions like schizophrenia, we can actually understand schizophrenia a bit more by looking at the brain.

As he spoke to me, he turned in his swivel chair, and nodded first at the monitors in front of him, through the glass of an internal window, out to the MRI itself where, at the time, no one was being scanned, and then back up the line. It's the standard layout of all imaging sites, and almost rivals pictures of the brain as iconic presentations of the technology: a chain of consequences and a structure of translations from subjects to objects. Nevertheless, it became clear that for him this trajectory was about gradually shifting research from finding brain abnormalities that could be directly related to existing diagnostic categories, to identifying pathology that could be identified, as he put it, "on its own terms"; that is, increasingly independent from existing psychiatric labels or evaluations made of actual behavior. Thus, the aim is not merely to find consistent biological markers—in terms of structural differences or regional function—for current definitions of illness, but to use the new knowledge to inform how illnesses are understood and eventually defined. It is clear that this is inextricably linked with the more general view that neuroscience needs ultimately to distance itself from its reliance on the mess of contemporary psychiatry, and find ways of establishing pathology independent of persons—a philosophy encapsulated by the line of vision he plotted from the empty scanner, through the glass, to the bank of computers and monitors upon which a specific "hot" region of the brain might perhaps be illuminated.

The project to disentangle neuroscientific research into abnormal brain function from current psychiatric practice is already requiring quite a high degree of critical evaluation amongst neuroscientists themselves. It is forcing them to make claims beyond the descriptive to the interpretative, which many are actually reticent to make at the present time.

> ME: What does a scan actually tell us?
> SCIENTIST: Well, it shows what specific regions of the brain are affected—and that the illness is truly biological—surely you can't argue with that?
> ME: Of course not—but then, I'm not "anti" biology … but what does this new information really tell us?
> SCIENTIST: Well, as well as showing once and for all that it is physically based, I suppose … what we're hoping to show is how different regions … which we are slowly beginning to understand and describe in terms of specific functions … and networks … like, how things relate together … where there is abnormal activity, and umm, where there is less.

A key aspect of the scientist's ontology is the logic that biology is, as he suggests, the "base" of illness and has to be the definitive, singular cause of disease. His argument implies that it has merely been the limitations of technology and detailed knowledge of the brain that have, up until this point, prevented this foundation of behavior and mental distress from being conclusively identified. But in addition, the apparent hesitancy in this exchange is perhaps also indicative of the fact that beyond merely demonstrating that an illness has a biological "base," the task will require a level of understanding not only of the extraordinary intricacies of brain function but equally its relationship to all the variations of human behavior and experience. Thus, paradoxically, though neuroscientific research is committed to escaping traditional psychiatry and its embedded nature to defining pathology and abnormality "objectively," at the present time it is virtually impossible to do so. As a result, current research invariably is forced to adopt a somewhat odd order of things towards the eventual material explanation of illness. So it is that patients diagnosed by traditional means are encouraged to voluntarily take part in new research that might identify biological abnormalities which not only endorse the old psychiatric logic, but potentially break free from it. The principle appears straightforward; features that can be statistically associated with these volunteers can, at some time in the future, become some of the key discriminating features of diagnosis itself.

The desire to extend psychiatry is reproduced uncritically through arguing that revealing the biology can only serve to endorse, not threaten, it. However, the subtle trepidation displayed by both the senior researcher mentioned above and the majority of others that I spoke to in the labs around making claims of an objective biology as the basis of observable behavioral symptoms, could also suggest that a new constraint might already be surfacing. As more and more neurochemistry, genetics, and brain function localization is eventually understood, the original disease entities might themselves be questioned and potentially lose their fragile integrity. In other words, as modern neuroscience progresses, the researchers may have to address how to maintain the fabric of current psychiatric classification that they see themselves refining. On being posed this paradox directly, few neuroscientists responded with much clarity:

> I'm not quite sure what you mean. We know people suffer from such things as bipolar or OCD ... they're manifestly different. It's true that as our understanding increases, some of the diseases may have to be further refined. Some might be shown to be related. But we're working to improve psychiatry, not dismantle it.

There is, then, an invisible curb to the unrestricted endeavor of moving upstream, in that neuroscience itself is reproducing many of the existing values and assumptions that underpin traditional psychiatry. The limitations inevitably imbue current research practice, shaping it, determining the kinds of questions being asked and the kinds of objects being sought. Incongruously, then, the faith in an object beneath and prior to the symptoms is the very thing that appears to ensure the new biology of mental disease will inherit much of the old social nature of mental illness.

Patients' Disruptions

For virtually all of the patients who agree to participate at the research labs as volunteers in the various experiments, the motive to have a scan, and their hope to take home a copy in the form of a printout or a set of image files, is ostensibly one driven by a search for legitimacy. This draws not only on the possibility of obtaining objective external markers of illness, but equally on the very subjective and hitherto largely concealed feelings, beliefs and experiences of living with a condition up until this time. All these varied dimensions tend to be encapsulated by the frequently proclaimed, if surreal, phrase that they hope the scans "prove it's not all in my head." Almost unanimously, they express the idea that evidence of something physical would not only demonstrate to themselves that the condition, as most put it, is "real," but more importantly that the neuroscientific confirmation could be used to address problems and anxieties that arise in relation to other people. Patients would talk to me about how they can "prove" their suffering to their relatives and friends, without "me having to say anything," and how more generally "no one will be able to say anymore that I have anything to do with it." As this phrase implies, it is frequently hoped that something from participating in the research will be able to stand firm, and demonstrate a fixed "reality" of suffering that is able to resist the skeptical views expressed by others.

The patients, however, while frequently secretive about these more personal motives for volunteering, nevertheless share a key feature with the neuroscientists. They too see the technology, and the evidence it might produce, as allowing them to revise their illnesses—indicating once and for all the physical basis to their symptoms, and so the pathology that had been causing their symptoms all along. Consequently, they also articulate a paradox, like the neuroscientists; that while they want the technology to revise and reveal their illness in a physical form they nevertheless require it to remain sufficiently stable and intact in order to ensure the scope of the newfound legitimacy is appropriate and sufficiently encompassing. Unsurprisingly, the image is uniquely useful to this process because it suggests the location of illness with a directness that only an image seems to have. Interestingly, the patients repeatedly insist that their enthusiasm is not driven by a crude desire to refute social stigma associated with their particular condition, since many don't actually want to lose this aspect of their identity. This frequently left me confused: on the one hand, they would willingly participate in imaging studies, enduring all the inconvenience that it entailed and aware it would not have any clinical consequence for them; yet, in the act of looking at their own brain on a screen, or taking a copy home with them, they would say that making their illness physical was about not wanting to completely divorce themselves from their condition. This apparent contradiction was put to me by one volunteer in the following way:

> I have bipolar, and I have done for years. It's who I am, and I can't imagine not suffering from it. So, you see, I don't want to suddenly wake up and not be a bipolar … What I want is to be able to say to people, "Look. This bit of my brain, that's why I am bipolar. But I am bipolar, so if I have to live with it, why can't you?"

Thus, the imperative for materiality is not necessarily straightforward and refutes the idea that what is sought is simply some fundamental physical "base" to their condition. What is really important is not a biological understanding of their conditions per se, in terms of a claim for a root cause, but rather the ways in which such new accounts of the concrete serve to disrupt their existing narratives of illness and experience of selfhood in their everyday lives.

> The scan is important because it shows just what has been wrong with me all these years … you don't have to listen to descriptions or anything, you can see it there before your very eyes.

Here, the illness is no longer determined "indirectly" through discursive words but is demonstrable in a physical form that is taken to be indisputable. Revisiting the idea that facts can be considered those things that are able to resist alternative explanations, it is evident that even if current neuroscience is only tentatively making claims of certainty, for patients the brain images already serve as solid intrusions into the circulation of ideas and beliefs about their condition. They consequently gain the standing of "facts" even if for scientists they have not yet gained this status. So, although the majority of neuroscientists I have spoken to are actually quite hesitant about the clinical relevance of their research, and unsure how it might be converted into actual psychiatric practice, this potential is readily taken up by patients, who utilize their brain scans as incontrovertible evidence and consequently transform them into "fact-like" things. For them, this capacity of resistance takes place in their social world rather than in the lab, where in the past illness could only be constructed through subjective description and personal account. Their use of scans to directly interject into conversations with others serves not merely to disrupt established perspectives, but presents something that, even if the specific science and technology is not understood, functions as an immutable object.

Across a range of illness, including schizophrenia and bipolar, conditions could not previously be established as incontrovertible precisely because there was nothing available to generate resistance. With the scan, this is created not by an alternative voice, but by the introduction of something that interrupts and refracts. In other words, it is this capacity to stand firm that establishes them as irrefutable, and as a consequence can demonstrate independently that their suffering is something that they, and others, can recognize as "real." In sum, the simple physicality of the scan, which is taken to be a direct representation of the physical nature of the illness, serves as a surface to create alternative encounters.

Making it Real

Let me draw on the story of Tommy as a vivid example of the role that brain images can have in the making of meaning for patients.[2] He was sitting on the toilet when it happened. A tiny aneurysm, a weakness in the wall of a blood vessel, ballooned and burst, allowing blood to invade the surrounding tissue of the brain. The attack is commonly called a stroke, but this wholly fails to convey any of the silent violence or possible damage in the assault on his health. The 54 year old was rushed to hospital, where a diagnosis was soon made, although no one was prepared to give any kind of prognosis. It all depended on what specific part of the brain may have been starved of oxygen, for how long, and the damage the escaping blood may have done. The extent to which there might be either partial or almost complete recovery was entirely unknown. It would all be a matter of time. After a number of scans, Tommy had to have a major operation, in which a metal clip was carefully applied across the tear in the blood vessel to seal it permanently, thus preventing it from continually hemorrhaging. He also had a catheter fed into the vessel to introduce lots of miniature coils to block off a second bulge that threatened to burst and risk a further stroke.

He did recover. Miraculously, the damage to his co-ordination and speech which initially incapacitated him, and which tend to be the two most common faculties permanently affected by such a catastrophic event, did not turn out to be permanent. But soon afterwards, he began compulsively to write poetry, desperately trying to make sense of his near-death experience. More than any intellectual searching, it was a creative compulsion that he had never experienced before. He had been a builder prior to the stroke, with no interest in literature. It is likely that functional regions of his brain reorganized themselves to ensure that key faculties were regained. But Tommy also suffered damage that has had a lasting effect on his personality and behavior. He did "recover," though he's not the same person he was.

From this moment on Tommy needed to write continuously, to do something expressive. As the days passed, his focus changed into a compulsion to paint and make sculpture, and now it includes any form of art. As he put it, the only painting he had ever done were the self-inscribed tattoos on his knuckles. Over the months his energy never ceased, as he endlessly applied paint to boards, canvas—even, one evening, the living room wall when he was seized once again by his obsessive compulsion. It's not

[2] I got to hear of, and to meet, Tommy through a friend and professional contact—Dr Mark Lythgoe of the Institute of Child Health. Tommy's story had already been reported in the local press of his home area of Birkenhead.

surprising that once his story was reported in the local newspapers Tommy's transformation was interpreted by some as a story of religious revelation, and proof of divine inspiration. For Tommy, though, the illness had taken on a very strange form—controlling him, and driving him to create. What had been a stroke was now, "this bloody madness. How can something like a stroke now make me be so obsessive? So out of control?"

Tommy knows that he's obsessed with his brain. He's got his own, growing collection of medical images—including CTs, MRIs, and fMRIs—and generally now only paints or models heads. He described creating one of his early clay sculptures of a head:

> I made this bust, about life size, more or less ... But then I found out you've got to hollow out the clay, so I just cut the top off, and scooped it out ... and then I thought, what could I do with the inside of the head? So I made smaller ones, little copies, out of the stuff I scooped out, in a row ... Lots of heads out of brains ... Like Russian dolls. I just want to find out what's going on inside, what went on inside my head.

Tommy sees himself as reshaped, remolded, and reconfigured—he even comments on how his body has changed shape since his attack. His paintings and sculptures wrestle with the status of his brain, and what kind of illness he now has. It's not all tragic; Tommy seems largely OK about who he is. But he's never at peace. As he says himself:

> I don't really think of myself as the same person. All my friends just thought I was acting, pretending to be someone else ... but now they realize that I really am different. And I do too ... I don't really think about the old Tommy at all. The scans show that.

While Tommy's story is unusual, in that it is so dramatic and his use of brain imagery so striking as a way of negotiating illness, virtually all the people I have spoken to echo his sense that the scan shows not only that the old sense of self has changed, but that it is also instrumental in elaborating a new one. On numerous occasions, while talking to people, we would both look at an image of their brain with an inevitable sense of significance. Neither I, nor they, would really know what we were looking at, or even whether the image demonstrated anything abnormal whatsoever. It usually would not, as volunteers are given impressively detailed structural scans, rather than the more hazy functional ones that constitute most of the current research. But the point is that, for them, just the idea of a new way of seeing their condition was regarded as not merely altering the illness that they suffered from, but even more importantly, the illness that they imagined they would always have.

The experience of many mental conditions is formed not only by how sufferers make sense of the present, but the extent to which this also envelops their sense of the future. The scans, and by implication the underlying interpretation that their illness is physical, potentially interrupt not only how they might currently view their illness, but perhaps even more crucially the ways in which they reconstruct their past and can equally imagine their future. Many talk about how, after having the scan, they hope things might change, and that their prospects might alter. No one talked about the possibility of new medical treatment, or even of no longer being ill; rather that the alternative neuroscientific representation might simply offer an opportunity to establish a different way of living with their illness. Because of this, they are taken home and treasured, shown to friends and loved ones. Many would tell me about how possessing a copy was important to them because it was an object invested with optimism and change. Far more than simply serving to define their illness as physical, in a literal sense, the neuroscientific promise is mobilized to interrupt the enduring sense of an ill self that many of them have developed, or have had developed through their interactions with others. As a result of the scan, not only is the past now something that might potentially be revised, but so too is the future which, until this point, had been imagined as merely a continuation of the present.

> For me, I just can't tell you how important it is. All these years, and now they can finally prove it. I'm sure that this will make a huge difference. I feel different already. Almost like new.

The point is that the volunteer patients' interpretation of the possibility of biological reductionism is something very different from simply demonstrating that their illness is physical, and from the scientists' commitment to finding the physical cause of their suffering. By investing the scan with the qualities of a revealing portrait, people see in them a new sense of who they can be. What is taken as "biological" by the patients is consequently not the possible causal base beneath their outward behavior that the scientists seek within the interior of the brain, but rather a definitive, overarching explanation of their diffuse experiences and intangible suffering that can be transposed as part of their experience of the condition. Consequently, in contrast to the scientists whose work is shaped by a need to maintain some degree of continuity with existing knowledge, and as a result generates uncertainty about its role in psychiatry, for the patients the notion of a biological representation of their illness is invested with a hope for discontinuity that is embraced as a means to interrupt the patterns and routines that constitute their living experience.

Discussion

A key concern that some raise is that the integration of this technology into clinical practice will ignore the fact that mental illness is as much a social and cultural as a clinical category. Further, that the widespread employment of a biological paradigm will simply serve to endorse existing psychiatric categories, and merely convert what were more obviously social and culturally determined evaluations in traditional psychiatry, into ones that appear neutral and arising solely from the physical. Others are concerned that with the rapid increase of brain imaging techniques the traditional basis of psychiatric diagnosis will be overshadowed, and that despite the extraordinary technology, it ultimately reflects a remarkably crude form of reductionism.

These remain valid concerns, in that the application of neuroscience to psychiatry might indeed serve to further mask or disguise the social nature of classification and difference. In practice, few neuroscientists state that what they are doing is directly "clinical," even though the majority willingly endorse the value of their work by saying that hopefully, one day, it will have a medical application. The result is that each study, each experiment, is taken to generate a highly limited and contained objective arena of certainty since it merely concerns this particular region of activity, or that threshold of a particular neuroreceptor. So, it is possible to see how the potential weight of responsibility for direct clinical relevance is averted through the sense that the field for each discrimination is only a contributing component of an overall association between illness and the brain. One leading neuroscientist professed that as their understanding of the living brain develops, through the amalgamation of research from a huge number of specialisms across the world, so eventually a range of different illnesses will be understood solely through the physiology and chemistry of the brain; "then," he said, "that's when we'll be able to regard psychiatry as a technique of the past, that dealt as best it could with not really understanding what was going on."

That developments in scientific knowledge challenge and sometimes breach classification boundaries is, of course, a truism; but how people address this, how they see new things, is a far more complicated and subtle process. There are undoubtedly some people with mental health problems who vehemently reject the current expansion of modern biological psychiatry and who argue that their illness can only be understood and treated as a social condition. In other words, they wish to resist attempts that claim to isolate the disease and divide it from the illness. Jake, who has had schizophrenia for many years, summed up his own feeling:

> I know who I am, and I know that sometimes I'm mad. I don't mind telling people, that's all there is to it. This whole stuff about scans and new drugs and everything … it's a waste of money and resources.

This position is nevertheless not as far removed from those sufferers who appear to support the development of a biological model as one might first assume. I have argued that rather than endorsing the reductionist model per se, the many patients who are enthusiastic about current developments in neuropsychiatry are positive precisely because, for them, it offers a potentially steadfast and compelling means to reconfigure the illness in their own social world, through the interjection of new ideas and objects able to resist doubts and shame instigated by others. For them, the promise of neuroscience is that it might actually demarcate their illness from the social, allowing a new way of negotiating stigma, responsibility and agency. Neither is this the end of the story, since they are also invested with the potential to alter people's own narratives of illness, and notions of the future. Thus, in contrast to critics who argue that as technological developments proceed the social dimension will increasingly become hidden, the cultural basis of the categories themselves may, in fact, be revealed.

I have tried to describe how the scans often become, quite literally, matters-of-fact that serve to disrupt the volunteer patients' sense of illness and self, and in so doing potentially offer revision and change. The point is that these images are both physical and representational—both material and immaterial. Alongside this, what seems most significant in their role for patients is the way that they potentially serve as a means to convey both disruption and resistance in their lives. Beyond a simple dialogue between patient and image, this is clearly determined by the manner in which the object of neuroscience (whether an actual scan or merely the notion of a material basis of illness) is negotiated in people's social lives. But, given the reality that the neuroscientific investigations actually rely far more on old models of psychiatric illness than might at first appear, the adoption and interpretation of neuroscientific claims by sufferers will also require a heightened and critical vigilance. Only in this way can neuroscientists chart how such promises will always be countered by the forces of continuity that are necessary to ground the new conceptualizations in existing knowledge and practices.

References

Beaulieu, A. (2002). Images are not the (only) truth: Brain mapping, visual knowledge, and iconoclasm. *Science, Technology, & Human Values, 27*(1), 53–86.

Carter, S. (1995). Boundaries of danger and uncertainty: An analysis of the technological culture of risk assessment. In J. Gabe (Ed.), *Medicine, health and risk: Sociological approaches.* Oxford: Blackwell.

Fleck, L. (1981). *Genesis and development of a scientific fact.* Chicago: Chicago University Press. (Original work published 1935).

Gibbons, S., & Novas, C. (Eds.). (2007). *Biosocialities, genetics and the social sciences: Making biologies and identities.* Oxon: Routledge.

Jack, A., & Roepstorff, A. (Eds.). (2003). *Trusting the subject?* Exeter: Imprint Academic.

Joyce, K. A. (2008). *Magnetic appeal: Mri and the myth of transparency.* Cornell: Cornell University Press.

Löwy, I. (2008). Ludwik Fleck on the social construction of medical knowledge. *The Sociology of Health and Illness, 10*(2), 133–55.

Phillips, K., First, M. B., & Pincus, H. (Eds.). (2003). *Advancing DSM: Dilemmas in psychiatric diagnosis.* Washington: American Psychiatric Association.

Rombouts S. A. R. B., Barkhof, F., & Scheltens, P. (Eds.). (2007). *Clinical applications of functional brain MRI.* Oxford: Oxford University Press.

Rose, N. (2007). *The politics of life itself: Biomedicine, power, and subjectivity in the twenty-first century.* Princeton: Princeton University Press.

Shorter, E. (1997). *A history of psychiatry: From the era of the asylum to the age of Prozac.* New York: John Wiley & Sons.

Vidal, F. (2009). Brainhood, anthropological figure of modernity. *History of the Human Sciences, 22*(1), 5–36.

Waldby, C., & Mitchell, R. (2006). *Tissue economies: Blood, organs and cell lines in late capitalism.* Durham, NC: Duke University Press.

9

Critically Producing Brain Images of Mind*

Joseph Dumit

We know we need hype to sell our research; let's try to keep it out of the results!
Louis Sokoloff, giving a plenary talk at a Society
for Neuroscience national meeting

As an anthropologist, I have observed and interacted with various facets of the brain imaging community for over six years, concentrating on PET (positron emission tomography) scanning. I feel PET to be an incredibly important and increasingly powerful technique for producing images of living human brains. On the basis of my research, I have identified an area of PET signification that I believe is critical in debates over the role of PET in the world today: the visual effect of PET brain images. By attending closely to PET images, I have chosen the most mobile aspect of PET experiments. These images travel easily and are easily made meaningful. Because they are such fluid signifiers, they can serve different agendas and different meanings simultaneously. While representing a single slice of a particular person's brain blood flow over a short period of time, one scan can also represent the blood flow of a type of human, be used to demonstrate the viability of PET as a neuroscience technique, and demonstrate the general significance of basic neuroscience research. As such, a critical neuroscience approach needs to take the production and circulation of the images, their captions and their reception seriously, as an inextricable part of the "doing" of neuroscience.

This chapter is excerpted from my book *Picturing Personhood: Brain Scans and Biomedical Identity*. It draws on fieldwork and interviews with PET researchers as they designed and carried out experiments, and as they debated the generation and publication of images in their articles.

* This chapter is a slightly revised and shortened version of chapter 3 of the author's monograph *Picturing Personhood: Brain Scans and Biomedical Identity*. Princeton University Press, 2004. The material is re-produced with kind permission from Princeton University Press.

Critical Neuroscience: A Handbook of the Social and Cultural Contexts of Neuroscience, First Edition.
Edited by Suparna Choudhury and Jan Slaby.

Brain images are produced for a variety of reasons, often contradictory. As with all natural human sciences, they contain assumptions from a whole apparatus but appear simple and represent types because of the imaging process. In most cases, PET brain-type research is triangulating between (1) groups of subjects selected according to often accurate but imprecise behavioral criteria; (2) the small sampling of the selected populations under study, usually between 4 and 20 people per group; and (3) a "functional" (flow rate) anatomy of the brain that is also imprecise and to some extent unknown at the millimeter level. The resulting PET images, generated at the intersection of these three imprecise referents, are thus paradoxically the most concrete, analytical data available as to whether a behavioral criterion (for example, a schizophrenia diagnosis) or task (for example, remembering a number) is reliably handled differently than by the brains of other subjects (for example, those not diagnosed with any medical condition) or by the same subjects doing a different task (for example, resting quietly). The miracle is that we are able safely and repeatedly to get any precise locational data at all about brain functions in living subjects. Historically, no other techniques except PET and similar tomographic imagers (functional magnetic resonance imaging [fMRI] and single-photon emission computed tomography [SPECT]) have given quantitative three-dimensional locational information about brain function.

Brain-imaging technologies like PET offer researchers the potential to ask a question about almost any aspect of human nature, human behavior, or human kinds and design an experiment to look for the answer in the brain. Each piece of experimental design, data generation, and data analysis, however, necessarily builds in assumptions about human nature, about how the brain works, and how person and brain are related. No researcher denies this. In fact, they constantly discuss assumptions as obstacles to be overcome and as tradeoffs between specificity and generalization. The aim of this chapter is to systematically outline how and where these assumptions are built in so they can be tracked as the images travel.

Properly representing results of these experiments is another balancing act. This time the balance is between the many kinds of audiences who will encounter these complex images: fellow brain-imaging researchers, other neuroscientists, science journalists, and the public. For those who publish brain images, the question is often how to balance the persuasiveness of the visual scans of simple difference with the desire for those images to also represent the significance of the experimental data.

This practice of actively constructing images for publication is neither surprising nor new. Similar issues have been observed concerning graphs, tables, digital astronomical images, and physics' images (Jones, Galison, & Slaton, 1998). Images are produced and selected for publication to make particular points and to illustrate the argument and other data presented, not to stand alone. They are, in other words, explicitly rhetorical. This is, one could say, the only way one can present images.

Researchers in the same field know this and read each others' images very critically. They go right to the data, methods, qualifications, and statistical results, and they adjust these depending on genre and audience: granting agencies, journals, inter-disciplinary forums, and the general public. Observing this practice, I am concerned with the ways in which brain images and their interpretations as referring to brain types are appropriated and transformed for further use at each stage of image

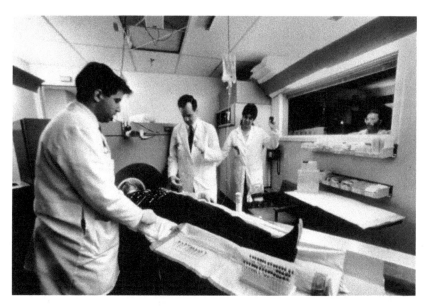

Figure 9.1 PET procedure in progress at Johns Hopkins University Medical Center. A research doctor, assisted by two technicians in the room and another one in the computer room behind shielded glass, draws blood and monitors the patient (Marcus, 1995).

production, selection, and dissemination, scientifically and popularly. Among scientists, this includes looking at how they design their machines and experiments, how they appropriate each other's work across disciplinary lines, and how they cooperate and compete. With each appropriation and subsequent translation, the content of the image, its qualifications, and brain-type referent, changes.

Despite its profound interdisciplinary complexity, brain-imaging data is presented in a particularly simple and compelling manner: PET images appear to be discrete, readable, and colorful. Similarly, because the process appears to produce clean pictures of functional brain activity, many simple diagrams of the PET process have been displayed as shorthand illustrations of it. Figure 9.1 makes PET seem almost as simple and as automatic as taking a snapshot. This leads not only to enthusiasm for brain imaging but to misplaced recruitment as well, as one researcher explained:

> It is kind of funny: I have had many people express an interest in using PET, typically established scientists in many fields who may be on a downhill curve of their career. Very overtly they express that PET is such a high road to science that they're willing to get involved now. They kind of held back before, but now they are willing to get involved because it is obviously so easy! They lack an understanding of what is entailed, I think, because the data comes out as pretty pictures. You put up these slides that show the brain turning on and turning off. They just don't understand the work that is involved in making these experiments happen (see Figure 9.1).

It is crucial, therefore, to unpack the kind of complexity required to produce and understand PET images as well as to understand the social function and efficacy of such simple diagrams.

The remainder of this chapter examines how the data produced in a PET experiment is visualized as an image of a living brain slice, and how those images are produced in the lab, selected and published to make meaningful, factual claims about the world. My thesis is that the visual nature of these images, their apparent familiarity and their transparency with regard to the brain all contribute to the potency of PET claims. PET researchers have acknowledged the difficulty of properly producing and understanding these images, and have warned that "we must understand our tools before we can hope to understand our results" (Perlmutter & Raichle, 1986). I am arguing that the processes of producing, selecting, and presenting images in both scientific articles and in public arenas require the same sort of understanding.

Creating Experiments: A Difficult Task

Creating experiments based on this work demands a tremendous team effort. The first thing one realizes when entering a PET lab is that the scanner is only one piece of a large-scale technical system. Technical descriptions of the scanning process only begin to define the work of conducting an actual experiment, however: they describe the stage and players, but not the play. Heuristically, we can break the whole process into four stages: design, measure, manipulate, and visualize.

1. **Experiment design:** The first stage of the process involves choosing participants for the study and designing their state and behavior in the scanner. Defining criteria for participant inclusion requires delimiting the boundaries of "normal human" for purposes of the study. Is a chronic smoker or coffee drinker normal enough? How about someone who had been found to have depression 10 years ago and has taken Prozac for 6 months—or someone whose brother is diagnosed with schizophrenia? Likewise, if the study is comparing two groups, the experimental group must also be characterized.

Because the purpose of the scan is to detect brain function, every part of the person's state of mind and brain needs to be controlled for. This includes what each subject eats or drinks beforehand, how rested or anxious the subject is, and what exactly the subject does inside the scanner. The more precise the state can be defined and calibrated, the easier it will be to compare results with those of other experiments.

2. **Measuring brain activity:** The second stage covers the scanning process proper. The radioactive molecules must be prepared and then injected into the person. The scanner must properly collect the data, and then a computer must algorithmically reconstruct the data into a three-dimensional map of activity, based on assumptions about the scanner and brain activity. The result is a dataset keyed to the individual's brain activity, a brainset.

3. **Making data comparable:** In stage 3, the individual brainsets are transformed and normalized so that the individual's brain locations can be correlated with those of others. With the use of MRI data and digital brain atlases, anatomical areas corresponding to the brainset can be found. Next, different brainsets can be combined and checked for statistical significance using subtraction, averaging, and other forms of data set manipulation. The result is a collective group brainset.

4. Making comparable data presentable: Finally, in stage 4, the brainsets are made visible. First, colors are used to substitute for the numbers in the dataset, and second, specific colored brainsets are selected, to be produced and published. Coloring involves transforming numeric variation into a contour map, highlighting some differences at the expense of others.

Turning then to the postproduction events for images, particular images are selected for publication and presented in journals. At the heart of this process is a common, standard, and often encouraged practice of selecting extreme images. This is an acknowledged, troubling practice, necessary for scientific work and yet increasingly problematic as these images travel outside expert circles and into popular culture, where new, less-qualified labels are applied.

Each of the stages and substeps within them is hotly debated, and along the way there are many assumptions about human anatomy, human physiology, and human nature. As discussed above, however, rather than exploding the coherence of the PET experiment, each assumption can become the grounds for a different discipline's article. The complexity and theory-ladenness of the PET experiment is thus incredibly productive of scientific results.

STAGE 1: EXPERIMENT DESIGN

Subject selection and injection

A senior PET researcher described subject selection:

> Collaboration is fine. Share data, collaborate, talk about it, work it out. But just having it in a base where somebody can pull it out, I think, creates a lot of chaos. One of the difficulties is [that] too many people have access to the databases and can make changes that you would never know about. So when I go for normal controls, I go to our normal control database, but I have to be very careful going through it. Just because they are labeled normal controls doesn't mean they are. I tend to use normal controls that I have generated myself in my own studies. I don't take the ones generated in other people's studies in the same group, because I don't really know what they did. But I don't think that that is a major impediment to the science.

Choosing people to be scanned for a study can be one of the most difficult procedures. In extreme cases, such as finding people diagnosed with schizophrenia who are drug naive (who have never taken medication or illegal drugs), the work of actually locating and validating proper subjects can constitute grounds for claiming first authorship on the published article! The problem, as I have come to understand it, involves group and individual definition of variability and constraint: to what extent is an individual representative of a group, and to what extent is the group well-characterized? These problems are exacerbated or exaggerated because PET often involves very small study sizes (4–20 subjects) because of cost, radioactivity, and time constraints, and because PET often provides information for which there is no independent verification. This means that often the only way to corroborate the findings of a PET study is with another PET study. There is no easy end to possible confounding variables.

Because there was no other way to verify the data that PET produces, one of the first tasks of PET researchers was to characterize "normals" (Mazziotta, Phelps, Plummer, & Kuhl,1981; Mazziotta & Phelps, 1985; Raichle, 1994). Only then could "non-normals" be compared. However, creating a baseline definition of "normals" is both a physiological and a social judgment. The following description provides a list of the tests used to characterize persons as "normals" in one study:

> The normal population consisted of 20 males aged 19–59 years. Inclusion in the study was determined by the absence of medical, neurological, and psychological pathology. Medical reasons for exclusion were a history of severe head trauma, chronic hypertension, significant vascular disorders, diabetes mellitus, thyroid abnormalities, and a history of psychiatric illness. Gross psychopathology was identified with the Structured Clinical Interview (SCI), an inventory of 17 yes-or-no items filled out by the examiner during a 20 to 30 min. interview. The SCI can also be scored for 13 overlapping scales: anger, hostility, conceptual dysfunction, fear, worry, incongruous behavior, incongruous ideation, lethargy, dejection, perceptual dysfunction, physical complaints, self-depreciation, and sexual problems. Any score significantly beyond the norms on any SCI component automatically excluded the candidate. Neurological and neurophysiological screening included medical history and testing for intelligence (WAIS [Wechsler Adult Intelligence Scale]), anterograde memory [Randt-NYU Memory Test], perceptual-motor function and structure (Bender Visual-Motor Gestalt Test), and handedness [Edinburgh Handedness Inventory]. Subjects also underwent a comprehensive laboratory work-up, a brief neurological screening examination, CT (to provide scans which could be correlated with PETT), computerized EEG [electroencephalography], and testing of visual and auditory evoked potentials.
>
> (Brodie et al., 1983, p. 201)

The tremendous amount of work put into finding such "normal" subjects was done with the intent of avoiding "noise" in the resulting data. Georges Canguilhem's book *The Normal and the Pathological* traces the history of the terms *normal, abnormal, pathological,* and *anomalous* through various sciences and medicines. Canguilhem noted that *normal* has been a polyvalent term that in different texts meant "typically healthy" (what the patient desires to be), "quantitatively average," "not anomalous," or "ideal" (in the sense of being not at all pathological or unhealthy) (Canguilhem, 1989). Medical characterizations of diseases are historically defined from a therapeutic perspective; one is diseased if one is not typically healthy and seeks therapeutic care. Initial studies with brain images are based on selections of "ideal" subjects, or "supernormals," who have no probable pathology.

> Normal age-matched controls have been studied in conjunction with this project. Healthy controls best consist of persons selected to minimize the possibility of covert pathology. These so-called "supernormals" are individuals who have been observed to be symptom-free for a number of years, have no personal or family history of psychiatric disorders, and are not users of substances known to influence mood.
>
> (Phelps & Mazziotta, 1985, p. 459)

The complexity of the project is part of the difficulty of mental-illness research and psychological research in general. Directly measuring the brain adds an additional factor. Possible confounders remain: are men sufficiently different from women to study separately, or are they sufficiently similar to women so that they can be averaged together? Such characteristics as age, ethnicity, handedness, culture (refugee status), sexuality, familial histories, past head trauma, and medical history are all still unknown confounders, raised as questions in meetings during presentation of results.

PET brain studies often use right-handed male subjects, unless gender is specifically being studied or a disease is being studied that is significantly more prevalent in females than in males. Although the reasons for this exclusion—cleaner data because of the lack of possible interference from gender or handedness differences—"may be viewed as practical from a financial standpoint, it results in ... a lack of information about the etiology of some diseases in women" (Rosser, 1994). By choosing only men for these studies, the researchers implicitly assume that gender matters. But by treating the results of the experiments as applicable to normal humans in general, they risk the consequence that a gender difference may appear as an abnormality.

For large-scale studies of schizophrenia, with over 50 people being studied, race is often recorded, though not consistently. In PET studies where the extreme expense of the procedure and the time involved results in very small samples, typically between 4 and 20, race has almost never been mentioned. Analogous to the circumstances for gender, the assumption is that there probably are significant population differences in brain chemistry and anatomy between different races. To eliminate this potentially confounding variable, however, race is often excluded from the sample altogether by using only Whites. Financially, for the experimenters, this is the only course of action that makes sense.

Even once these lifelong or *trait* characteristics are accounted for, temporary or *state* characteristics remain. For some studies, "normals" are only those who have not had caffeine that day. Use of nicotine, vitamins, or other drugs must also be monitored. Debates go on about proper cooling-off periods for drugs and medication. Also, questions remain as to the value of "normal databases" based on such exclusionary definitions of normal.

Because there are so many different definitions of normal, of who could be included as a normal control, and how explicitly their attributes should be noted, attempts to standardize a database have so far failed (see Beaulieu, 2000, for more).

At the 1995 meeting of the Society of Nuclear Medicine, a new confounder was introduced: one lab reported that the time of day during which the scan took place significantly and regionally affected PET results (Diehl & Mintun, 1995). This means that a scan taken of a person in the morning, when compared with a scan of the same person taken under the same conditions but in the afternoon, seemed to show a difference in certain areas of the brain and not others. The authors suggested that time-of-day differences might account for specific differences among labs. Certainly, this finding adds to the difficulty of replicating PET findings.

Yet, because the assumption behind this decision to exclude population differences is that these differences probably matter, the production of generically unmarked images with labels such as *normal* and *schizophrenic* (rather than, for example, *White US right-handed males diagnosed with schizophrenia*) means that we should assume that non-Whites will probably not look normal. When we combine this analysis with the practice of choosing "extreme" images for publication (where normal is chosen because it is farthest away from that particular group of subjects with schizophrenia), we can see yet another reason why the non-sampled non-White could more easily be found to be not normal.

Task design (types of scans and confounders)

Once the subjects have been selected, they must be injected with the radioisotope. What the subjects do or think, once injected, makes task selection fundamental to the PET data produced, even when the task is not the object of the study. This is one area where PET (and fMRI) are completely different from CT or MRI, which image structure. Structure does not change from moment to moment. PET scanning maps rates of flows of molecules in the brain over a relatively small period of time. Consequently, correctly characterizing and understanding a person's behavior, mood, and cognitive activity is essential to understanding the meaning of the flows.

Once injected with the radiotracer, the patient is now "on display." His or her body is emitting radioactivity. During this time, especially for brain studies, what the patient does—moving, thinking, hearing—bears greatly on the final PET scan data. For instance, one classic study compared *seeing* words versus *hearing* words. During the seeing-words task, subjects watched video screens where words were flashed up. During the hearing-words study, subjects listened to different words. PET has proved to be sensitive to different cognitive activities, and discovering the regional differences in brain activity during these activities is often the aim of these studies.

Even if the aim of the study is to characterize disease states, however, the behavior of the subjects must still be controlled for. "Resting" turns out to be a complicated task (Mazziotta, Phelps, Plummer, & Kuhl, 1981). Should one rest with eyes closed or open? With ears blocked, in silence, or listening to music? Does having an injection in one arm focus attention there? Anxiety has been studied, for instance, in part because the PET scan procedure itself might cause anxiety (for example, at being motionless in a scanner for 30 minutes or being injected with a radioactive substance) (Reiman, 1988; Reiman, Fusselman, Fox, & Raichle, 1989; Wu et al., 1991). Anxiety levels are usually measured before and after studies. With PET, in other words, one is always performing a task. Baseline states are all confounding variables to consider in designing a task to be studied.

Depending on the half-life of the tracer used, the subject will carry out the task either before getting into the scanner or while strapped inside. With FDG, for example, the critical uptake time is the first 40 minutes after injection. During this time, the brain traps almost all of the radiotracer in different cells and keeps it there, emitting radioactivity, for about another hour. After the 40 minutes, the subject is placed in the scanner and a picture of the trapped, still-radioactive glucose analog is taken. With oxygen, which has a two-minute half-life, the subject must already be in the scanner when injected. Scans are performed during the first two to five minutes, while the subject is performing the task.

Task design is itself one of the most active areas of studies. Studies include cognitive task comparisons (looking at words), states comparisons (such as anxiety or sadness, or cued-state studies such as showing cocaine addicts a video of drug use), resting trait comparisons (patients diagnosed with Huntington's disease versus those without it), task–trait comparisons (patients diagnosed with schizophrenia who are hallucinating versus at rest), neurotransmitter binding studies (dopamine, serotonin, for example), and challenge studies (where a drug is given and the brain's reaction to it is studied). The key problem in designing a particular study of any one of these types is finding a way to keep the other types from interfering (Frith, 1991). For cognitive neuroscientists, used to large sample sizes, PET added a new challenge: how to control for as many dimensions of variability as possible. A simple-sounding task like recognizing words might reveal a host of confounding variables, each correlating with a different set of brain regions: the size of the displayed words (how much of the visual field is consumed in the recognition process), the brightness of the word, the rate of presentation (which in fact turned out to produce very different brain activations), the language of the word (is a more ideographic language like Chinese processed differently from English?), educational level, effects on attention, novelty, and learning. (Are there effects from simply having to repeat a very simple task over and over that are different from purposeful recognition of words? Is proofreading a different activity from reading?)—these are in addition to designing the series of tasks so that a particularly desired component of language is being isolated. If the underlying presumption of modularity is correct and the task correctly isolates the component simple mental operation, then "from such data emerges a map of the distributed modular organization of the brain underlying normal human cognition and emotion" (Raichle, 1990).

There are debates in psychology over whether the modularity hypothesis itself can be tested with PET at all, or whether it must just be assumed (Kossyln, 1994; Szasz, 1996; Uttal, 2001; Wilson, 1998). Philosopher Jerry Fodor summarized one of the issues as a very serious struggle over limited resources and the value of different lines of questioning:

> I quite see why anyone who cares how the mind works might reasonably care about the argument between empiricism and rationalism; and why anyone who cares about the argument between empiricism and rationalism might reasonably care whether different areas of the brain differ in the mental functions they perform ... But given that it matters to both sides whether, by and large, mental functions have characteristic places in the brain, why should it matter to either side where the places are? ... what is the question about the mind–brain relation in general, or about language in particular, that turns on where the brain's linguistic capacities are? And if, as I suspect, none does, why are we spending so much time and money trying to find them?
>
> (Fodor, 1999)

Another form of dispute concerns the significance of individual variability. PET researcher Richard Haier calls himself an "individual differences psychologist," which means he is interested specifically in tasks for which people differ in their performance. If this is the case, then he can look for a correlation between performance and some brain measure. He begins by comparing his work to cognitive psychology.

HAIER: You know what cognitive psychologists do? They ask you to press a button when you see an *M* or an *N*. Either it is presented in this visual field or that visual field. They use very simple stimuli to get at complex processes. The idea of using something like Raven's Advanced Matrices [a measure of general intelligence] is just outrageous. Even the idea of individual differences in cognitive psychology is not a very big idea. Cognitive psychologists almost by definition are not interested in individual differences.

DUMIT: They are interested in how people share certain characteristics.

HAIER: That is right. So the variance in people's reaction times is regarded as error variance by cognitive psychologists. They want a task that minimizes that. They don't want a task that has a wide range of performance, they want everyone to do about the same, so they can discover "the" process. We took a completely different point of view. It is not that our point of view is better or worse—it is just a different starting point. This is common in psychology.

Haier is describing yet another dimension of brain function. In this "individual differences" dimension, humans vary in their performances on tasks and in their brain activation during performances. Video game experiments, for instance, use a task that people get better at. Most PET studies in the last 20 years have minimized this dimension, concentrating on tasks presumed to be relatively similar in performance and brain activity across humans and in time.

In sum, during the design stage, the basic terms of human nature are already built into the experiment. Subject selection defines a concept of the "normal" human being in the form of an ideal (super)normal. Abnormal categories, such as mental illness, are likewise normed as ideals. This process takes types of humans (or the generalized human as a type) as given, not to be discovered through the experiment but only to be correlated with brain activity. Similarly, task design must assume that the specific task behaviors correspond to discrete mental "functions." It might be suspected that if results are found indicating that different brain activity is correlated with each task or group, this verifies the human or task typology. This assumes, however, that the contrary—finding no significant difference—would be meaningful. Instead, the finding of no significance is interpreted as a need for better equipment. Psychiatrist and neuroscientist Nancy Andreasen stated this very clearly in a 1997 review of the field:

There are, at present, no known biological diagnostic markers for any mental illnesses except dementias such as Alzheimer's disease. The to-be-discovered lesions that define the remainder of mental illnesses are likely to be occurring at complex or small-scale levels that are difficult to visualize and measure

(Andreasen, 1997, pp. 1586–1587).

STAGE 3:   MAKING DATA COMPARABLE

Measuring brain activity is no simple task. Choices must be made among tracers with different half-lives and behavior in the body, among scanner architectures and head position with different areas of the brain that are more resolvable, and among reconstruction algorithms that privilege different theories of brain activation. Several levels of analysis are thus required to develop an interpretation of what is visualizable, between the flow of the tracer molecule and the final set of data that scientists work with. For further discussion of this "stage 2," see *Picturing Personhood*, chapter 3 pp. 68–81. Meaning then has to be made of these data, against some sort of reference— which brings us to stage 3.

> Using methods called "image processing" ... the computer acts as an extension of the eye and the brain by selecting information the scientists cannot see.
>
> (Blumenthal, 1982)

The scanner has now produced a brainset, an apparently stable set of numbers that represents the flow rate of the tracer and apparent activation. The next stage of the process of producing a brain image consists in first adjusting and transforming the dataset so that it corresponds to some other brainset, either the subject's own MRI, for instance, or a reference brainset. In the first case, the PET data is computationally combined, or "registered," with the MRI information so that the activity voxels can be given anatomical locations. Often this is combined with the process of then transforming or warping the subject's brainset into a standardized human brainset or "atlas." As Anne Beaulieu describes in *The Space inside the Skull*, this process presumes the meaningful and practical possibility of a *generalized human brain*, and then produces it (Beaulieu, 2000).

The following discussion of different brain atlases by MRI imager Matthew Brett, illustrates some of the difficulties:

> The MNI [Montreal Neurological Institute] defined a new standard brain by using a large series of MRI scans on normal controls. Recall that the Talairach brain is the brain dissected and photographed for the famous Talairach and Tournoux atlas. The atlas has Brodmann's [anatomical] areas labelled, albeit in a rather approximate way. In fact, what the authors did was to look at pictures of the Brodmann map and estimate where the same place was on their brain. To quote from the atlas, p. 10: "The brain presented here was not subjected to histological studies and the transfer of the cartography of Brodmann usually pictured in two-dimensional projections sometimes possesses uncertainties."
>
> The MNI wanted to define a brain that is more representative of the population. They therefore did a large number of MRI scans on normal subjects (305 of them), and did a simple linear match of each brain to the brain in the Talairach atlas ... The problem introduced by the MNI standard brains is that the MNI linear transform has not matched the brains completely to the Talairach brain. As a result, the MNI brains are slightly larger (in particular higher, deeper and longer) than the Talairach brain. The differences are larger as you get further from the middle of the brain, towards the outside, and are at maximum in the order of 10 mm.
>
> (Brett, 1999)

There are many techniques for transforming and mapping the three-dimensional PET and MRI data onto each other. Various warping techniques include (1) finding standard landmarks and "stretching" the dataset; (2) registering the data to the subject's magnetic resonance image and then transforming the image to match the atlas; (3) warping on the basis of the surface of the brain; and (4) performing a nonlinear three-dimensional warping of brain structures. Each of these methods, of course, has tradeoffs, and is still being debated and adjusted. The resulting compound image (MRI + PET) combines high-resolution anatomical information with quantitative physiological data. Each of these methods trades off precision in one realm for accuracy in another.

The net result is that all the brainsets are rendered comparable to each other and each activity voxel can be located within the atlas and given a more or less precise anatomical location (for instance, in the basal ganglia). Unfortunately, there is disagreement between many labs over the proper reference brain—the Talairach atlas, for instance, was generated from a woman in her 60s who died shortly after having an MRI. Consequently, brain data located on one atlas are not easily comparable with other atlases without significant work (Beaulieu, 2000; Talairach, 1957; Talairach & Tournoux, 1988).

Brainsets often must be normalized to each other in activity levels. In some people, the overall flow in each hemisphere is slightly different. To assist in comparing regions between the two hemispheres, they are often adjusted so that they are of the same average overall activity. Then, because voxel activity measures are dependent on total isotope emission activity, people with higher metabolisms will tend to have higher overall brain blood flow. Because most labs are interested in regional activity and the relative difference in activity in one voxel compared with another, the total overall amount of voxel activity is usually adjusted so that comparisons can be made across individuals. In this case, the absolute activity is defined as not relevant to the study.

Once the brainsets are made into comparable brainsets, the work of extracting significance from them can begin. Significance in PET brain imaging is usually defined as regional differences in activation between two brainsets—for example, the set of voxels corresponding to the basal ganglia are more active in the brainset of an anxious person than in the brainset of the same person when calm. As discussed earlier, these differences can be between the brainsets of an individual doing one task and the same individual doing another one, between two individuals doing the same task, between an individual with a condition (like schizophrenia) and one without, or between two groups of individuals.

In each case, the emphasis is on determining which voxels of activity differ enough between the two brainsets to suggest that the anatomical location of these voxels—a specific brain region—is "involved" in whatever defines the comparison. For example, brainsets of an individual looking at a colored pattern compared with brainsets of the same individual looking at the same pattern in black and white reveal that a set of voxels identified as located in part of the visual cortex had 10% more activity. The suggestion of this data is that part of the visual cortex is "correlated with seeing color" or with "color processing." Because all that can be determined is correlation, this kind of study cannot prove that the brain region is involved or responsible for the function of color processing. Instead, PET scanning is often described as "hypothesis generating" (suggesting brain regions that might be involved in an activity) rather than "hypothesis confirming."

This example also demonstrates that PET must conceptually assume that activation change is significant and represents the "participation" of the "area" (set of voxels) differentially activated in the correlated task. Activation is also conceptual, understood as linear: more is better—more activation means more participation in the function. The corollary of this assumption is that voxels that do not differ between two brainsets are not involved in the task or comparison. In the living brain, all areas are "constantly" active, except the areas that are dead due to, for example, a stroke. All of the neurons are in use, oxygen and glucose are being consumed and neurotransmitters are being released and taken in.

When images are colored (discussed later) only the voxels that differ are given colors, and the other voxels are often rendered black. Comments made to the effect that "no other areas were active" point to the visual and conceptual acceptance of the brainset as the brain. These are shorthand phrases that fill in for "were differentially more active," but they act to reinforce the notion that the other areas of the brain *could* be uninvolved, because they were "off."

Methods of comparing images and determining significance vary from lab to lab. (It may seem tedious to repeat yet another reminder of process differences between labs, but there is no other way to demonstrate the complexity of the interacting layers of assumptions underlying a PET image and how each of these assumptions is not standard within the PET field but contested.) The example just described involved *subtraction*. The value of each voxel in the black-and-white brainset was subtracted from the corresponding voxel in the color brainset. Ideally, most corresponding voxels will have been equal in value, subtracting to zero, implying that the brain activity in that voxel was not affected by the difference between the tasks. The resulting brainset thus highlights those voxels that differed (see color plate 1, top row).

Conceptually, subtracting one image from another assumes that regions of the brain that show no overall change in activity are not directly involved in the task or condition. This is an assumption similar to that with a computer's hardware, where the math coprocessor heats up only when algorithms needing certain functions are run. However, in computers without a math coprocessor, the same functions can be programmed into regular RAM (random-access memory). In this case, there is no overall difference in any particular component when the computer performs the algorithms. The RAM is critically involved in, and directly responsible for, those functions, but it is also involved in database manipulations and Internet surfing at the same intensity. Consequently, an "image" of the latter computer would not detect the role of the RAM program in performing the algorithms because the functional difference is "hidden" as a difference in coding within a constant-use unit, not "present" in a specific, dedicated unit.

A second analogy will further constrain this concept of activation. Assume that we want to detect the top tennis players in a country but are able to measure only general muscle intensity of its inhabitants. We might try to correlate the intensity of activity with tennis tournaments and hypothesize that the top tennis players will be more active during tournaments than not. But what if these tennis professionals also spend every day practicing at great intensity? Then even if they do the work of playing tennis for the country, they will not be detectable through correlation with tennis events. Analogously, we might wonder about regions of the brain that "practice" analyzing patterns of color during the time that they are not actually analyzing new color input.

A different paradigm that competes with the concept of participative activation is that of individual differences and learning. Richard Haier, for instance, designed a study of people playing the computer game Tetris in which scans were done of people (1) just learning to play; (2) as they were becoming more skilled; and (3) when they could play the game consistently at its highest level. Correlating these images, he claims to have found that some specific regions of the brain were more active when learning, then got less active as the person became more skilled, and finally were less active than at rest when the person was playing as an expert. He described this data as conforming to an "efficiency hypothesis," in which a brain region is very active when adapting to a new task and then over time the region becomes very streamlined or efficient at that task and, therefore, needs less and less activity to carry it out. In the final instance, one can imagine the region on a kind of autopilot, less active even than when it is not performing the activity at all (and perhaps participating obliquely in other tasks).

The efficiency hypothesis is useful to highlight the particular design of most cognitive science tasks. They are specifically chosen as those kinds of tasks that people do not tend to get better at. Thus, they are suited to repeating over and over with the same person and—it is hoped—to causing the same response behaviorally and neurologically each time. Equally, they allow many different people to be tested without worrying about how good they are at the task. The functional brain map of cognitive neuroscience tends to be a map of those functions for which there is little or no learning. This abstraction of the range of human functions is common to much of psychology today and has captured much of PET scanner research.

Having clarified the paradigms of isolation of tasks in brainsets of individuals through subtraction, we can now attend to how these results can be combined with each other to produce results in groups. The basic method is one of averaging (see color plate 1, middle and bottom rows). In the case of the color-seeing task, the subtracted brainsets of each of five individuals are normalized to each other as described: their average activity level is altered to the same average, and the brainsets are deformed to the same absolute reference brain atlas. Now the same voxel value in each normalized brainset can be added together and divided by the total number of brainsets to provide the average group voxel value. Repeating for each voxel, the end result is a new "average group brainset."

This average brainset is intriguing because it has conceptualized significant activity as only the subtracted activity that is most common to the set of individual brainsets. Subtracted activity that is common to only one or two is redefined from being potential "individual participative activity" to "noise." This individual variability is often not represented at all in the resultant average brainset, being rendered black. This is intentional. Individual differences are treated as noise in cognitive psychology, whose mission is to discover the baseline mental functions that are common to (most) normal people. What are retained as significant in the averaged brainsets are those regions that can be said to participate in the task in most individuals.

The study of patients, to investigate the recovery of language functions, raises further problems, in particular whether it is appropriate to average patient data. The answer in many cases is likely to be that it is inappropriate ... mixing the results from patients reveals only common features, and individual differences of great potential interest are obscured. However, the comparison of an individual patient's results with grouped normal data, to look for significant regional differences, is a relatively insensitive technique, and one that is open to problems of interpretation—for example, does the patient show a regional difference from normal subjects because of an adaptive change in the neural networks processing the task, or because of an irrelevant stimulus such as discomfort from a full bladder of which the investigator was unaware at the time the patient was studied? Irrelevant stimuli are likely to be randomly distributed amongst a group of normal subjects, and therefore conveniently "lost" during inter-subject averaging.

(Wise, Hadar, Howard, & Patterson, 1991)

Turning back now to the example figure (color plate 1), another conceptual abstraction can be discovered. The five subtracted brainsets each have a fairly lateralized activation in the visual cortex, meaning that the left side is significantly more active than the corresponding right side, or vice versa. The average brainset, however, is prominently bilateral, with both the left and right side of the visual cortex showing high (white) subtracted activity. Thus, the process of averaging here produces a new *quality* in the average brainset that is not present in any of its source brainsets. When I have discussed this image with other brain-imaging researchers, the most common response has been, "Yes, that is right, but if you think that is bad, let me tell you a story ...." The point of their stories is that there are many such inherent but well-known risks in every algorithm. The key is keeping them from ending up in the results section of the journal article, not in keeping them out of the images (see below under "Extreme Images").

Averaging can also be done before subtracting images. A group of brainsets of patients with schizophrenia might be averaged together, and then an averaged brainset from a group of controls without schizophrenia can be subtracted from it. In this case, the differences between the brainsets of the subjects without schizophrenia and the differences between the schizophrenia diagnosed subjects' brainsets are filtered out as noise first, and only the group-shared intensities are subtracted. This result is then interpreted as potentially specific to brains of those diagnosed with schizophrenia.

This is a two-step process. First, the selected (super)schizophrenic patients are scanned and their brainsets averaged, creating an "average schizophrenic subjects–group brainset." Already, the presumption to be able to meaningfully average together a group of subjects with schizophrenia is sliding into the notion of a "schizophrenic brainset." This is to be compared with the "average (super)normals-group brainset," interpreted as a "normal brainset." In the second step, the brainset of subjects without schizophrenia is subtracted from the brainset of subjects diagnosed with schizophrenia, with the result suggesting a "brainset of schizophrenia itself"—that is, the disease is presented as the "only" difference between the two groups, all other difference, it is hoped, having been eliminated as noise. *Difference* between brain images is another one of those words, such as *significance*, whose multiple meanings often ambiguously and productively play off each other. Here the difference (as non-similarity) between the two groups is layered on top of the difference (as the result of an arithmetic subtraction) between the two brainsets. Identifying areas of the averaged brainset that are significant is the province of computational algorithm writers who debate the relative merits of each system.

There are different assumptions built into each kind of statistical algorithm. Most algorithms always highlight one or more brain regions—they choose the highest peaks, for example. As such, they cannot be used to disconfirm the premise that there are active brain regions (Uttal, 2001, p. 185). The fundamental point of contention between different approaches is that there is no other method of proving what significant brain activation should look like. Should a set of voxels be interpreted via a center-of-mass algorithm as Fox described, or using SPM as Friston does, or via a field activation approach as Per Roland argues in *Brain Activation?* (Roland, 1993). At the present time, these are all competing approaches to analyzing brainset data for significance.

Finally, data on individuals from different machines and different institutions can be combined into a large database of "human brain anatomy and function." The Institute of Medicine set up a National Neural Circuitry Database Committee in October 1989 to evaluate how such a database might be constructed. This committee's difficulty with levels of analysis of brain data led to the publication of a set of priorities and recommendations for pilot studies, *Mapping the Brain and its Functions: Integrating Enabling Technologies into Neuroscience Research* (Pechura & Martin, 1991), a book which features four PET scans on its cover. The Human Brain Project is another project funded as a result of this effort, and it includes grants for BrainMap and the Probabilistic Atlas (National Institutes of Health, 1993). BrainMap is a distributed computer program (distributed across several physical locations and connected through the Internet) that integrates information from peer-reviewed studies of the functional brain so they may be cross-referenced by anatomical location (Beaulieu, 2000; Fox & Woldorff, 1994). The Probabilistic Atlas is an attempt to correlate scales of information about the brains of subjects without a medical condition matched for handedness, age, and gender with variability across different populations (Mazziotta, Toga, Evans, Fox, & Lancaster, 1995).

These techniques of averaging, subtracting, and creating a database are both very powerful and very tricky in terms of evaluation and significance. These techniques emphasize similarities across individuals and treat differences between them as "noise" (irrelevant information). They necessarily presume that there is no significant anatomical variability in the functions being studied (Fox & Pardo, 1991). These techniques have been successfully and prominently used in the study of language, for instance, in spite of studies that have shown widespread individual variability:

> Mapping of cortical language sites by stimulation studies of the surgically exposed dominant hemisphere demonstrates that there is tremendous inter-individual variability in the location of essential language areas … Many of these areas fall outside the classically delineated Wernicke's and Broca's areas. Furthermore, any specific zone within Wernicke's or Broca's area was found to be essential for language in less than half of the cases … It is apparent that the variability of language organization is so great that a mapping procedure must be carried out in each individual for whom language localization is important.
>
> (Martin et al., 1990, citing Ojemann, 1979 and G.Ojemann,
> J.Ojemann, Lettich, & Berger, 1989)

The issue of variability is not unaddressed within most PET articles, but it is subordinated to PET's ability to generate statistically significant results. Calling attention to this subordination, one editorial was entitled, "Can Statistics Cause Brain

Damage?" (Ford, 1983). This is not the place to examine critiques of PET statistics but to simply acknowledge that these issues are undergoing lively debate within the corridors, discussions, and appendixes of the PET community. The most significant conceptual concern seems to be whether, and where, PET should be used as inferential (hypothesis confirming) or exploratory (hypothesis generating). Rapoport reports on the "heated" discussion of this issue at a 1989 workshop on PET data analysis, relating to "whether it is better to avoid type I errors (where a statistically significant positive finding proves erroneous) rather than type II errors (where statements of statistical insignificance prove erroneous)" (Rapoport, 1991, A142). Rapoport appeared to lean toward the exploratory use of PET, where results are presented that may be wrong but that can spark further studies.

STAGE 4: PRODUCING INTERPRETED IMAGES

Inferences drawn from qualitative in vivo measurements ... must be viewed with extreme caution despite their intuitive visual appeal. Unfortunately, this sort of inference is the rule rather than the exception

(Perlmutter & Raichle, 1986).

Parenthetically, the [PET scan] pictures that are particularly attractive that you have seen in general are fairly heavily doctored, in the sense of making them more attractive than they should be.

(Michel M. Ter-Pogossian, interview; cited in Dumit, 1995).

Significant, correlated difference, having been determined in the form of voxels, must now be made visible. This dataset of quantitative results can now be mapped onto a spatial coordinate system and displayed on a computer screen as a brainset (Wolf, 1981). Although the resulting image is two dimensional, the brainset is actually three dimensional, where the third dimension is typically represented using color or brightness (see color plate 2).

Peter Galison describes a historical process in which mechanical objectivity—the insistence on the natural transfer of the real objects to image—gives way to an improved object: the interpreted image (Galison, 1997, p. 349). The interpreted image is seen as a more "realistic" process because it can be recognized by non-specialists. "For the image to be purely "natural" was for it to become, *ipso facto*, as obscure as the nature it was supposed to depict" (p. 351).

DUMIT: One of the strengths of PET is that it gives you quantitative data. And at the same time you produce visual, qualitative images. How do these two things work together? Can you read images?

WAGNER: There is a tremendous amount of data. When you say *quantification*, you are talking about numbers, and these spatially oriented studies, these four-dimensional studies, three dimensions in space and one dimension in time, can only be abstracted and displayed in a meaningful way in the form of images. Otherwise there are too many numbers. Your brain can't really handle more than a couple of variables at one time if they are quantitative, so you have to have abstractions. And images are a very, very nice way of abstracting quantification.

(a)

(b)

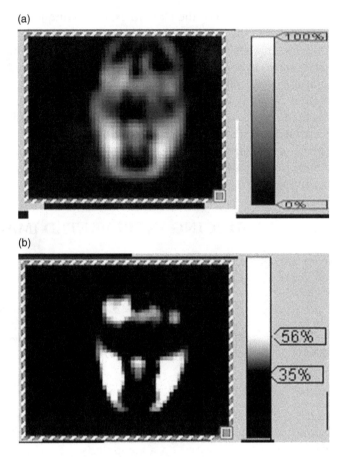

Figure 9.2 Gray scale differences. Figures (a) and (b) have the same numerical data set behind them, but they are colored according to two different tables of black, gray, and white rules. (Screen capture of the Image Viewer applet (ePET) developed by Val Stambolstian, Ph.D., Interactive Media Group, Crump Institute for Molecular Imaging).

Starting with the "normalized" brainset or with the averaged subtracted brainset, the primary problem is how to make sure the reader can understand both the location of the voxels of significance and the meaning of the (relative) activity values. The "simplest" method is to assign each number a shade of gray, starting with black for 0 and ending with white for 100, assuming that the values range from 0 to 100. Because it is not always possible or desirable, however, to present 100 shades of gray-scale, decisions have to be made as to how to group different values together into different shades. This process of grouping is called "windowing," meaning that one range of values (for example, 0–10) will be assigned to black, another range (11–20) to dark gray, and so on (Figure 9.2a). If most of the variation between two images takes place between 40 and 50, however, this will render the two images nearly identical. In this case, the windows can be adjusted so that perhaps 1–35 are black, 35–40 are darkest gray, and most of the variation in color takes place along the bands 41–42, 43–44, 45–46, 47–48, and 49–50, with 50–55 being lightest gray and 56–100 being white (Figure 9.2b). This windowing scheme makes the difference between the two images stand out clearly, and conceptually it makes the close similarity of the two brainsets appear not to be very similar at all. Similar to the way that voxels define a specific scale of spatial resolution and invent brain regions, here the different windows define *activity resolution* and invent a set of discrete *activation levels*, visually eliminating the variability with the levels. Voxels have become *pixels*.

A more elegant solution to the windowing problem is to use colors rather than grayscale. The use of color scales to display differences in intensities in brain images was pioneered by Louis Sokoloff at the National Institutes of Health (NIH). He explained that in digital autoradiography (one of the precursors of PET), the researcher's eye could not see all the shades of gray that could be displayed (Sokoloff, 1986; Sokoloff et al., 1977). Color was introduced to make subtle distinctions visible. This consists of assigning to each subrange of numbers (the full range of which varies from, say, 1–100) a specific color (e.g., 1–10 are black, 11–40 are blue, 41–60 are green, 61–70 are red, 71–100 are yellow). Now the brainset can be presented as a picture, either three dimensionally or by slice. The coloring process is very important, as the final images look very different depending on how they are colored, even if they are based on exactly the same brainset. The data is thus dynamic even after all of the transformations have been accounted for. One effect of colorizing is that new areas appear as discrete and sharply bounded, rather than diffuse.

The effect can be profound. Color is not a simple linear or even two-dimensional array of values. It is best represented by some form of three-dimensional model. Choosing a set of colors to represent linear activity values is therefore an arbitrary choice. Because these colors do not correspond to the real colors of the brain, they are known as pseudo colors. Michel Ter-Pogossian explained it this way in an interview:

> Pseudo-color exaggerates and may distort the information that is in the imaged data. There are a number of color scales, like the heated-object color scale or color mapping that the visual system knows enough about the relation of different colors to each other to be able to say, "Well, that color represents a hotter object than the other colors." So you can order the colors. Whereas how do you order a pseudo-colored object? You can't—you can't tell whether blue is more or less than green. It is a two-dimensional color space anyway, and how you wander around in that color space is not well defined.

As with other aspects of PET, different labs have different preferred color schemes. The ePET applet is revealing in this because in addition to descriptive names for various color schemes such as "Black on White," "Hot Metal," and "Rainbow 1," there is also a location name "UCLA" (see color plate 3).

The debates over various color schemes concern clarity versus a notion of fidelity. Many color schemes, such as the rainbow one, shift from bright to dark to bright again while changing colors. This can create a significant visual shift, rendering a small change in numerical value as a solid boundary between what now appears to be two distinct regions. In this case, the spatialized brain regions of the brainset combine with the activity resolution of the windowing to create a visible "functional anatomy," regions defined as contiguous voxels all having the same color of activity. The arbitrariness of the colors reinforces the sense that these regions are internally coherent, separate from their neighbors, and therefore able to adequately represent the "functioning of the task" in question.

It must be emphasized that the criticism here is part of the aporia of visual representation of data: to make the activity visible in itself to readers, and not simply a representation of activity in general (the way that electroencephalograms often appear), there is a necessary addition of supplementary meaning. PET researchers readily describe their struggles with this problem.

> DUMIT: One of the things that I am interested in is the color pictures in terms of the different things that they can signify. In one case, they can signify that there is a lot of activity going on here.
>
> TER-POGOSSIAN: Well, yes, they signify whatever you want them to signify. This is the pitfall, of course. You can emphasize, for example, a given phenomenon very artificially, if you want to do it with color. It is misleading, too. You have to be very careful when you are using it.
>
> DUMIT: Now when there is purple; that is going outside of the boundaries of the person's head there. Is there any significance to the mottle that is going on?
>
> TER-POGOSSIAN: No, this is noise. That purple, that is noise; this is reconstruction noise. The reason why you see lines is that they are really reconstruction artifacts. And you see that in any reconstruction scheme, including CT scanning. However, very often you erase that by just windowing it out. In other words, these represent very low values, as seen on this scale. So all you have to do is put a cutoff limit and it is removed. But that is what it is. And this, you see, this is a reconstruction artifact. Parenthetically, the pictures that are particularly attractive that you have seen in general are fairly heavily doctored, in the sense of making them more attractive than they should be.
>
>   So to answer your question, no, to the best of my knowledge there is no standardized scale. People have a tendency, of course, to use the scales that emphasize what they like to emphasize.
>
> DUMIT: Yes, I have been struck that each different institution's pictures tend to look very different from each other. It seems very difficult to compare PET scans from different institutions.
>
> TER-POGOSSIAN: It is very difficult. It is very, very difficult indeed. It is misleading to just use purely aesthetic values.

Ter-Pogossian here describes one of the more surprising aspects of the brainset. Despite having fixed numbers for each pixel, the ability to choose a coloring and windowing scheme allows one to use them to "signify whatever you want them to signify." The brainset is thus highly dynamic—so dynamic, in fact, that Brian Murphy, the director of computing and the PET clinical physicist in the Department of Nuclear Medicine at the State University of New York, at Buffalo produced the visually stunning set of images in color plate 4 as a cautionary visual explanation for PET physicians.

> What's the difference between the 40 images [in color plate 4]? Which is normal, which has a tumor, and which has indications of stroke? Actually they're all the same image of a healthy normal volunteer—just displayed with different colour scales. The effects

created by various colour scales may be visually dramatic but may also cause one to see distinct boundaries where there are none. With so much image analysis occurring on the computer, where dialing up any colour scale you like is relatively easy, it is possible to make almost any feature stand out with the right tweaking (affectionately referred to as "dialing a defect"). For this reason, it is important to include a colour scale legend somewhere on these images if they're going to be shared with others so that viewers will have some idea of how the underlying image intensity is being represented (first and last image are presented with a linear ramp gray scale).

Note: The full series of images below appeared on the December 1996 cover of the *Journal of Nuclear Medicine Technology*. One of the motivations for creating these images (aside from their artistic merit) was to illustrate that different "interpretations" are possible for the same image under the simple artificial manipulation wrought by adjusting the colour scale. An additional potential source of interpretation error was added at the time of publication—image orientation. One must be extremely careful when viewing images in an artificial colour scale, especially when they are upside down and left/right reversed.... Pay particular attention to the hot spot at the base of the image and note how it can appear "hot," "cold" or "disappear" depending on the colour scale used.

(Murphy, 1996)

Murphy and Ter-Pogossian both describe the danger in attempting to actually read a PET image out of context. Their discussions highlight the tension between what semiologist Jacques Bertin has referred to as *elementary* (and intermediate) readings of graphic images—in which the image is analyzed internally for relations between elements or groups of elements—and the overall reading, in which the image is apprehended as a whole, a gestalt impression, or in gross comparison to another image. An elementary reading of a PET image, for example, would involve attempting to determine the flow rate for a particular anatomical area, by attempting to read the value for a particular pixel as the flow rate for the voxel. An intermediate reading would involve comparing one hemisphere with another, or the value of an ROI in one image with the same ROI in another. These distinctions in reading practices concern how "technologies of representation" are deployed by scientists and others to build persuasive accounts about the structure of natural and social worlds. This is what Lynch and Edgerton called aesthetics, "the very fabric of realism: the work of discriminating difference, ... and establishing evident relations" (Lynch & Edgerton, 1988). PET is a particularly good case to examine in this regard because the data it provides are so interdisciplinary and expert, yet its images also appear quite convincing to non-experts as well. In addition to color schemes, there are also completely different conventions for representing the data as brain images. The examples in the color plates give an idea of the difficulty of reading images across labs.

Your brain on ecstasy

Especially when these images travel outside of the laboratory and journal articles to the wider public, the dangers of misreading and reinforcing stereotypes is further amplified. For example, four kinds of escalating rhetoric describe the same study of MDMA (3,4-methylenedioxy methamphetamine, or ecstasy) users. In 1998, a PET study was done of 14 people who used MDMA heavily compared with 15 non-users.

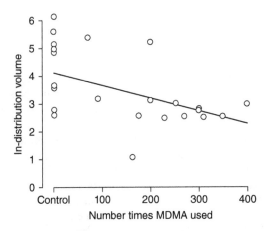

Figure 9.3 Ecstasy user's brain graph (from McCann et al., 1998)

The results were significant both mathematically and socially. The study concluded: "these data suggest that human MDMA users are susceptible to MDMA-induced brain 5-HT neural injury ... Our data do not allow conclusions about reversibility or permanence of MDMA-in-duced changes in brain 5-HT transporter." (McCann, Szabo, Scheffel, Dannals, & Ricaurte, 1998).

Its publication in the *Lancet* was followed by the appearance of a series of letters calling into question the parameters and generalizability of the illustration of how the range of neurotransmitter activity differed among the two groups. Among other problems, the range overlapped so much that if the data were in fact generalizable, one might be able to make a guess but certainly not a diagnosis of someone's past drug use on the basis of a scan (Figure 9.3). Nonetheless, the images published with the study were of one individual from each group, each looking extremely different. It appeared as if they were, in fact, the extremes (see color plate 5).

As the study results traveled outside the medical journal, the stated implications intensified. In the newsletter NIDA (National Institute on Drug Abuse) Notes, staff writer Robert Mathias described the broad outline of the study and discussed the overall results in a competent manner as a correlation. In the caption to the image included with the article, though, he claimed it showed causality: "Dark areas in the MDMA user's brain show damage due to chronic MDMA use."

When the images travelled to the US Senate Caucus on International Narcotics Control, however, the director of the NIDA used them to proclaim not only absolute causality but diagnostic ability as well: "Through the use of positron emission tomography (PET), we can actually see that the brain images on top belongs to an individual who has never used MDMA ... Clearly the brain of the MDMA user on the bottom has been significantly altered." At this point, the extreme images of two people were being used to ground a strong biosocial claim, not to illustrate a weak one, as they did in the *Lancet* article. It should come as no surprise, then, that the NIDA 25-year poster took the further step of creating a single didactic image about the ecstasy brain images (see color plate 6). Building on the "your brain on drugs" campaign, this poster visually argued that twisted drugs lead to twisted brains (and therefore to a twisted self). The poster also went a step too far. As if the choosing of color scales, windowing ranges, and extreme images is not enough, in combining the right half of a "normal brain" with the left half of a "brain on ecstasy," the graphic artists actually inverted the color scale of purple and black. The result is an even more stunning and tragic looking drug-ravaged brain, but at the expense of putting forth a visual lie.

The ease with which the Brain on Ecstasy images could be misinterpreted was a source of discomfort amongst most PET researchers. They hated this fact, which at best often led to misconceptions, at worst getting them dragged into court. I therefore concentrated my attention on the difficult decisions that had to be made in presenting images.

Extreme images

Once the data have been condensed into a series of images and analyzed, the researchers must decide which images to publish. In the following discussion, a researcher comments on the process of using PET images in his own articles. The image is one part of an argument that necessarily includes a textual component.

> DUMIT: When you do an article on PET for a journal …
>
> TER-POGOSSIAN: I'm working on one right now!
>
> DUMIT: … and you are trying to select images for the article to demonstrate one way or another what is going on. It looks to be the case—and I can't tell because often there is not that much information presented about why these two images were chosen—it looks to be the case that the most extreme images are chosen.
>
> TER-POGOSSIAN: Sure
>
> DUMIT: I'm curious about this. Is this a kind of heuristic idea, that these images display the difference that is being talked about? Are they representative?
>
> TER-POGOSSIAN: Well, it varies. It depends on how you show the images. For example, if indeed you want to emphasize a difference, you show the extreme cases. However, in any responsible article, it behoves you to emphasize also the overlapping areas— and these are in any kind of study that involves, say, the comparison of something or another—it behoves one to use a statistical analysis. In most instances we have a statistician on the staff and we [ask] him, "How do we present the data?" And he in general has his own approach; I'm not a very good statistician myself. And he gives you that data.
>
> But to get back to what you are saying, very often indeed, in most instances you are going to select images that emphasize your case, sure. But also, you might, if you so wish, show images that on the contrary show a false positive. It depends on what you want to do. But yes, you select the images that prove your case. However, the case is also proven, supposedly, in your text.

Ter Pogossian emphasizes how extreme images—images that look the most different from each other—may be used to imply that there are significant differences that are demonstrated in the text. Alternately, an image may be used to imply that, in spite of a significant finding, there remains a strong possibility of mistaking a normal case for an abnormal one. In spite of this, as he indicates, extreme images are often used.

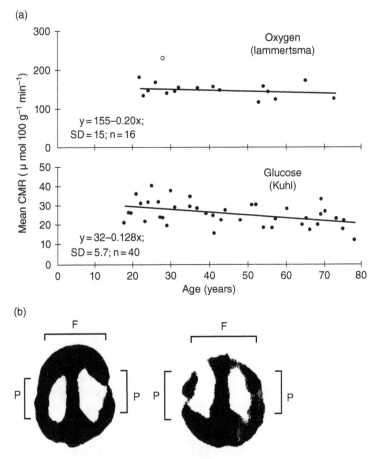

Figure 9.4 Aging Graph. Graph showing decline in cerebral glucose utilization (CMRglu) with age is the same in mean overall cortex, caudate-thalamus, and white matter. Each data point represents the average measurements from five normal subjects. Error bars represent one standard deviation. (From Kuhl et al., 1982)

For example, in looking at PET images in a scientific article, I was struck by the way extreme images were presented as iconic proof of significance in an experiment. In this experiment, an attempt to measure the effects of ageing on the brain, forty normal volunteers, aged 18–78 years, were scanned with PET (Figure 9.4a). A series of graphs accompanied the article produced as a result of this experiment. The caption reads: "The degree of metabolic hyperfrontality varies considerably among normal subjects, but on average declines gradually with age" (Kuhl, Metter, Riege, & Phelps, 1982). The graphs show that although the averages for groups of five subjects does decline, the typical variation for any age category actually overlaps the averages of every other category. In other words, given another PET scan of an unknown subject near any of the averages, there would be no basis for deciding which age category that person belongs in.

In spite of this constraint, exactly two PET images are presented in the text that look quite different from each other; one is of a 27-year-old and the other is of a

75-year-old (Figure 9.4b). They were chosen, not because they represented the youngest and the oldest in the set, nor because they were the average, but because they were the most extreme cases, "the extremes of [the] ratio" (Kuhl et al., 1982). In this case, the two most visibly different images of a set are presented as if representative of two different types of brains. I asked one of the researchers about this:

> DUMIT: In the article, there are only two images shown, and it says underneath that these images were chosen because they were the most extremely different. Is that a standard practice, to choose the most extreme images rather than, say, the average for each?
>
> PHELPS: Yes. What is maybe not so common a practice is to point out that you did that ...
>
> DUMIT: Right.
>
> PHELPS: Well, yes. If you are honestly and forthrightly trying to show something in the article, you try [to] take the data and the images and process them to point that what you know to be true you can see. So we take the extreme cases for the readers to be able to see them. You have the tabulated data to look at all cases. It is fine.

Embedded in his explanation is a twofold critique: on the one hand, having carried out the experiment, the expert knows that there is a significant finding in the data. He or she can see it in the numbers, yet others, non-experts, cannot. The expert, however, can produce a picture using some of the data, that illustrates what the data as a whole show—an ideal to represent a statistical trend. On the other hand, this researcher is careful to note a potential abuse lurking in this practice; that is that the part may be taken for the whole. In this case, without the careful caption and without the accompanying data graphs, it would be easy to conclude that younger brains are simply quite distinguishable from older ones. It should be noted that though choosing to print extreme images appears to be standard practice, in practice such a choice is almost never stated. Researcher Richard Haier concurred:

> We always publish group statistical data—usually analysis of variance, sometimes multiple T-tests. That is always reported in detail in the paper. Our conclusions are based on the statistics. Most of the time, although not all of the time, we include a color picture, because journals like color pictures, everybody likes color pictures—and that is what they remember. When we do that, we select images that illustrate the group statistical finding. It is not the other way around. So the picture that was in *Newsweek*, I just took the person with the highest score and the person with the lowest score [see color plate 7]. And it looked so compelling, but that's what the effect was, that is why it was so compelling. I took the best exemplar, I took the best pair, to exemplify that. That is true. But I don't see anything wrong with that.

The images presented in these popular and scientific articles are not then to be carefully interpreted pixel by pixel. The displayed images should not be measured; they are not meant to be. Rather, they are consciously selected to enhance the textual argument. They are crafted to underpin, teach, and illustrate the process of discursive and statistical persuasion. One researcher has commented that:

> Functional information is communicated very approximately by images and requires quantification to be meaningful. Thus the imaging capabilities of PET, which derive

from the mode of data collection, can at best serve as an aide memoir, or illustration, of much more detailed data pertaining to a variety of cerebral functions.

(Frackowiak, 1986, p. 25)

Despite such qualifications, however, it is precisely these simplified "illustrations" that are valorized when these images travel from the laboratory into articles and into popular culture. In textbooks, as well, extreme images can have cultural effects. Used as illustrations of types of brains, these images become "classic" expressions of pathology, or "textbook images." In the "Chairman's Corner," an editorial spot in the journal *Investigative Radiology*, Melvyn Schreiber comments on social conditioning with which medical students learn to identify "beautiful" pictures:

> We don't mean that it's pretty but rather that it is exquisitely representative of the classic expression of the disease. When our mental conception of the textbook picture of an abnormality is reproduced perfectly in life, we describe the image as beautiful, largely out of appreciation for its verisimilitude and partly out of recognition of the ease with which the diagnosis can be made when all of the necessary elements are present and recognisable, as they so rarely are.
>
> (Schreiber, 1991, p. 771)

The verisimilitude that Schreiber refers to is the fidelity of the observed image to the textbook image. The practice of producing extreme images is also encouraged by regulatory agencies and pharmaceutical companies. One researcher who had worked at a large pharmaceutical company, for instance, told of how he had to search "until the ends of the earth" to find two images, one normal and one pathological, which could clearly show the difference to the FDA. Another researcher commented on these and other popular uses of difference images:

> Well, we put a lot of emphasis in trying to get pharmaceutical contracts when we started. And I think it was our experience in general that they weren't terribly interested. They were only interested at certain stages of development. If they thought it would help them get through the FDA, then they would be interested, but we found that most of our pharmaceutical contracts really came through the PR departments, the advertising departments, not through the science departments. And they were after pretty pictures to put in the ads, which apparently worked, and worked well.

I cite these examples in order to demonstrate the persuasiveness of this visual practice of exemplary images whose purpose is easy recognizability (in spite of the rarity of such recognition in practice), yet whose function is often one of proof of difference.

The risk that these pictures pose, I am arguing, lies in their multivocal readings. They are both veridictory (evidentiary) graphs and emphatic illustrations (see Greimas & Courtes, 1982). This risk appears in stark outline in courtrooms (see Dumit, 2000; Dumit, 2004, chapter 4), where the exemplary images of the most normal and most abnormal can be transformed into types, into typical representatives of normal and abnormal, to "make clear the difference." Such a process, although scientifically and legally sanctioned, risks making it appear as if one could go from single scan to diagnosis, from picture to text.

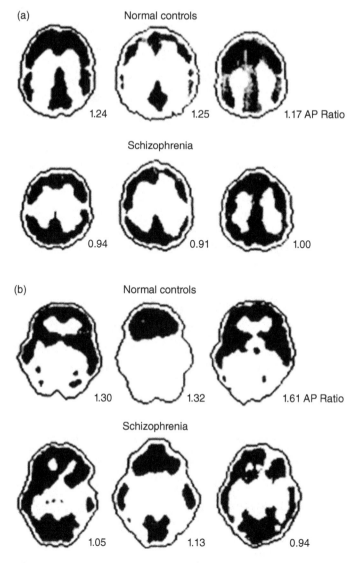

Figure 9.5 Schizophrenia extremes. PET supraventricular slices (a), and PET intraventricular slices (b), for three subjects without schizophrenia and three patients with schizophrenia. (From Buchsbaum et al., 1985)

PET as a difference engine

In the presentation of the search for biological correlates of schizophrenic diagnoses, this collapse of scan to diagnosis seems to predominate when the correlates are located in the brain. Even though research since the 1970s has shown many relationships between this diagnosis and symptom relief through pharmacological treatment, visual presentation of "schizophrenia" seems to promise much more (Buchsbaum, DeLisi, Holcomb, Hazlett, & Kessler, 1985). In Figure 9.5, taken from a book chapter on functional imaging, again the brain images shown are the most extreme, leaving a visual sense of clear differentiation between people with schizophrenia and people without, even though there are many people diagnosed with schizophrenia whose brains look like those of people without, and people without schizophrenia whose brains look like those of people with it (on this sort of problem, see Nelkin & Tancredi, 1989). Significant in terms of the virtual community of images is the way in which, though the brain scans of the volunteers without schizophrenia are labeled *normal controls*, the brain scans of the those diagnosed with schizophrenia are labeled *schizophrenia*. The image is thus labeled as showing the "disease" itself, rather than a correlate symptom of someone found to have schizophrenia. Hence, the symptom has been collapsed into the referent.

The collapse of symptom into referent in this article should give us pause. Just because we think schizophrenia is in the brain does not mean that an experimentally discovered correlation in the brain is the cause and, therefore, the thing itself. This is an example of what critical neuroscience, following Honneth, calls a "social pathology of reason." Our sense of the truth, our cultural contexts, allows us to short-circuit careful scientific reasoning and declare that we do not need to ask more questions, such as: what might cause that correlation? Briefly consider a counterexample: even if we found a large correlation between schizophrenia and bloodflow in the big toe, we would not state that schizophrenia was in the toe, we would immediately get to work discovering if the correlation was spurious, and if not, what caused it!

As facts "loop" between labs and the wider culture, this particular process of extreme images has the effect of reinforcing rather than challenging our assumptions. A critical neuroscience therefore must attend not only to making sure that the text is careful but that the images are too. One way this might be done is not only to show two extreme images, but to put next to them average or mean images from the two groups. In most cases, the group average images are almost indistinguishable from each other. Showing this might go a long way towards making it clear to readers that though there is a statistical result worth pursuing, that the two groups nonetheless overlap quite a bit.

References

Andreasen, N. C. (1997). Linking mind and brain in the study of mental illnesses: A project for a scientific psychopathology. *Science, 275*(5306), 1586–1593.

Beaulieu, A. (2000). *The space inside the skull: Digital representations, brain mapping and cognitive neuroscience in the Decade of the Brain.* Amsterdam: University of Amsterdam.

Blumenthal, D. (1982). Image processing advanced techniques enable scientists to pursue new studies. *NIH Record, 24*(8), 11.

Brett, M. (1999). *The MNI brain and the Talairach atlas,* 2001. Retrieved from http://www.mrc-cbu.cam.ac.uk/Imaging/mnispace.html.

Brodie, J. D., Wolf, A. P., Volkov, N., Christmann, D. R., DelFina, P., DeLeon, M. et al. (1983). Evaluation of regional glucose metabolism with positron emission tomography in normal and psychiatric populations. In W.-D. Heiss & M. E. Phelps (Eds.), *Positron emission tomography of the brain.* Berlin: Springer-Verlag.

Buchsbaum, M., DeLisi, L., Holcomb, H., Hazlett, E., & Kessler, R. (1985). Cerebral glucography in schizophrenia. In T. Greitz, D. Ingvar & L.Widen (Eds.), *Metabolism of the human brain studied with positron emission tomography* (p. 471ff.). New York: Raven Press.

Canguilhem, G. (1989). *The normal and the pathological* (C. R. Fawcett & R. S. Cohen, Trans.). New York: Zone Books.

Diehl, D. J., & Mintun, M. A. (1995). Morning versus midday differences in baseline regional cerebral blood flow in healthy volunteers demonstrated by PET (abstract). *Journal of Nuclear Medicine, 38*(5), 128P.

Dumit, J. (2004). *Picturing personhood: Brain scans and biomedical identity.* Princeton: Princeton University Press.

Dumit, J. (2000). When explanations rest: "Good-enough" brain science and the new sociomedical disorders. In M. Lock, A. Young & A. Cambrosio (Eds.), *Living and working with the new biomedical technologies: Intersections of inquiry* (pp. 209–232). Cambridge: Cambridge University Press.

Dumit, J. (1995). Twenty-first-century PET: Looking for mind and morality through the eye of technology. In G. E. Marcus (Ed.), *Technoscientific imaginaries: Conversations, profiles, and memoirs* (Vol. 2, pp. 87–128). Chicago: University of Chicago Press.

Fodor, J. (1999, Sep 30). Diary. *London Review of Books, 21.*

Ford, I. (1983). Can statistics cause brain damage? (Editorial). *Journal of Cerebral Blood Flow and Metabolism, 3,* 259–262.

Fox, P., & Pardo, J. (1991). Does inter-subject variability in cortical functional organization increase with neural "distance" from the periphery? *Ciba Foundation Symposium, 163,* 125–140; discussion 140–124.

Fox, P., & Woldorff, M. (1994). Integrating human brain maps. *Current Opinion in Neurobiology, 4*(2), 151–156.

Frackowiak, R. S. J. (1986). An introduction to positron tomography and its application to clinical investigation. In M. R. Trimble (Ed.), *New brain imaging techniques and psychopharmacology* (pp. 25–34). Oxford: Oxford University Press.

Frith, C. (1991). Positron emission tomography studies of frontal lobe function: Relevance to psychiatric disease. *Ciba Foundation Symposium, 163,* 181–191; discussion 191–187.

Galison, P. L. (1997). *Image and logic: A material culture of microphysics.* Chicago: University of Chicago Press.

Greimas, A.-J., & Courtes, J. (1982). *Semiotics and language: An analytical dictionary.* Bloomington: Indiana University Press.

Jones, C. A., Galison, P. L., & Slaton, A. E. (Eds.). (1998). *Picturing science, producing art.* New York: Routledge.

Kosslyn, S. M. (1994). *Elements of graph design.* New York: W. H. Freeman.

Kuhl, D., Metter, E., Riege, W., & Phelps, M. (1982). Effects of human aging on patterns of local cerebral glucose utilization determined by the [18F]fluorodeoxyglucose method. *Journal of Cerebral Blood Flow and Metabolism, 2*(2), 163–171.

Lynch, M., & Edgerton Jr., S. Y. (1988). Aesthetics and digital image processing: Representational craft in contemporary astronomy. In G. Fyfe & J. E. Law (Eds.), *Picturing power: Visual depiction and social relations* (pp. 184–220). London: Routledge.

Martin, N., Grafton, S., Vinuela, F., Dion, J., Duckwiler, G., Mazziotta, J., & Becker, D. (1990). Imaging techniques for cortical functional localization. *Clincal Neurosurgery: Congress of Neurological Surgeons, 38.*

Mazziotta, J., & Phelps, M. E. (1985). Human neuropsychological imaging studies of local brain metabolism: Strategies and results. *Research Publications: Association for Research in Nervous and Mental Disease, 63,* 121–137.

Mazziotta, J., Phelps, M., Plummer, D., & Kuhl, D. (1981). Quantitation in positron emission computed tomography: 5. Physical–anatomical effects. *Journal of Computer Assisted Tomography, 5*(5), 734–743.

Mazziotta, J., Toga, A., Evans, A., Fox, P., & Lancaster, J. (1995). A probabilistic atlas of the human brain: Theory and rationale for its development. The International Consortium for Brain Mapping (ICBM). *Neuroimage, 2*(2), 89–101.

McCann, U. D., Szabo, Z., Scheffel, U., Dannals, R. F., & Ricaurte, G. A. U. (1998). Positron emission tomographic evidence of toxic effect of MDMA ("Ecstasy") on brain serotonin neurons in human beings. *The Lancet, 352*(9138), 1433–1437.

Murphy, B. (1996). *Colour scales: Dialling a defect.* Retrieved from http://www.nucmed. buffalo.edu/nrlgy1.htm.

Nelkin, D., & Tancredi, L. (1989). *Dangerous diagnostics: The social power of biological information.* New York: Basic Books.

National Institutions of Health (NIH). (1993). *The human brain project: Phase I feasability studies* (Vol. PA-93–068).

Ojemann, G. (1979). Individual variability in cortical localization of language. *Journal of Neurosurgery, 50,* 164–169.

Ojemann, G., Ojemann, J., Lettich, E., & Berger, M. (1989). Cortical language localization in left, dominant hemisphere: An electrical stimulation mapping investigation in 117 patients. *Journal of Neurosurgery, 71,* 316–326.

Pechura, C. M., & Martin, J. B. (Eds.). (1991). *Mapping the brain and its functions: Integrating enabling technologies into neuroscience research.* Institute of Medicine (IOM). Committee on a National Neural Circuitry Database. Washington, DC: National Academy Press.

Perlmutter, J., & Raichle, M. (1986). In vitro or in vivo receptor binding: Where does the truth lie? (Editorial). *Annals of Neurology, 19*(4), 384–385.

Phelps, M., & Mazziotta, J. (1985). Positron emission tomography: Human brain function and biochemistry. *Science, 228*(4701), 799–809.

Raichle, M. E. (1994). Visualizing the mind. *Scientific American, 270*(4), 58–64.

Raichle, M. (1990). Anatomical explorations of mind: Studies with modern imaging techniques. *Cold Spring Harbor Symposia on Quantitative Biology, 55,* 983–986.

Rapoport, S. (1991). Discussion of PET workshop reports, including recommendations of PET Data Analysis Working Group. *Journal of Cerebral Blood Flow and Metabolism, 11*(2), A140–146.

Reiman, E. (1988). The quest to establish the neural substrates of anxiety. *Psychiatric Clinics of North America, 11*(2), 295–307.

Reiman, E. M., Fusselman, M. J., Fox, P. T., & Raichle, M. E. (1989). Neuroanatomical correlates of anticipatory anxiety. *Science, 243*(4894), 1071–1074.

Roland, P. E. (1993). *Brain activation.* New York: Wiley-Liss.

Rosser, S. V. (1994). *Women's health–missing from U.S. medicine.* Bloomington: Indiana University Press.

Schreiber, M. H. (1991). Ugly organs. *Investigative Radiology, 26*(8), 771.

Sokoloff, L. (1986). Basic principles in the imaging of rates of biochemical processes in vivo. In O. Hayaishi & K. Torizuka (Eds.), *Biomedical Imaging* (p. 183). Tokyo: Academic Press, Inc.

Sokoloff, L., Reivich, M., Kennedy, C., Des Rosiers, M., Patlak, C., Pettigrew, K., Shinohara, M. (1977). The [14C]deoxyglucose method for the measurement of local cerebral glucose utilization: Theory, procedure, and normal values in the conscious and anesthetized albino rat. *Journal of Neurochemistry, 28,* 897–916.

Szasz, T. S. (1996). *The meaning of mind: Language, morality, and neuroscience.* Westport, CT: Praeger.

Talairach, J. J. (1957). *Atlas d'anatomie stéréotaxique: Repérage radiologique indirect des noyaux gris centraux des régions mesencephalo-sous-optique et hypothalamique de l'homme.* Paris: Masson.

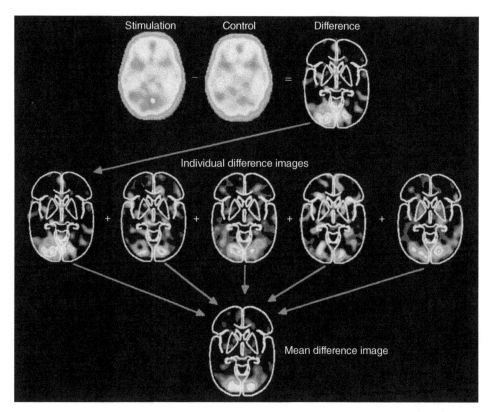

Plate 1. PET scans illustrating the subtraction and averaging processes (Posner & Raichle 1994).

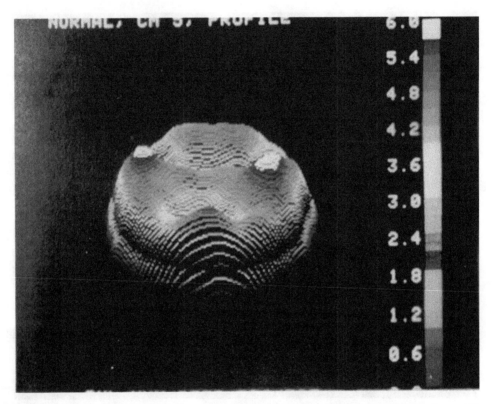

Plate 2. Three-dimensional PET scans of "normal" and "schizophrenic brains". See p. 90 (Wolf 1981a).

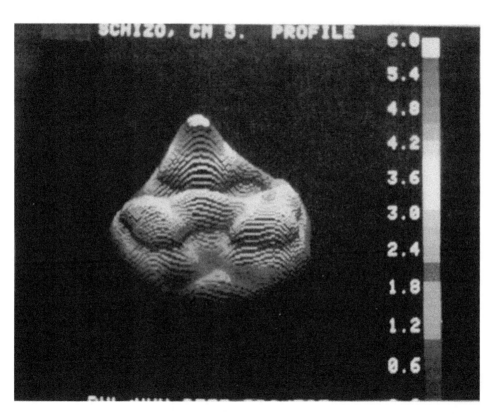

Plate 2.   *(cont'd)*

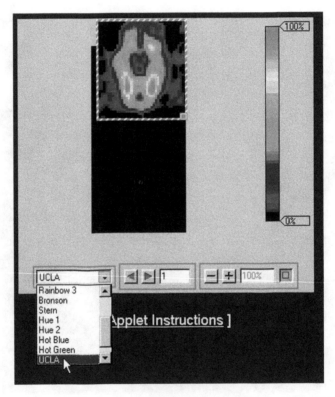

Plate 3. Screen capture of the Image Viewer applet (ePET) developed by Val Stambolstian, Ph.D., Interactive Media Group, Crump Institute for Molecular Imaging).

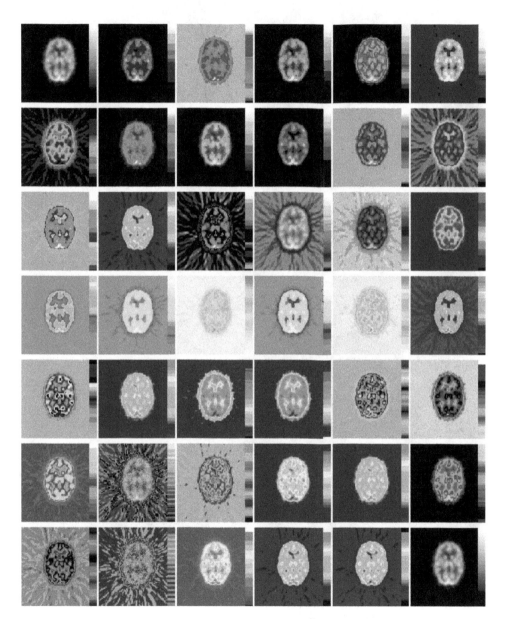

Plate 4. Identical PET scans illustrating pseudo-color choices (courtesy of Brain Murphy).

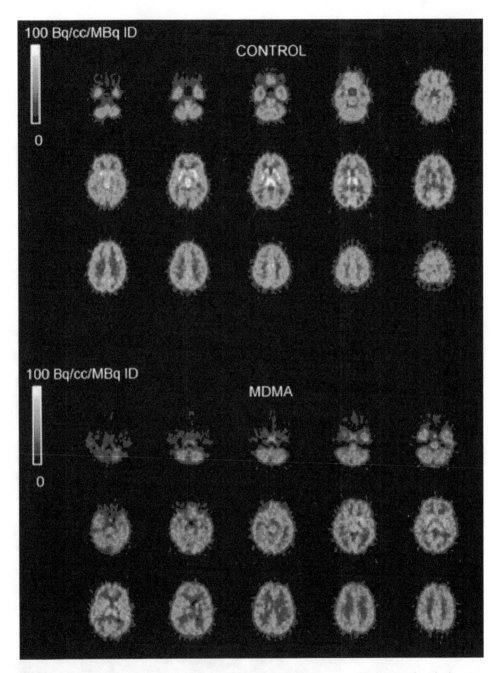

Plate 5. PET scan of the brain of a heavy user of MDMA ("ecstasy") compared with the scan of a normal control subject (McCann, Szabo, Scheffel, et al. 1988).

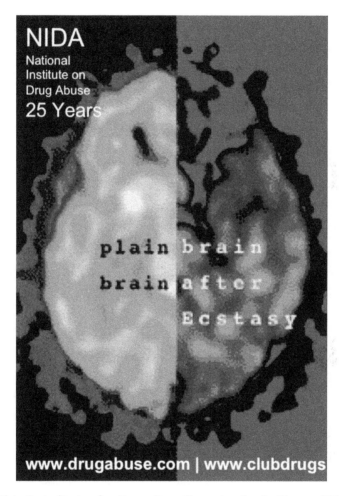

Plate 6. "Plain Brain/Brain after Ecstasy"; an illustration for the Twenty-fifth Anniversary Poster NIDA (National Institute of Drug Abuse).

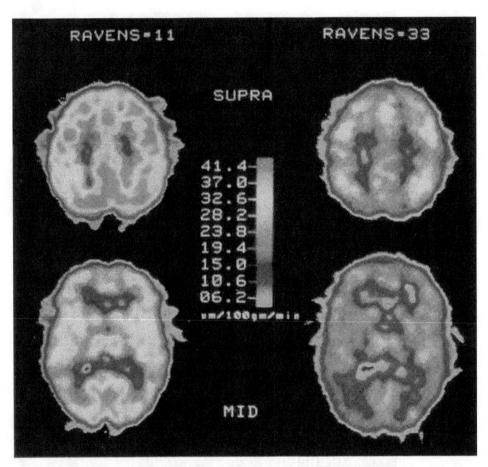

Plate 7. PET scans of the Ravens Advanced Matrices intelligence test, from Sharon Begley's "How to Tell if You're Smart – See Your Brain Light Up," in Newsweek 1988 (vol. 64).

Talairach, J. J., & Tournoux, P. (1988). *Co-planar stereotaxic atlas of the human brain: A 3-dimensional proportional system, an approach to cerebral imaging.* Stuttgart: G. Thieme Medical Publishers.

Uttal, W. R. (2001). *The new phrenology: The limits of localizing cognitive processes in the brain.* Cambridge, MA: MIT Press.

Wilson, E. A. (1998). Neural geographies: Feminism and the microstructure of cognition. New York: Routledge.

Wise, R., Hadar, U., Howard, D., & Patterson, K. (1991). Language activation studies with positron emission tomography. *Ciba Foundation Symposium, 163,* 218–228; discussion 228–234.

Wolf, A. P. (1981). PETT–Quantitative in vivo measurement of human function and metabolism. In C. G. Lunsford & Philip Morris Incorporated (Eds.). *Frontiers of analytical techniques and their application: Proceedings of the fourth Philip Morris Science Symposium, Richmond, Virginia, October 29, 1981.* New York: Philip Morris.

Wu, J. C., Buchsbaum, M. S., Hershey, T. G., Hazlett, E., Sicotte, N., & Johnson, J. C. (1991). PET in generalized anxiety disorder. *Biological Psychiatry, 29*(12), 1181–1199.

10

Radical Reductions

Neurophysiology, Politics and Personhood in Russian Addiction Medicine

Eugene Raikhel

Several years ago a Russian research group carrying out an experimental treatment for opiate addiction caught the attention of English-language readers, both in the popular media (Whitehouse, 1999) and in professional publications (Orellana, 2002). Since 1998 this group led by Svyatoslav Medvedev, director of St. Petersburg's Institute for the Human Brain, had carried out several hundred stereotactic neurosurgeries on addicts. The procedure involved the ablation of the cyngulate gyrus, a brain region which has been associated by biological psychiatrists and neuroscientists with Obsessive Compulsive Disorder (Medvedev, Anichkov, & Polyakov, 2003). The practice caused significant controversy in Russia. When a patient filed suit against the institute in 2002, charging that after he had paid $6,000 for the surgery, he had developed headaches, the Russian state temporarily stopped the procedures (Orellana, 2002).

Writing in the journal *Addiction*, University of Queensland bioethicist Wayne Hall (2006a), condemned the stereotactic surgery, along with similar ones taking place in China, as a misguided attempt at a "great and desperate cure," and as an extension of a punitive system of drug addiction treatment. Hall argued that "The addictions field will need to speak with a united voice if we are to ensure that neurosurgical treatment of addiction is not introduced into developed countries by enthusiastic private practitioners without formal evaluation, as purported 'cures' for heroin addiction all too often are," (2006a, p. 2). In a separate editorial Hall used the Russian example as an illustration of the "potentially less welcome social consequences of the 'brain disease' model of addiction," (2006b, p. 1529; see also Hall, Carter, & Morley, 2004). While ostensibly the object of Hall's critique, the neurosurgical procedures taking place in St. Petersburg served more of an illustrative and cautionary purpose than anything else. Hall's argument was geared towards the potential ethical and social pitfalls and consequences of neuroscientific research into addiction in what he called "developed countries."

Critical Neuroscience: A Handbook of the Social and Cultural Contexts of Neuroscience, First Edition.
Edited by Suparna Choudhury and Jan Slaby.
© 2012 John Wiley & Sons, Ltd. Published 2016 by John Wiley & Sons, Ltd.

Underlying Hall's argument is the common assumption that the different forms which neuroscience and its interventions have taken in different countries consist primarily in institutional conditions such as levels of research funding, the influence of commercial interests in science, government policy and regulation of clinical research, and the state of the infrastructure for ethical oversight. While such differences are extremely important, in this chapter I suggest that they are inextricable from a broader set of distinctions and particularities. I am referring to the multiple elements—including, but not limited to, institutional conditions—which shape what have been called styles of reasoning or epistemic cultures of science and clinical medicine.[1] A critical neuroscience which seriously hopes to bring the tools of the social sciences to bear on issues addressed by neuroethics, would do well to pay close attention to the ways in which distinct styles of reasoning take shape, as well as to how knowledge is translated between them.

While I attempt just such an analysis in this chapter, I should add that my aim is not to provide a thorough explanation for the phenomenon of stereotactic surgery in contemporary Russia per se. Such an account would require one to trace the history of neurosurgery in Russia (for example, Lichterman, 1998), to examine the development of the technical means which have made such surgery feasible, and so on. However, because stereotactic surgery remains a relatively unusual and largely experimental intervention for drug addiction in Russia, I take a somewhat different approach, focusing more broadly on the development of Russian addiction medicine. My account draws on ethnographic fieldwork carried out in a number of clinical institutions in St. Petersburg, Russia, devoted to the treatment of substance dependence, as well as on readings of the Soviet and Russian medical literatures on addiction.[2] I examine how Russian addiction medicine has been shaped by a clinical style of reasoning specific to a Soviet and post-Soviet professional psychiatry, itself the product of contentious Soviet intellectual and institutional politics over the knowledge of the mind and brain. I argue that whereas psychosocial explanations and interventions played a central role in governing addiction in Western Europe and North America for

---

[1]   Throughout this article, I use the notion of "style of reasoning" drawn from the work of Ludwik Fleck (1979) by Ian Hacking (1992). As Allan Young describes it concisely, a style of reasoning "is composed of ideas, practices, raw materials, technologies and objects … It is a characteristically self-authenticating way of making facts, in that it generates its own truth conditions," (2000, p. 158).

[2]   Although the account given in this chapter draws heavily on textual sources, this project was not prompted by research questions emerging from the history of medicine literature, but by an ethnographic investigation into contemporary Russian addiction medicine. The fieldwork was conducted over 14 months between 2002 and 2004 in a number of addiction treatment facilities in St. Petersburg, Russia. In addition to interviewing and interacting informally with patients and physicians in the municipal Narcological Service, I conducted fieldwork at one commercial addiction clinic and a charitable 12-step-based rehabilitation center. I also interviewed several narcologists and psychiatrists in private practice, sat in on a series of training lectures on narcology for physicians, attended open sessions of Alcoholics Anonymous, and observed séances conducted by a self-proclaimed "Orthodox psychotherapist." Finally, I conducted extensive textual research in the Russian-language scientific and medical literature on addiction and its treatment.

The project was approved by the IRB of my home institution at the time (Princeton University) as well as by the St. Petersburg Department of Public Health. In order to ensure the confidentiality of informants, all of the patients and physicians whom I interviewed in the Narcological Service have been given pseudonyms or general appellations, and some identifying details have been changed.

much of the twentieth century, Soviet addiction medicine was based on a very particular biomedical model, which claimed its origins in Pavlov's physiology of reflexes. This model helped to shape the prominence of therapeutic methods for alcoholism based on mechanisms of aversion and suggestion. Finally, in light of recent arguments about the relationship between neurobiological knowledge and personhood, I examine how the treatment methods which emerged from Russian addiction medicine affect (or more precisely, fail to significantly affect) patient's self-identifications.[3]

From "Diseased Wills" to "Hijacked Brains"

As many of the chapters in this volume attest, the past 20 years have seen not only the biologization of psychiatry, but its reframing as a "clinical neuroscience discipline" (Insel & Quirion, 2005). To some, this has meant that an increasingly prevalent assumption in psychiatry is that "a successful theory of the mind will be a solely neuroscientific theory" (Gold & Stoljar, 1999, p. 809). Moreover, coinciding as it has with the neoliberal transformation of health care in many countries and a consequent search for "cost-effective" therapies, this cardinal shift in psychiatry has also facilitated a growing emphasis on pharmaceutical interventions for mental illnesses (Healy, 1997; Luhrmann, 2000; Shorter, 1998). While some critics have interpreted these developments as examples of normalization or social control, others have argued that in order to adequately assess the changes associated with the biosciences of the late twentieth and early twenty-first centuries, scholars in the human sciences must develop new categories for description, new concepts for analysis, and new modes of intellectual intervention (Rabinow & Rose, 2006). For example, Nikolas Rose and others have argued (2003; 2007; Vidal, 2009) that as neurobiological ways of thinking about and acting upon human beings diffuse beyond the laboratory, a somatic understanding of the self is increasingly displacing—or at least being layered onto—the psychological or identity-based subject of the twentieth century. Further, this "neurochemical personhood" or "brainhood" is associated with a characteristically neoliberal way of governing pathological behavior, the model of which is the individual who internalizes functions once carried out by a sovereign state or social institutions and assumes responsibility for the management of his or her own susceptibilities and desires (Rose, 2003).

Such arguments—which ascribe an epochal shift in personhood and self-governance to neuroscience and psychiatry largely in relation to contemporary Anglo-American societies—need to be tempered through close attention to particular cases. Ethnographic studies have shown the multiple ways in which the meanings, uses, and effects of psychiatry's diagnostic categories and clinical interventions change, as they are translated from bench to bedside or from one cultural setting to another (Gaines, 1992; Kleinman, 1982; Young, 2000). Nosological categories (most often those of the DSM), psychopharmaceuticals, and their attendant ways of understanding mental

---

[3] In making this argument I neither intend to equate accounts of addiction produced by mid-twentieth century Soviet neurophysiology and those of contemporary Anglo-American neurobiology or to suggest that Soviet addiction medicine was entirely isolated or disconnected from English-language literatures on addiction.

distress, have moved far beyond the North American and European settings in which they were developed, articulating in varying ways with differences in national and local styles of psychiatric reasoning (Kitanaka, 2008; Lakoff, 2006; Lee, 1999; Lloyd, 2008), domestic economies (Biehl, 2004) and institutional and political economic conditions (Ecks & Basu, 2009; Jain & Jadhav, 2009). Even in the post-industrial societies where brain-based modes of conceptualizing and managing distress have become ubiquitous, they have continued to mesh in unforeseen ways with lay explanatory models, forms of identification and sociality (Dumit, 2003; Martin, 2007), and longstanding patterns of marginalization (Oldani, 2009). Moreover, while neurobiology is often viewed as providing a basis for an integrative framework in psychiatry, evidence suggests that most psychiatrists continue to think about clinical cases in largely dualistic terms (Miresco & Kirmayer, 2006).[4]

In the case of pathological consumption of psychoactive substances, the story has been somewhat more complex, at least in the English-speaking world. Social scientists and historians have observed that even as many other forms of human suffering have been medicalized and increasingly biologized, practices such as problematic drinking and drug use have resisted complete subsumption under the aegis of biomedicine or psychiatry (May, 2001; Valverde, 1998). Complementary and often competing strategies of governance have been applied to the practices themselves (drinking, drug use), the actors ("addicts"), the related substances, and the settings of their use. In part, this has had to do with the multiplicity of competing theories (both professional and lay) which ascribe the etiology of addictive behaviors to aspects of these various sites.

To be sure, the conceptualization of alcoholism as a "disease" has a long history. Historians and sociologists of medicine have generally identified the disease concept of alcoholism—the notion of a chronic, progressive compulsion to consume a particular substance, of which the subject eventually "loses control"—as an invention of the early industrial age in Anglo-American countries, linked to the behavioral strictures imposed by a valorization of self-reliance, independence and productivity (Ferentzy, 2001; Levine, 1978). However, in the case of psychoactive substances, medical or disease models have also historically encountered significant resistance, in part because of their tension with notions of human beings as rational choice-making actors, and with widespread assumptions about volition or the will. While proponents of medical models of addiction have long seen their frameworks as humane alternatives to interpretations of addiction as a moral failing, others have seen in disease models a behavioral determinism which threatens Enlightenment notions of the free-willing subject (Hyman, 2007).[5] Not surprisingly, in North America, those arguing against disease conceptualizations of addiction in recent years have generally offered some notion of "choice" as an alternative (Heyman, 2009; Satel, 2001). And although there have also been numerous critiques based on social, psychological, or contextual

---

[4] Ian Hacking makes a related–though somewhat different—argument in suggesting that rather than overcoming mind/body or mind/brain dualism, the capabilities of contemporary medical technologies to manipulate bodies and brains are in fact fostering a revival of Cartesian thinking (2007).

[5] While the medicalization of addiction in North America is often viewed as incomplete or unsuccessful because of the ubiquity of criminalizing and supply-control policies, it has been suggested that the notion of addiction as a chronic relapsing brain disease is actually congruent with such policies (Reinarman, 2005).

interpretations of addictive behaviors (for example, Peele, 1989), expert arguments that addiction "is a choice" reflect ideas which are very widespread in lay discourses.

Moreover, while the general notion of addiction as a disease has circulated in Europe and North America at least since the late eighteenth century, it has taken many radically different forms, invoking different loci and mechanisms of addictiveness and linked to different forms of intervention and lines of scientific research. For example, nineteenth-century notions of inebriety emphasized the inherently addictive qualities of alcohol—a notion which meshed with the prohibitionist arguments of many in the North American temperance movement. Following the failure of Prohibition in the US, Alcoholics Anonymous (AA) articulated a conception of alcoholism-as-disease according to which some aspect of the particular drinker (rather than the substance itself) leant itself to pathological consumption (Gusfield, 1996, pp. 247–256). This general assumption about alcoholism, in turn, informed decades of studies on "predisposition" and "risk-factors"—which sought to identify the specific aspect of the drinker (whether social, psychological, hereditary) or his or her environment, to which addiction might be ascribed (Valverde, 1998). Early researchers working on narcotics, on the other hand, were equally interested in the psychoactive and addictive properties of particular drugs, and their efforts were often institutionally linked to a search for non-addicting analgesics (Campbell, 2007). While the 1970s saw a shift to research on "alcohol and drugs" the popularization of "addiction" as a supra-category encompassing not only problematic use of substances but also pathological manifestations of behaviors such as gambling, was shaped as much by the burgeoning mutual help and addiction recovery movement as it was by biomedical research (Schull, 2006; Sedgwick, 1993; Valverde, 1998).[6] In short, the field of addictions has long been characterized by a multiplicity of models and conceptual frameworks or, in the words of one leading researcher, "conceptual chaos" (Shaffer, 1997) which has precluded it (in the eyes of many) from being considered a "normal" or paradigmatic science.

However, since the mid-1990s a number of researchers—led by the National Institute on Drug Abuse (NIDA)—have forcefully promoted the notion of addiction as a "chronic, relapsing brain disease:" a dysfunction of normal brain systems involved in reward, motivation, learning, and choice (Hyman, 2005; Kalivas & Volkow, 2005; Leshner, 1997). Based in research conducted since the 1970s on the neurochemical underpinnings of craving and pleasure, this model was further bolstered and legitimated by the powerful imaging technologies which became widely available during the 1980s and 1990s (Campbell, 2007; Vrecko, 2006a). The chronic, relapsing brain disease model replaces an earlier biomedical narrative in which withdrawal and growing physical tolerance lead people to consume increasing quantities of a substance to maintain the same level of intoxication, with one in which substances and particular behaviors "hijack" endogenous systems evolved to reward behaviors necessary for survival—the so-called "dopamine hypothesis" (Hyman, 2005; Kalivas & Volkow, 2005). Moreover, as a revised edition of the DSM takes shape, the chronic, relapsing

---

[6] While this therapeutic arena is itself vast and highly varied in North America, it is largely unified in accepting the notion of addiction as a "disease of denial," itself traceable to psychoanalytic ideas about denial as a type of ego defence (Carr, 2010).

brain disease model has fuelled debate among psychiatrists about whether to recognize as addictive disorders behaviors such as pathological gambling, Internet use, and overeating which do not involve psychoactive substances but nonetheless may correlate with dysfunctions in the same brain circuits (Block, 2008; O'Brien, Volkow, & Li, 2006; Petry, 2006; Volkow & O'Brien, 2007). Finally, in addition to shifting much public discussion of addiction's roots from psychology, family dynamics and social factors to neurotransmitters and brain functions, this research has resulted in the development of several pharmacological treatments for addiction, including drugs which dampen the neurochemical effects of opiates or alcohol and reduce sensations of craving (Lovell, 2006; O'Brien, 2005; Valverde, 2003).

For many social scientists interested in the effects of such clinical interventions, the significant question does not rest on a polarized distinction between free will/choice and compulsion/determination (as has often been the case in bioethics discussions), but on an empirical examination of the ways in which neurobiological explanations and pharmacological treatments presume or foster certain behaviors or self-conceptions on the part of patients. For example, some observers have argued that whereas psychotherapeutic interventions and the 12-step program alike employ what Summerson Carr calls a linguistic "ideology of inner reference" (2006) to teach patients to conceive of themselves as particular types of persons—namely alcoholics and addicts—pharmacological treatments seem to operate in a radically different way. Scott Vrecko suggests that people who use the opiate-antagonist naltrexone as a means of managing their problematic drinking are encouraged to conceive of themselves as "targeting and controlling specific elements of neurochemistry" rather than resisting cravings through force of will or with the aid of fellow sufferers (2006b, p. 302).[7] Thus, the argument is often made that the "neurochemical self" encouraged by pharmaceutical interventions is (or will be) radically different from the "psychological self" which underlies psychotherapeutic discourse and practice (see Rose, 2003; 2007).

While these potential distinctions are useful in helping to structure research questions and agendas on the social effects of neuroscience, it is also possible to emphasize the continuities and similarities between psychosocial and pharmacological clinical interventions in the addictions. For example, as Jamie Saris points out, both psychosocial and pharmacological interventions are "techniques and technologies of transformation"—even if they accomplish this transformation in different ways (2008, p. 266). Here I emphasize a somewhat different continuity between these clinical technologies. What I have in mind is the relationship between models of addiction (or for that matter, mental illness more generally) held by researchers or clinicians, and the self-identification of patients. I suggest that in both "talk therapies" broadly construed (including 12-step programs) and pharmacological therapies—as they are practiced in North America—there seems to be a homology between clinicians'

---

[7]   It is important to note that these arguments—as well as the one made in the present chapter—focus on issues of personhood, and are thus concerned with whether or how clinical interventions and neurobiological concepts shape shared assumptions about what it means to be a person. They are thus related to, but clearly distinct from, psychological questions such as how the neurochemical effects of psychopharmaceuticals may or may not affect a particular person's sense of herself (see Gold & Olin, 2009).

models of addiction or mental illness and the self-conceptions of patients undergoing successful treatment. Or, perhaps, more than a homology—it is possible to say that patients are encouraged to self-identify with counselors' and clinicians' models of addiction.

This is somewhat more obvious in the case of talk therapy and 12-step therapy, in which the techniques either demand that patients conceive of themselves as persons of a certain sort, or teach them to do so. It is not simply that 12-step programs require people to publicly self-identify as "alcoholics" and "addicts" and to narrate their life-stories appropriately. Rather, as Laurence Kirmayer has argued, on a more basic level, 12-step therapy and various systems of psychotherapy "depends on implicit models of the self, which in turn, are based on cultural concepts of the person," (Kirmayer, 2007, p. 232). Such models of the self are implicit in certain capacities which most psychotherapies demand of patients. Sometimes grouped under the rubric of "psychological mindedness" these include a capacity to articulate one's life story according to particular narrative conventions; an awareness of oneself as possessing an unconscious, into which repressed or denied portions of one's mental life are pushed; a capacity for self-reflection; an ability to recognize and identify one's emotional responses to experiences; and a desire "to accept and handle increased responsibility for the self," (Kirmayer, 2007, p. 236).[8] An understanding of oneself in such terms is a precondition for many modes of psychotherapy, while other modes of talk therapy—including 12-step-based programs—often seek to develop such capacities in patients. For example, patients in an alcoholism rehabilitation program I once observed were presented with lists of emotion terms to aid them in reflecting upon and naming their present emotional states during group therapy.

In the case of pharmaceutical interventions, patients' self-identification and their conceptualization of their illness plays a somewhat different role in the treatment process. While it is not intrinsically necessary for patients to conceive of their problems as originating in their brains in order for psychopharmaceutical interventions to work effectively, patients are in fact encouraged to think this way for a number of reasons: to produce demand for pharmacological products; to increase compliance with pharmacological treatment regimens; and to ensure a relationship of trust and mutual understanding between physician and patient under an ethical regime of patient autonomy. Thus, while many actual patients may continue to think about themselves and their distress in a variety of different terms, it might be fair to say that the ideal patient implicit in the imagination of many biologically-oriented psychiatrists is one who conceptualizes his or her symptoms as stemming from a neurochemical imbalance (Dumit, 2003).

In what follows I suggest that Russian narcology—as addiction medicine is known locally—and its treatments stand in a rather different relationship to the personhood of patients. However, first I will examine how narcology's clinical style of reasoning took shape within the context of the Soviet sciences of the mind and brain.

---

[8] Congruent as they were with many older Euro-American epistemological assumptions, such constructions of personhood entered the general public discourse in North America in the early to mid-twentieth century and deeply influenced the conceptual tools which people had at hand for conceiving of themselves, whether or not they actively participated in psychotherapy (Kirmayer, 2007).

Reductionism as Politics

To understand how Pavlovian theory came to dominate Soviet psychiatry—and early understandings and treatments of alcoholism—it is important to sketch the genealogy of the links between materialist physiology and politics in Russia. As early as the 1860s, materialist physiology became associated with political radicalism in the Russian popular imagination. When Ivan Sechenov, the "father" of Russian physiology, published *Reflexes of the Brain* in 1863, it was the subtitle which simultaneously struck a chord with radicals and aroused the suspicion of the Tsarist censors: *An Attempt to Introduce Physiological Foundations for Psychic Processes.* Sechenov's argument for the reduction of psychological phenomena to material processes, largely informed by the mechanism of German physiologists, sparked debate with introspective psychologists and theologians alike (Janousek & Sirotkina, 2003; Joravsky, 1989). While most academic physiologists were liberal rather than radical in their politics, the image of the radical physiologist became popularized by literary depictions such as those in Chernyshevsky's *What is to be Done?* and Turgenev's *Fathers and Sons.* It was political thinkers and activists, rather than scientists, who tended to see in the new physiology a materialist reductionism which seemed to both undercut Orthodoxy, a central pillar of Tsarist autocracy, and lend itself to utopian ideas about the transformational power of science (Joravsky, 1989, p. xiv).

For the Bolsheviks, who inherited these assumptions, the relationship between physiology and psychology was deeply contentious and ideologically significant, because it was in this sphere of knowledge that Marxists hoped to link their understanding of human beings as historical actors with an objective science of humans as material beings (Joravsky, 1989; Smith, 1992, p. 191). This project could not be achieved by a simple reduction of psychology to physiology; instead, it was attempted through the concepts and language of dialectical materialism. Ivan Pavlov's reflex theory was taken up in this context, not simply as an explanation of a particular type of learning (as it was largely interpreted outside the USSR), but as a way of framing the relationship between human biology and the environment as "dialectical" (Graham, 1987, p. 163; Joravsky, 1989). Thus, although Pavlov's own politics were viewed as "reactionary" (until his rapprochement with the Soviet regime during the mid-1930s), his doctrine was embraced by the Bolsheviks and praised in 1924 by Nikolai Bukharin (at the time a leading Party ideologist) as a "weapon from the iron arsenal of materialism," (quoted in Joravsky, 1989, p. 212).

Despite these assumptions on the part of leading Bolsheviks, Pavlovian theory came to dominate the Soviet sciences of the mind and brain somewhat gradually, following a brief period in the early 1920s when a variety of schools and research traditions coexisted—including psychoanalysis. When mass industrialization and the collectivization of agriculture were instituted late in the decade, Stalin and other Party leaders shifted away from their previously conciliatory policy towards professionals and initiated the project of creating a cadre of specialists whose primary allegiance would be to the party-state rather than to their professional group (Fitzpatrick, 1992). This shift in policy set the stage for the creation of a Soviet psychiatry which would, in its broad contours, persist at least until the late 1980s (Calloway, 1992; Skultans, 2003;

1997). Psychoanalysis and various Russian psychological schools were increasingly condemned as "idealist," while Pavlov's theory of conditioned reflexes was promoted in increasingly forceful terms (Etkind, 1997a; Miller, 1998; Todes, 1995). Soon after the official endorsement of Trofim Lysenko's anti-Mendelian theories of heredity, a series of conferences on physiology, psychiatry, and psychology were held (between 1950 and 1952) at which Pavlov's doctrine was declared the objective foundation for the Soviet sciences of the mind and brain, and scientists who had previously dissented publicly "confessed" their errors (Joravsky, 1989, p. 413; Windholz, 1997). While a resurgence of interest in psychology and theories of consciousness took place following Stalin's death, the influence of Pavlovian doctrine on clinical psychiatry extended well past the early 1950s, in part due to the prominent institutional posts held by its adherents, in part simply to the persistence of certain clinical interventions (Segal, 1975).[9]

For example, the dominance of Pavlov's theories had a number of infrequently discussed—and presumably unintended—consequences in that it legitimated hypnosis and other suggestion-based practices by providing a coherent explanation for them in scientific terms, as forms of inhibition (Chertok, 1981, p. 11). Pavlov described inhibition, along with excitation and equilibrium, as a fundamental process, taking place in the nervous system. Inhibition encompassed all processes which weakened conditioned reflexes, and could be distinguished into the categories of "external inhibition," "internal inhibition," and the inhibition associated with sleep (Smith, 1992, pp. 200–201). Hypnosis resulted when the inhibitory process that led to sleep occurred to a less extensive degree; it was also a state of consciousness which facilitated suggestibility (Pavlov, 1925/1994, p. 84; Platonov, 1959).

Whether one interpreted such accounts as reduction to physiology or as "dialectical," Pavlov's theories helped to render hypnosis scientifically legitimate, allowing it to be incorporated into mainstream psychiatry (Babayan & Gonopolsky, 1985; Babayan & Shashina, 1985; Hoskovec, 1967; Rozhnov & Burno, 1987). As one Soviet textbook argued, "The strictly objective Pavlovian method of investigating higher nervous activity dispelled the fog of mystery and the subjective psychological conceptions that had for so long wrapped the problem of hypnosis in darkness," (Babayan & Shashina, 1985, p. 99). In helping to legitimate hypnosis, the Pavlovian dominance in psychiatry led to the development of multiple suggestion-based interventions categorized as "psychotherapy."[10] These clinical applications of hypnosis emerged partly from the

---

[9] While the extent to which Pavlovian reflex theory came to dominate the Soviet psy sciences during the 1940s and 1950s was clearly driven by Party officials' shaping of the discursive field, it also paralleled contemporaneous scientific and cultural developments in North America and Europe. For instance, the mechanistic metaphors of Pavlovian reflex theory shared much figuratively, with the images of aspiring modernity used by European intellectuals with increasing frequency since the Industrial Revolution—but particularly during the period immediately preceding and following World War I (Rabinbach, 1992). Additionally, the period of the 1920s–1950s, also saw the rise of a radically reductionist and Pavlov-influenced behaviorism in North American psychology, exemplified by the work of B.F. Skinner and others (Moore, 2005). And yet in much of Europe and North America, the robust psychoanalytic movement offered a conception of mind and meaning (as well as clinical techniques) which vied with those informed by behaviorism.

[10] Suggestion was also central to the research of Vladimir Bekhterev, the eclectic psychiatrist who developed a theory of "associative reflexes" in many ways parallel to Pavlov's, and who is often described in the Soviet literature as the "founder of Russian psychotherapy," (Platonov, 1959, p. 11).

theory itself; for Pavlov, sleep and hypnosis were "protective" forms of inhibition—an idea which also facilitated the widespread use of sleep therapy in Soviet psychiatry (Wortis, 1950, p. 161).

Indeed, with a few notable exceptions, as well as a strong tradition of "rational psychotherapy," most interventions identified as psychotherapy during the Soviet period were employed hypnosis and suggestion (Etkind, 1997b; Kirman, 1966; Lauterbach, 1984; Segal, 1975; Wortis, 1950, p. 88). These methods included various types of individual and group hypnosis, "direct suggestion," in which the patient remains in a waking state and is aware of the procedure, "indirect suggestion," (which included the use of placebos) and techniques of autosuggestion and the "autogenous training" developed by German therapists (Lauterbach, 1984, p. 81). As I describe below, such methods made up the majority of long-term interventions used to treat alcoholism in the Soviet Union. However, in order to understand how this came to be, I first briefly trace how narcology conceptualized alcoholism and addiction, and then examine how these disease models meshed with styles of clinical reasoning.

Disease States

When narcologists and psychiatrists during the late Soviet period wrote about chronic alcoholism as a disease, they typically elided etiological arguments and focused on physiological mechanisms; chronic alcoholism referred to the pathological conse-quences of regular, long-term heavy drinking, and not to a phenomenon linked to a bodily or psychological predisposition (Galina, 1968; Strel'chuk, 1954).[11] However, expert conceptualizations of alcoholism in the Soviet Union had not always been so narrow, nor had they always been monopolized by psychiatry.

After the Bolshevik Revolution and through the 1920s, the Soviet Commissariat of Public Health promoted investigations and interventions that examined the social etiology of illness, under the rubrics of social- and psycho-hygiene. For Soviet social hygienists, alcoholism was a "social disease," in that the social component of its development and "transmission" was held to be the primary one (Solomon, 1989, p. 257). Specifically, social hygienists focused on the role played by such factors as stress, family background, wage level, and the drinker's "level of culture [*kul'turnnost'*]." The negative political implications of widening these "environmental factors" to a point where they might have constituted a critique of Soviet society, may have led many hygienists to focus on the "micro-social environment" of the family or immediate community (Solomon, 1989).

While social hygienists' views of disease were not inherently mutually exclusive to those of psychiatrists, their drastically divergent object of study led them to recommend very different forms of intervention, fostering a professional rivalry over so-called lifestyle alcoholics (Joravsky, 1989; Solomon, 1989). Thus, social hygienists argued that lifestyle alcoholics were to be resocialized and their habits transformed

---

[11] While I discuss the distinctions which specialist discourses on narcology drew between such categories as "alcoholism" and "chronic alcoholism" in the following section, popularizing texts of the 1960s and 1970s (even those authored by physicians) often failed to distinguish between "drunkenness" and "alcoholism" altogether (Zenevich 1967).

through a series of measures focused on public education (Solomon, 1989). Additionally, social hygienists and sympathizing psychiatrists advocated the treatment of alcoholics through out-patient dispensaries, an approach which clashed with that of most psychiatrists who viewed alcoholics as a subset of their broader contingent of mental patients, whom they preferred to treat in hospitals or isolated psychiatric colonies (Solomon, 1989, p. 266). Ultimately, social hygienists' claim to produce authoritative knowledge and treatment for lifestyle alcoholics was short lived. In April 1927, the Soviet government issued a decree which allowed drinkers categorized as "socially dangerous" to receive treatment without consent, effectively creating a legal equivalence between mental illness and alcoholism, and by 1930 the party-state had stopped funding social research on alcoholism (Solomon, 1990, 1989).[12]

The eclipse of social hygiene by psychiatry in the management of alcoholism came at the same time that the latter was becoming increasingly dominated by a Pavlovian neurophysiology—as described above. The dominance of neurophysiology in psychiatry—and the politicization of genetics—facilitated a focus on the functional mechanisms of addiction and away from its etiology, whether social or hereditary. This avoidance of etiological arguments was also shaped by the political sensitivity of alcoholism itself. Whether they were criminologists or psychiatrists, Soviet writers on alcoholism found themselves constrained in similar ways. On the one hand, they drew on Marxist arguments to bolster their contention that various forms of "deviance" were fundamentally social phenomena; on the other, the risk of articulating an overt critique constrained the sphere of "the social" to which this etiology could be ascribed. At least until the 1970s, popular and specialist texts alike continued to describe alcoholism as a "capitalist survival" or "birthmark."[13]

Without recourse to etiological arguments, Soviet psychiatry focused more closely on the mechanisms underlying alcoholism—and particularly on alcoholic psychoses— than on the motivations or explanations of the behaviors leading to alcohol consumption. At the peak of Pavlovian orthodoxy in the sciences of the mind and brain it was

---

[12] Susan Gross Solomon argues that while social hygienists avoided making the kind of broad social critiques which their researches may have facilitated, their emphasis on gradual adaptation, moderation and voluntary resocialization clashed with the ethic of impatience and speed, as well as the belief in the unlimited possibilities of sheer will-power, championed during the 1930s industrialization. At a more fundamental level, whereas hygienists' depiction of alcoholism as a "social disease" conferred primary responsibility onto the Soviet state, psychiatrists' definition of it as a mental pathology placed this responsibility more fully onto the patient (Solomon, 1989, 1990).

[13] The dominant argument was simply that alcoholism was one of many social ills (poverty, prostitution and crime among them) which were inevitably fostered by capitalist relations of production (Galina, 1968, p. 6). By radically transforming these social roots of alcoholism, many texts argued, the construction of socialism would, by definition, eradicate such phenomena. While such claims were patently absurd by the 1950s and 1960s (after—according to official definitions—more than two decades of life under socialism), the questions which they begged were avoided in public discourse. By the 1970s and 1980s some Soviet commentators were distinguishing the "primary social roots" (exploitative relations of production) which fed alcoholism under capitalism from the "secondary" ones which explained its persistence under socialism: "people's habits and norms" [*privichki i nravy*] (Beisenov, 1981, p. 12). Thus, perhaps, it was not surprising that the broad conclusions reached by these researchers were similar to those of the social hygienists: heavy drinking or alcohol abuse [*zloupotreblenie*] was depicted simply as a learned behavior or habit born out of the drinker's relationships in his "micro-social environment" [*mikrosotial'naia sreda*] (Galina, 1968, pp. 50–58; Tkachevskii, 1974, p. 37; Zenevich, 1967).

sometimes argued that "chronic alcoholism" was simply a conditioned set of reflexes. In other words, if heavy drinking or alcoholism was a learned behavior, the theory of conditional reflexes was used to explain how that behavior was learned, but not why it took place in the first place (Janousek & Sirotkina, 2003, p. 438). Even well into the 1960s chronic alcoholism was often referred to in somewhat tautological terms as in this popularizing medical text entitled *Harmful Habit or Disease?* [*Vrednaia Privychka ili Bolezn?*]:

> At the root of chronic alcoholism lies lifestyle drunkenness, conditioned by various fac-tors, fed by traditions and customs. The systematic consumption of alcoholic drinks which emerges on this basis leads to a singular passion for alcohol, accompanied by numerous disturbances to one's health, that is, to chronic alcoholism.
>
> (Zenevich, 1967, p. 20)

The legacy of Pavlovian thinking in narcology extended to diagnostics as well. In fact, it was in this sphere that I encountered it during my conversations with narcologists in 2004. For instance, when I asked him about the symptoms or signs which distinguished alcohol dependence, the medical director of the Municipal Addiction Hospital explained:

> The dependence syndrome: there are several criteria according to which you can clearly tell that a person can't live without alcohol or some psychotropic substance, that he needs systematic use ... One of the signs, is when the so-called defensive reflex has been lost. If you drink too much you have a hangover, and if you try to drink a little more you feel nauseated. If you are nauseated, that means your defensive vomiting reflex is working. That means you're not an alcoholic. If the reflex is lost, then this is already alcoholism. A person in this state just needs to drink and he's fine. A person who's not an alcoholic—even if people put pressure on him and say—comrade, just drink a little beer—just from the sound of it he gets nauseated. This is one of the signs, the symptom—the vomiting reflex, through which you can categorize all people into alcoholics and non-alcoholics.

Like other practicing narcologists I spoke to, this physician identified the "vomiting reflex," as the primary criterion for a diagnosis of alcohol dependence. In part because of its congruence with Pavlov's theory of reflexes, this marker was mentioned in late Soviet textbooks as coinciding with the beginning of the first stage of alcoholism (Babayan & Gonopolsky, 1985, p. 100).[14] Textbooks on narcology written throughout the Soviet period generally referred to a progressive "three-stage" schema of alcoholism. In his *Lectures on Narcology*, Nikolai Ivanets, the country's head narcologist during the 1990s and 2000s, argues that this schema, which employed diagnostic criteria such as

---

[14] For example, the use of the vomiting reflex as a diagnostic criterion was congruent with the psychiatric tradition favoring longitudinal diagnoses and was written into a particular Soviet diagnostic code. Throughout the late Soviet period, psychiatrists and narcologists used a version of the World Health Organization's International Classification of Diseases (ICD) amended to reflect the differences in Soviet conceptions of mental illnesses, most notably schizophrenia (Calloway, 1992). While a commitment to switch to the international version of the ICD was made by representatives of Soviet psychiatry during the late 1980s, (partly as a condition for their re-admittance to the World Psychiatric Association), the new diagnostic criteria were not instated until the late 1990s (Smith, 1996).

the loss of a vomiting reflex, defined alcoholism significantly more broadly than does contemporary Russian narcology, by including under its aegis the phenomenon of "alcohol abuse" [*zloupotreblenie*] or "problem drinking," (here Ivanets borrows and literally translates the Anglo-American terminology). In other words, more recent diagnostic criteria—namely those of the ICD-10—draw a bright line between biological "dependence" and "abuse," the latter a category which, to some degree maps onto the popular notion of drunkenness [*pianstvo*] (Ivanets, 2001, p. 48).

Moreover, Ivanets emphasizes that under the new set of diagnostic criteria, the loss of the vomiting reflex, as well as the entire "first stage" of alcoholism, belong to the "pre-clinical stage of the illness" (2001, p. 47). Along with an increased tolerance for alcohol, the loss of the vomiting reflex is simply a sign that the patient has ingested high levels of alcohol over a long period of time. Neither indicates physical dependence, the cardinal sign of which is "alcohol abstinence syndrome," described in Anglo-American literature as "withdrawal syndrome," (2001, p. 48).[15] While Ivanets writes that the "abstinence syndrome" was not widely accepted as the primary diagnostic marker of dependence in the Soviet Union until the 1950s or 1960s (2001, pp. 48–49), it was clear from my conversations that even in 2004 some narcologists referred to the earlier diagnostic criteria. Indeed, many narcologists seemed to continue to translate the new diagnostic terms back into the old ones on an everyday basis. One physician I spoke to compared the shift in diagnostic terminology to a currency change or devaluation:

> Officially, [the terminology of] the three stages of alcoholism is not used any longer. But really we often think in terms of the old system. My grandmother told me that after the monetary reform of the 1950s, she always used to mentally convert prices back into pre-reform rubles. In the same way, we diagnose patients according to the old rules and then translate these into the new syndromes.

To some degree such disjunctures between the officially promoted categories of Russian narcology—which are increasingly oriented away from Soviet models and toward those of international biological psychiatry and addiction medicine—and those used by narcologists in everyday clinical practice, mirrored those between medical research and clinical medicine. While Soviet studies in neurophysiology were, by the 1970s, increasingly moving away from the terminology and ideas of Pavlovian reflex theory and focusing increasingly on the neurochemistry of addiction, their conceptions were slow in diffusing to clinical practice (Shabanov & Shtakelberg, 2001). This is not to say that practicing narcologists were not aware of such research: many clearly were. However, new treatment methods were slow in coming, and in the meantime therapeutic methods based on Pavlovian theory had

---

[15] Despite its somewhat confusing name, many Russian narcologists refer specifically to "alcohol abstinence syndrome," rather than "withdrawal," indexing the Soviet scientific origins of the former. As described during the late 1920s and early 1930s by Soviet neurophysiologist S. G. Zhislin, "abstinence syndrome" [*abstitentnyi sindrom*]—sometimes referred to as "hangover syndrome" [*pokhmel'nyi sindrom*]—is characterized by tremors, sweats and difficulty sleeping (Zhislin, 1959). Ivanets argues that Zhislin's work went unnoticed overseas, until it was essentially replicated as "withdrawal syndrome" by American researchers during the 1950s (2001).

became part of an institutional logic with its own inertia. Indeed, the most commonly used methods in Russian narcology in the present day are still those that rely on the mechanisms of suggestion and aversion, and are resonant with the language and concepts of early to mid-twentieth-century neurophysiology.

Condition and Suggest: Narcology's Therapies

The thought style prevalent in Soviet psychiatry helped to give rise to a number of clinical interventions specifically addressing alcoholism. For much of the Soviet period, a dominant treatment for alcoholism was conditional-reflex therapy—sometimes referred to as "apomorphine treatment"—an aversive treatment developed by Soviet medical researchers working within the framework of Pavlov's theory of reflexes (Zhislin & Lukomskii, 1963). In Pavlovian terms the idea was to condition a reflex to the taste, smell, sight, or mention of alcohol based on the unconditional reflex to an unpleasant stimulus. In the mid-1930s Soviet researchers worked out a technique that used emetics like apomorphine as an unconditional stimulus, so that subjects developed a gag or nausea reflex upon tasting or smelling alcohol (Sluchevsky & Friken, 1933). Between the 1940s and 1980s in the Soviet Union, this conditional-reflex therapy was recommended in textbooks as a first-line clinical therapy for use after detoxification, and its use was mandated in certain penal institutions (Babayan & Gonopolsky, 1985; Shtereva, 1980).[16] Although the 1960s saw a post-Stalin liberalization of science which allowed Soviet physiologists to move well beyond Pavlov's formulations, conditional-reflex therapy maintained its currency in Soviet psychiatry and narcology because it complemented both the official medical knowledge about alcoholism and the institutional conditions in which that knowledge was put into practice.[17]

During the late 1940s and early 1950s, the alcohol antagonist disulfiram was introduced in the Soviet Union.[18] Often referred to in Russia as teturam, Esperal, or Antabuse, disulfiram prevents the body from fully processing alcohol. By blocking the action of aldehyde dehygrogenase (ALDH)—an enzyme in the metabolic pathway of ethanol—the drug causes a build-up of the toxic by-product acetaldehyde, with deeply unpleasant consequences for patients: flushing, nausea, and high blood pressure,

---

[16] It should be noted that the use of apomorphine treatment was not confined to the Soviet Union, although it was less widespread elsewhere. During the twentieth century, apomorphine therapy was employed against alcohol and drug addiction by British physician John Dent, whom William Burroughs credited with breaking his addiction to heroin (Burroughs, 1957; Dent, 1949). During the 1930s, Walter Voegtlin, a gastroenterologist who had studied with Pavlov, established a sanitarium in Seattle for the treatment of alcoholism by conditional-reflex therapy. Along with several other clinics throughout the United States and United Kingdom, this institution—now known as the Schick-Shadel Hospital—continues to treat addiction using a type of aversion therapy (Lemere, 1987; White, 1998, pp. 106–108).

[17] The use of apomorphine treatment was largely discontinued during the late 1980s and early 1990s, a period of political reform during which many therapies viewed as punitive were put to rest. While many narcologists and patients in 2003–2004 described apomorphine treatment as an inherently punitive practice— in effect, a punishment—others emphasized the coercive conditions under which it had been carried out, arguing that, as a consequence, the method's tarnished reputation had profoundly undercut its efficacy.

[18] For a more detailed account of disulfiram treatment in Soviet and Russian addiction medicine see Raikhel (2010).

referred to in the medical literature as a disulfiram-ethanol reaction (DER) (Mann, 2004). When disulfiram was introduced in the Soviet Union, its use was modeled after that of conditional reflex therapy.[19] The potential effects of drinking alcohol with disulfiram in one's body were demonstrated to patients in physician-administered "tests" (Babayan & Gonopolskii, 1985; Strel'chuk, 1952). Moreover, a clinical researcher, who was among the first to publish on disulfiram treatment in the Soviet Union, described it as fostering a "negative conditioned reflex to alcohol" in patients, adding that this was observed even in "patients who had not taken antabuse in nearly a year," (Strel'chuk, 1952, p. 49). In other words, the idea that patients might experience the negative effects of disulfiram in absence of the drug itself was present almost from the inception of its use in the Soviet Union. By the late 1960s Soviet researchers were reporting clinical experiments with the use of placebo-therapy: the replacement of the drug with a saline solution or vitamins (Ialovoi, 1968). Originally intended for patients for whom the drug was contraindicated, such placebo-therapy became increasingly widespread over the following decades. By the 1990s (and perhaps earlier) it was entirely commonplace (Fleming, Meyroyan, & Klimova, 1994).

During the 1970s and 1980s sub-dermal implantations of depot disulfiram developed in France began to be commonly used in the Soviet Union (Fleming, Meyroyan, & Klimova, 1994; White, 1998, p. 228). While these were meant to gradually release the chemical into the patients' bloodstream over the course of an extended period (such as a year), clinical studies have shown that no disulfiram or ALDH inhibition is detectable in patients shortly after the insertion of commercially available implants—and thus no threat of a DER exists (Johnsen & Morland, 1992).[20] Nevertheless, although other placebo therapies were subsequently developed, such as the "tablet" and the "torpedo" (these were represented to patients as oral and intravenous forms of "long-acting" disulfiram, respectively) implantation remained by far the most popular (Chepurnaya & Etkind, 2006). Laypeople called this therapy an "implant," [*podshivka*] and patients would commonly say "I was implanted," [*menia podshili*]. Narcologists referred to all of these variants of disulfiram or placebo therapy as *khimzashchita*. Such treatment remained extremely common among patients I spoke to in 2004, and some returned regularly for repeat implantations.

While narcologists depicted *khimzashchita* to their patients as a pharmacological therapy, among themselves they spoke about it as a form of psychotherapy and emphasized its parallels with a type of hypnosis known as emotional-stress psychotherapy

[19] Indeed, many of the earliest Soviet publications about disulfiram were written by Ivan Vasil'evich Strel'chuk, who also developed widely-employed protocols for apomorphine treatment, some using methods of hypnotic suggestion (Miroshnichenko, Pelipas, & Ivanets, 2001, p.139; Strel'chuk, 1951, 1952).

[20] Thanks to Colin Brewer for pointing out a minor error made in the discussion of disulfiram implants in Raikhel (2010). Namely, the duration of disulfiram's chemical action does not depend on its presence in the bloodstream, as I suggested in that publication. Brewer points out (personal communication) that individual ALDH molecules are *permanently* inhibited by disulfiram—and thus the normal metabolism of ethanol does not resume until new ALDH molecules are produced. This process can take anywhere from one or two days to over a week, during which period a patient can still experience a DER even if disulfiram is no longer in the bloodstream. Despite this pharmacological detail—it is clear that ALDH inhibition would not continue for anything like six months or a year–the lengths of time which narcologists often claim *torpedos* and implants to work.

or coding [*kodirovanie*], perhaps the most commonly used method of addiction treatment in contemporary Russia (Mendelevich, 2005). While this therapy had antecedents in mainstream Soviet psychiatry (for example Rozhnov & Burno, 1987), it was developed during the 1970s by Alexander Dovzhenko, a physician working in Crimea, and became popular as a "rapid" form of therapy during the 1980s and 1990s (Dovzhenko et al., 1988; Miroshnichenko, Pelipas, & Ivanets, 2001). Frequently depicted as a magic-bullet cure—and criticized for this very reason—coding quickly became a deeply commercialized type of therapy, with "séances" carried out on auditoriums full of patients. Like *khimzashchita*, coding is a therapy meant to keep patients from drinking, seemingly by convincing them that their brains or minds have been altered in a way that makes the consumption of alcohol harmful or fatal. In his original protocol for coding, Dovzhenko wrote, "It is suggested to patients that the efforts of the doctor, with the help of a set of 'hypnotic' and physiogenic [sic] actions on their brains, will create a stable centre of excited nerve cells, which from the moment of 'coding' will block their craving for alcohol for a given length of time—1 year, 5, 10, 25 years or longer," (Dovzhenko et al., 1988, p. 94). As in the case of *khimzashchita*, patients were to sign a document stating they understood that the consequences of drinking during the duration of the "code" could cause severe illness or death (Fleming, Meyroyan, & Klimova, 1994). While it has remained controversial, Dovzhenko's method has gained legitimacy among many narcologists, and it continues to be widely used in the state-funded addiction treatment service (Finn, 2005). Variations on the method—particularly its hybridization with Russian Orthodox or "occult" imagery and symbolism—have become popular among self-styled alternative practitioners and spiritual healers (Grigoriev, 2002).

While the clinical styles of reasoning through which narcologists understood and enacted these interventions also influenced patients' experiences with the treatments, a number of additional factors played a significant role as well. As I have describe at greater length elsewhere (Raikhel, 2010), a particular patient's experience with these methods often had to do both with his or her prior treatment experiences and with the broader social context of his or her life. While patients' descriptions of methods like *khimzashchita* and coding ranged from confused to compliant, to defiant to desperate to cynical, it was most often the patients with jobs and intact families— those likely to be viewed as potentially compliant—who spoke about the treatment positively. These were also the patients whom physicians were most likely to characterize as "believing" in the efficacy of their methods.

A factory worker in his 50s whom I met at St. Petersburg's Addiction Hospital—I will call him Vyacheslav—was typical of such compliant patients. As he explained, Vyacheslav's hospitalization was part of a yearly cycle which began with a drinking binge [*zapoi*] and ended with his wife successfully persuading him to return to the Hospital. There he would undergo the usual month-long detoxification, after which he always received a *torpedo* injection. "There's also a special injection they can give you in your vein," he explained, adding, with a deference to the authority of medical professionals: "It's all figured out by the professors so that it gradually dissolves." Frightened of the potential negative effects of drinking with the substance in his body, Vyacheslav explained that he always waited until the course of the *torpedo* was finished before beginning another drinking binge. Then the cycle would repeat itself again. Acknowledging that his abstinence from alcohol dampened his social life, Vyacheslav also insisted that he and his family were successfully managing his tendency to indulge in drink.

Of course, many others at the Hospital were not such "good" patients: they spoke about the narcological treatments cynically and swapped stories meant to illustrate the inefficacy of the methods. Not surprisingly these were—most often—patients who had already lost the most and had been socially marginalized. However, other patients who were similarly categorized as "hopeless" cases were able to construct new relationships and build for themselves a kind of sociality at the Hospital—albeit one typically based on their dependence on and deference to their clinicians. In other words, patients' ideas about and commitments to treatment were often mediated (in varying ways) by their social capital, itself linked to the—often downward—trajectories of their lives.

However, more significant to the present argument, is how Vyacheslav spoke about himself—or just as importantly, how he did not speak about himself. Not surprisingly, he did not speak about his drinking problem as a chemical imbalance to be modulated. Nor did he articulate an illness-based addict identity—as advocated by 12-step programs—and speak of himself as "an alcoholic;" or for that matter speak about "alcoholism" as an illness or all-encompassing category at all. Rather, to the degree that he wanted to speak about the treatment at all, Vyacheslav described himself simply as someone who was managing his drinking binges. While it would be easy to view such an attitude through the analytic of "addiction as a disease of denial," I would suggest that the fact that these therapies make such few claims on patients' selves or identities only increases their appeal to those post-Soviet people wary of totalizing frameworks of self-transformation. While some patients found methods such as the *torpedo* or coding useless as means of achieving even temporary sobriety, and others passed through cycles of increasingly brief remission, at least for some, like Vyacheslav, narcology's methods worked as pragmatic aids for the care of the self which bolstered personal motivations for sobriety. The reason for these differences between patients had less to do with anything specific to the treatment protocol, than with the broader configuration of institutions and relationships (both inside and outside the clinic) within which any particular instance of the treatment took place.

Conclusion: Radical Measures

As I have suggested throughout this chapter, the narrative of Soviet and Russian addiction medicine calls into question the argument that a somatic model of the self follows from a thoroughly biologically-based psychiatry. In what may seem a paradox from the purview of Anglo-American psychiatry, in Russia a neurophysiological style of reasoning facilitated the dominance of treatments that relied on largely psychological mechanisms. Moreover, these methods were rather indifferent to the self-identifications of patients and their conceptions of illness. In other words, unlike most psychotherapies and even many contemporary psychopharmacological treatments, the effectiveness of methods like *khimzashchita* did not depend on patients aligning their understanding of addiction with that of their therapists.[21] If anything, these techniques required clinicians to conceal certain facts from their patients, widening—rather than closing—the divergence between patients' and therapists' models of illness and therapy.

What about the stereotactic surgery with which I opened this chapter? Many narcologists I spoke to were highly skeptical of this procedure. Putting aside the ethical considerations, most emphasized that the rate of success claimed by the neurosurgeons was an artifact of the process by which they selected their patients. In other words, only highly motivated patients were accepted. A recovering alcoholic and 12-step activist I knew was not so quick to dismiss the ethical dimension. "I know a man who went through this procedure," he told me, "They turned him into a freak."

At the same time, many narcologists have expressed great interest in the procedure. A colleague who attended a recent conference on addiction treatment and management in Kazan reported that Russian narcologists in attendance expressed far more interest in the presentation on stereotactic surgery than in those focused on harm reduction. Thus, it was not altogether surprising that some of the younger narcologists I spoke to articulated something of a disdain for the treatments, such as *khimzashchita*, which they used on a daily basis, and a hope for future therapies which would "cure" addiction through biological means. A particularly striking example was Vyacheslav's doctor, Anton Denisovich. Described by the hospital's medical director as a young star, Anton Denisovich had been appointed the head of his ward only four years after completing his MD in 2000. When I asked him which methods of treatment or rehabilitation he found most effective, he described the methods currently used in narcology as "palliative measures, not radical ones." The future, Anton Denisovich explained, lay with psychopharmacology, or with even more invasive biological interventions: "Either neurosurgery or genetics, I don't know, but with some kind of radical measures."

While it would be easy to interpret Anton Denisovich's hope for "radical measures" which might completely cure addiction only in light of the Soviet emphasis on biological mechanisms, I would argue that his statement echoes many North American biological psychiatrists' hopes for an effective treatment for addiction. In other words, what is striking both about therapies such as the stereotactic neurosurgery and statements such as this one is not simply that they represent an extreme of biological interpretations of addiction, but that they suggest the dovetailing of post-Soviet neurophysiological narcology and the dominance of a biological paradigm in the transnational psychiatric community.

After all, the collapse of the USSR and the subsequent reintegration of narcologists into transnational professional networks, brought them into contact not only with psychosocial models of addiction treatment, but also with the biologizing trends sweeping global psychiatry. Even as multiple forms of talk therapy have flourished in post-Soviet Russia (Matza, 2009), broadly biological styles of reasoning in Russian psychiatry have arguably been reinforced by this confluence of geopolitical rupture and disciplinary shift. Thus, while the stereotactic surgeries taking place in Russia were understandably viewed as ethically questionable both locally and in the broader

---

[21]   Of course, the treatment methods prevalent in Russian addiction medicine are by no means unique in this regard. A similar argument could be made of a number of so-called behavioral methods of treatment, particularly those which are conceptualized primarily as technologies and make little recourse to underlying theories of illness or pathology.

psychiatric community, this method is not entirely dissimilar in its assumptions about addiction and its treatment than the interventions becoming prevalent in North American and Western European psychiatry. These include the various pharmacological treatments for addiction mentioned earlier, but also recent attempts to use deep brain stimulation to treat alcoholism (Heinze et al., 2009). In their understanding of addiction not only in primarily biological terms, but also as—at least potentially—fully curable, these methods stand in direct tension with both the identity-based treatments of 12-step methods and with the assumptions of harm reduction advocates (for example Elovich & Drucker, 2008).[22]

Clinical interventions emerging from biological psychiatry and neuroscience are often portrayed by their critics as *inherently* problematic in their supposed reduction of personhood to neural mechanisms. However, such a perspective is only tenable if we can claim that clinical interventions are themselves reducible to therapeutic protocols, a reduction which depends on the assumption that clinical technologies are discrete, portable, and transposable between contexts with little transformation. For example, anthropologists have emphasized that the replacement (rather than the supplementation) of other types of mental health care with the prescription or distribution of medication—which has occurred in some resource-poor settings—is not simply a result of the availability of psychopharmacology, but its confluence with the goals of a neoliberalizing health sector (Biehl, 2004; Jain & Jadhav, 2009). Similarly, a critical neuroscience which hopes to examine the global circulation of neurobiological and psychiatric knowledge, substances, and techniques will have to grapple with the ways that these objects intersect with clinical performances and relationships, clinician's styles of reasoning and local research traditions, and institutional and political economic settings of treatment to shape the trajectories of patients' lives.

Acknowledgements

Thanks to Suparna Choudhury, Jan Slaby, Laurence Kirmayer, and Rob Lemelson for inviting me to participate in the workshop on Critical Neuroscience at which this chapter was first presented, and thanks to all of the workshop participants for their comments and suggestions. This chapter—as well as various versions of the material contained herein—has benefited greatly from the readings and suggestions of John Borneman, Joao Biehl, Colin Brewer, Suparna Choudhury, William Garriott, Laurence Kirmayer, Anne Lovell, Tobias Rees, Ian Whitmarsh, and Allan Young. Thanks to Katrin Maclean and Rachel Sandwell for their help with copyediting. The fieldwork described in this chapter was generously funded by a Fulbright-Hays Doctoral Dissertation Research Abroad Fellowship. Further research and writing was supported by the Fellowship of Woodrow Wilson Scholars at Princeton University and by a Postdoctoral Fellowship funded by the CIHR Strategic Training Program in Culture and Mental Health Services Research in the Division of Social and Transcultural Psychiatry at McGill University.

---

[22]   Elovich and Drucker argue that narcology's focus on "cures" for addiction is a central factor underlying the opposition to methadone and buprenorphine substitution treatment for opiate addiction in Russia (2008).

References

Babayan, E. A., & Gonopolsky, M. K. (1985). *Textbook on alcoholism and drug abuse in the Soviet Union*. New York: International Universities Press.

Babayan, E. A., & Shashina, Y. G. (1985). *The structure of psychiatry in the Soviet Union*. New York: International Universities Press.

Beisenov, B. S. (1981). *Alkogolizm: Ugolovno-pravovye i kriminologicheskie problemy*. [*Alcoholism: Legal and criminological problems*]. Moscow: Iurid. Literatura.

Biehl, J. (2004). Life of the mind: The interface of psychopharmaceuticals, domestic economies, and social abandonment. *American Ethnologist, 31*, 475–496.

Block, J. J. (2008). Issues for DSM-V: Internet addiction. *American Journal of Psychiatry, 165*(3), 1529–2401.

Burroughs, W. (1957). Letter from a master addict to dangerous drugs. *The British Journal of Addiction, 53*(2), 119–132.

Calloway, P. (1992). *Soviet and western psychiatry: A comparative study*. Keighley, Yorkshire: Moor Press.

Campbell, N. D. (2007). *Discovering addiction: The science and politics of substance abuse research*. Ann Arbor: University of Michigan Press.

Carr, E. S. (2010). *Scripting addiction: The politics of therapeutic talk and American sobriety*. Princeton: Princeton University Press.

Carr, E. S. (2006). "Secrets keep you sick": Metalinguistic labor in a drug treatment program for homeless women. *Language in Society, 35*(5), 631–653.

Chepurnaya, O., & Etkind, A. M. (2006). Instrumentalizatsiia smerti: Uroki antialkogol'noi terapii. [The instrumentalization of death: Lessons of an anti-alcohol therapy]. *Otechestvennyie Zapiski, 2*(27).

Chertok, L. (1981). *Sense and nonsense in psychotherapy: The challenge of hypnosis*. London: Pergamon Press.

Dent, J. Y. (1949). Apomorphine treatment of addiction. *British Journal of Addiction, 46*(1), 15–28.

Dovzhenko, A. R., Artemchuk, A. F., Bolotova, Z. N., Vorob'eva, T. M., Manuilenko, Y.A., Minko, A. I., & Dovzhenko, V. A. (1988). Stressopsikhoterapiia bol'nikh alkogolizmom v ambulatornykh usloviiakh [Outpatient stress psychotherapy of patients with alcoholism]. *Zhurnal nevropatologii i psikhiatrii imeni S. S. Korsakova, 88*(2), 94–7.

Dumit, J. (2003) Is it me or my brain? Depression and neuroscientific facts. *Journal of Medical Humanities, 24*(1), 35–47.

Ecks, S., & Basu, S. (2009). The unlicensed lives of antidepressants in India: Generic drugs, unqualified practitioners, and floating prescriptions. *Transcultural Psychiatry, 46*(1), 86–106.

Elovich, R., & Drucker, E. (2008). On drug treatment and social control: Russian narcology's great leap backwards. *Harm Reduction Journal, 5*(1), 23.

Etkind, A. M. (1997a). *Eros of the impossible: The history of psychoanalysis in Russia*. Boulder, CO: Westview Press.

Etkind, A. M. (1997b). There are no naked thoughts: Psychoanalysis, psychotherapy and medical psychology in Russia. In E. Grigorenko, P. Ruzgis & R. Sternberg (Eds.), *Psychology in Russia: Past, present, future*. Hauppauge, NY: Nova Publishers.

Ferentzy, P. (2001). From sin to disease: Differences and similarities between past and current conceptions of chronic drunkenness. *Contemporary Drug Problems, 28*, 363–363.

Finn, P. (2005, October 2). Russia's 1-step program: Scaring alcoholics dry. *Washington Post*.

Fitzpatrick, S. (1992). *The cultural front: Power and culture in revolutionary Russia*. Ithaca: Cornell University Press.

Fleck, L. (1979). *Genesis and development of a scientific fact.* Chicago: University of Chicago Press. (Original work published 1935).

Fleming, P. M., Meyroyan, A., & Klimova, I. (1994). Alcohol treatment services in Russia: A worsening crisis. *Alcohol & Alcoholism, 29*(4), 357–62.

Gaines, A. D. (1992). Medical/psychiatric knowledge in France and the United States: Culture and sickness in history and biology. In A. D. Gaines (Ed.), *Ethnopsychiatry: The cultural construction of professional and folk psychiatries* (pp. 171–202). Albany: State University of New York Press.

Galina, I. V. (1968). *Alkolizm razrushaet sem'yu.* [*Alcoholism destroys the family*]. Moscow: Meditsina.

Gold, I., & Olin, L. (2009). From Descartes to Desipramine: Psychopharmacology and the self. *Transcultural Psychiatry, 46*(1), 38–59.

Gold, I., & Stoljar, D. (1999). A neuron doctrine in the philosophy of neuroscience. *Behavioral and Brain Sciences, 22*(5), 809–830.

Graham, L. R. (1987). *Science, philosophy, and human behavior in the Soviet Union.* New York: Columbia University Press.

Grigoriev, G. (2002). "Batyushka," In *Istselenie slovom* [*Healing with the word*]. St. Petersburg: Alexander Nevskii Temperance Society.

Gusfield, J. R. (1996). *Contested meanings: The construction of alcohol problems.* Madison: University of Wisconsin Press.

Hacking, I. (2007). Our neo-Cartesian bodies in parts. *Critical Inquiry, 34*(1), 78–105.

Hacking, I. (1992). " 'Style" for historians and philosophers. *Studies in History and Philosophy of Science, 23*, 1–20.

Hall, W. (2006a). Stereotactic neurosurgical treatment of addiction: Minimizing the chances of another "great and desperate cure." *Addiction, 101*(1), 1–3.

Hall, W. (2006b). Avoiding potential misuses of addiction brain science. *Addiction, 101*, 1529–1532.

Hall, W., Carter, L., & Morley, K. I. (2004). Neuroscience research on the addictions: A prospectus for future ethical and policy analysis. *Addictive Behaviors, 29*(7), 1481–1495.

Healy, D. (1997). *The antidepressant era.* Cambridge, MA: Harvard University Press.

Heinze, H.-J., Heldmann, M., Voges, J., Hinrichs, H., Marco-Pollares, J., Hopf, J.-M., & Münte, T.F. (2009). Counteracting incentive sensitization in severe alcohol dependence using deep brain stimulation of the Nucleus accumbens: Clinical and basic science aspects. *Frontiers in Human Neuroscience 3*. Retrieved from http://www.frontiersin.org/human-neuroscience/paper/10.3389/neuro.09/022.2009/

Heyman, G. M. (2009). *Addiction: A disorder of choice.* Cambridge, MA: Harvard University Press.

Hoskovec, J. (1967). A review of some major works in Soviet hypnotherapy. *International Journal of Clinical and Experimental Hypnosis, 15*(1), 1–10.

Hyman, S. E. (2007). The neurobiology of addiction: Implications for voluntary control of behavior. *American Journal of Bioethics, 7*(1), 8–11.

Hyman, S. E. (2005). Addiction: A disease of learning and memory. *American Journal of Psychiatry, 162*(8), 1414–1422.

Ialovoi, A. I. (1968). [Substitution of the alcohol-antabuse test with a placebo in the treatment of alcoholism]. *Zhurnal nevropatologii i psikhiatrii imeni S. S. Korsakova, 68*, 593–596.

Insel, T. R., & Quirion, R. (2005). Psychiatry as a clinical neuroscience discipline. *JAMA, 294*(17), 2221–2224.

Ivanets, N. N. (2001). Sovremennaia kontseptsiia terapii narkologicheskikh zobolevanii [Contemporary conceptions of therapy for addictive disorders]. In N. N. Ivanets, (Ed.) *Lektsii po narkologii* [*Lectures on addiction medicine*]. Moscow: Medica.

Jain, S., & Jadhav, S. (2009). Pills that swallow policy: Clinical ethnography of a community mental health program in northern India. *Transcultural Psychiatry, 46*(1), 60–85.

Janousek, J., & Sirotkina, I. (2003). Psychology in Russia and central and eastern Europe. In T. Porter (Ed.), *The Cambridge history of science. Vol. 7: The modern social sciences,* (Chapter 24, pp. 431–449). Cambridge: Cambridge University Press.

Johnsen, J., & Morland, J. (1992). Depot preparations of disulfiram: Experimental and clinical results. *Acta Psychiatr Scand, 86*(S369), 27–30.

Joravsky, D. (1989). *Russian psychology, a critical history.* London: Blackwell.

Kalivas, P. W., & Volkow, N. D. (2005). The neural basis of addiction: A pathology of motivation and choice. *American Journal of Psychiatry, 162*(8), 1403–1413.

Kirman, B. (1966). Psychotherapy in the Soviet Union. In N. O'Connor (Ed.), *Present-day Russian psychology.* New York: Pergamon Press.

Kirmayer, L. J. (2007). Psychotherapy and the cultural concept of the person. *Transcultural Psychiatry, 44*(2), 232–257.

Kitanaka, J. (2008). Diagnosing suicides of resolve: Psychiatric practice in contemporary Japan. *Culture Medicine and Psychiatry, 32*(2), 152–176.

Kleinman, A. (1982). Neurasthenia and depression: A study of somatization and culture in China. *Culture Medicine and Psychiatry, 6*(2), 117–190.

Lakoff, A. (2006). *Pharmaceutical reason: Knowledge and value in global psychiatry.* Cambridge: Cambridge University Press.

Lauterbach, W. (1984). *Soviet psychotherapy.* London: Pergamon Press.

Lee, S. (1999). Diagnosis postponed: Shenjing Shuairuo and the transformation of psychiatry in post-Mao China. *Culture Medicine and Psychiatry, 23*(3), 349–380.

Lemere, F. (1987). Aversion treatment of alcoholism: Some reminiscences. *Addiction, 82*(3), 257–258.

Leshner, A. (1997). Addiction is a brain disease, and it matters. *Science, 278* (October 3), 45–47.

Levine, H. G. (1978). The discovery of addiction: Changing conceptions of habitual drunkenness in America. *Journal of Studies on Alcohol, 39,* 143–74.

Lichterman, B. L. (1998). Roots and routes of Russian neurosurgery (from surgical neurology towards neurological surgery). *Journal of the History of the Neurosciences: Basic and Clinical Perspectives, 7*(2), 125–135.

Lloyd, S. (2008). Morals, medicine and change: Morality brokers, social phobias, and French psychiatry. *Culture Medicine and Psychiatry, 32*(2), 279–297.

Lovell, A. (2006). Addiction markets: The case of high-dose buprenorphine in France. In A. Petryna, A. Lakoff, & A. Kleinman, (Eds.), *Global pharmaceuticals: Ethics, markets, practices.* Durham, N.C.: Duke University Press.

Luhrmann, T. (2000). *Of two minds: The growing disorder in American psychiatry.* New York: Knopf.

Mann, K. (2004). Pharmacotherapy of alcohol dependence - A review of the clinical data. *CNS Drugs, 18*(8), 485–504.

Martin, E. (2007). *Bipolar expeditions: Mania and depression in American culture.* Princeton: Princeton University Press.

Matza, T. (2009). Moscow's echo: Technologies of the self, publics, and politics on the Russian talk show. *Cultural Anthropology, 24*(3), 489–522.

May, C. (2001). Pathology, identity and the social construction of alcohol dependence. *Sociology, 35*(02), 385–401.

Medvedev, S. V., Anichkov, A. D., & Polyakov, Y. I. (2003). Physiological mechanisms of the effectiveness of bilateral stereotactic cingulotomy against strong psychological dependence in drug addicts. *Human Physiology, 29*(4), 492–497.

Mendelevich, V. D. (2005). Sovremennaia rossiiskaia narkologiia: paradoksal'nost' printsipov i nebezuprechnost' protsedur [Contemporary Russian narcology: The paradoxicality of principles and imperfection of procedures]. *Narkologiia, 5*(1).

Miller, M. (1998). *Freud and the Bolsheviks: Psychoanalysis in Imperial Russia and the Soviet Union.* New Haven: Yale University Press.

Miresco, M., & Kirmayer, L. (2006). The persistence of mind-brain dualism in psychiatric reasoning about clinical scenarios. *American Journal of Psychiatry, 163*, 913–918.

Miroshnichenko, L. D., Pelipas, V. E., & Ivanets, N. N. (2001). *Narkologicheskii entsiklopedicheskii slovar* [*The encyclopaedic dictionary of narcology*]. Moscow: Anakharsis.

Moore, J. (2005). Some historical and conceptual background to the development of B. F. Skinner's 'radical behaviorism.' *Journal of Mind and Behavior (Part 1) 26*, 65–94.

O'Brien, C. (2005). Anticraving medications for relapse prevention: A possible new class of psychoactive medications. *American Journal of Psychiatry, 162*, 1423–1431.

O'Brien, C., Volkow, N., & Li, T.-K. (2006). What's in a word? Addiction versus dependence in DSM-V. *American Journal of Psychiatry, 163*(5), 764–765.

Oldani, M. (2009). Uncanny scripts: Understanding pharmaceutical emplotment in the aboriginal context. *Transcultural Psychiatry, 46*(1), 131–156.

Orellana, C. (2002). Controversy over brain surgery for heroin addiction in Russia. *Lancet Neurology, 1*(6), 333.

Pavlov, I. (1994). Relations between excitation and inhibition, delimitation between excitation and inhibition, experimental neuroses in dogs. In *Psychopathology and Psychiatry.* New York: Transaction Publishers. (Original work published 1925).

Peele, S. (1989). *Diseasing of America: How we allowed recovery zealots and the treatment industry to convince us we are out of control.* San Francisco: Jossey-Bass.

Petry, N. (2006). Should the scope of addictive behaviors be broadened to include pathological gambling? *Addiction, 101*(Suppl. 1), 152–160.

Platonov, K. I. (1959). *The word as a physiological and therapeutic factor: The theory and practice of psychotherapy according to I. P. Pavlov.* Moscow: Foreign Languages Publishing House.

Rabinbach, A. (1992). *The human motor: Energy, fatigue, and the origins of modernity.* Berkeley: University of California Press.

Rabinow, P., & Rose, N. (2006). Biopower today. *Biosocieties, 1*(2), 195–217.

Raikhel, E. (2010). Post-Soviet placebos: Epistemology and authority in Russian treatments for alcoholism. *Culture, Medicine and Psychiatry, 34*(1), 132–168.

Reinarman, C. (2005). Addiction as accomplishment: The discursive construction of disease. *Addiction Research & Theory, 13*, 307–320.

Rose, N. (2007). *The politics of life itself: Biomedicine, power, and subjectivity in the twenty-first century.* Princeton: Princeton University Press.

Rose, N. (2003). The neurochemical self and its anomalies. In R. Ericson & A. Doyle (Eds.), *Risk and morality.* Toronto: University of Toronto Press.

Rozhnov, V., & Burno, M. (1987). Sistema emotsional'no-stressovoi psikhoterapii bol'nykh alkogolizmom [A system of emotional-stress psychotherapy for alcoholism patients]. *Sovetskaia Meditsina, 8*, 11–15.

Saris, A. (2008). An uncertain dominion: Irish psychiatry, methadone, and the treatment of opiate abuse. *Culture, Medicine and Psychiatry, 32*(2), 259–277.

Satel, S. (2001). Is drug addiction a brain disease? In P. Heymann & W. Brownsberger (Eds.), *Drug addiction and drug policy: The struggle to control dependence.* Cambridge, MA: Harvard University Press.

Schull, N. (2006). Machines, medication, modulation: Circuits of dependency and self-care in Las Vegas. *Culture Medicine and Psychiatry, 30*(2), 223–247.

Sedgwick, E. K. (1993). Epidemics of the will. In *Tendencies* (pp. 130–142). Durham: Duke

Segal, B. (1975). The theoretical bases of Soviet psychotherapy. *American Journal of Psychotherapy, 29*(4), 503–23.

Shabanov, P., & Shtakelberg, O.Y. (2001). The Formation of narcology: Clinical and biological tendencies of development. [Russian]. Medline.ru *2*(53), 314–318. Retrieved from http://www.medline.ru/public/art/tom2/art53.phtml.

Shaffer, H. (1997). The most important unresolved issue in the addictions: Conceptual chaos. *Substance Use and Misuse, 32*(11), 1573–1580.

Shorter, E. (1998). *A history of psychiatry: From the era of the asylum to the age of Prozac.* New York: John Wiley & Sons.

Shtereva, L. V. (1980). *Klinika i lechenie alkogolizma* [*The treatment of alcoholism*]. Leningrad: "Meditsina," Leningradskoe otdelenie.

Skultans, V. (2003). From damaged nerves to masked depression: Inevitability and hope in Latvian psychiatric narratives. *Social Science & Medicine, 56*(12), 2421–2431.

Skultans, V. (1997). A historical disorder: Neurasthenia and the testimony of lives in Latvia. *Anthropology & Medicine, 4,* 7–24.

Sluchesvky, I., & Friken, A. (1933). Lechenie khronicheskogo alkogolizma apomorfinom [On the treatment of chronic alcoholism with apomorphine]. *Sovetskaia vrachebnaia gazeta, 12,* 557–561.

Smith, R. (1992). *Inhibition: History and meaning in the sciences of mind and brain.* Berkeley: University of California Press.

Smith. T. C. (1996). *No asylum: State psychiatric repression in the former USSR.* London: Macmillan.

Solomon, S. (1990). Social hygiene and Soviet public health, 1921–1930. In S. Solomon & J. Hutchinson (Eds.), *Health and society in revolutionary Russia.* Bloomington: Indiana University Press.

Solomon, S. (1989). David and Goliath in Soviet public health: The rivalry of social hygienists and psychiatrists for authority over the *bytovoi* alcoholic. *Soviet Studies, 41*(2), 254–275.

Strel'chuk, I. V. (1954). *Alkogol': Vrag cheloveka.* [*Alcohol: The enemy of mankind*]. Moscow: Znanie.

Strel'chuck, I. V. (1952). Dal'neishie nabliudeniia za lecheniem chronicheskogo alkogolizma antabusom (tetraetiltiuramdisul'firadom) [Continued observations on the treatment of chronic alcoholism with antabuse (Tetraethylthiuram disulfide)]. *Zhurnal nevropatologii i psikhiatrii imeni S. S. Korsakova, 52*(4), 43–50.

Strel'chuck, I. V. (1951). Novyi metod lecheniia alkogolikov antabusom [Treatment of alcoholics with antabuse; Preliminary report]. *Nevropatologia i Psikhiatriia, 20*(1), 80–83.

Tkachevskii, Y. M. (1974). *Pravovye mery bor'by s pianstvom* [*Legal means in the struggle with drunkenness*]. Moscow: Moscow University Press.

Todes, D. (1995). Pavlov and the Bolsheviks. *History and Philosophy of the Life Sciences, 17*(3), 379–418.

Valverde, M. (2003). Targeted governance and the problem of desire. In R. Ericson & A. Doyle (Eds.), *Risk and morality.* Toronto: University of Toronto Press.

Valverde, M. (1998). *Diseases of the will: Alcohol and the dilemmas of freedom.* Cambridge: Cambridge University Press.

Vidal, F. (2009). Brainhood, anthropological figure of modernity. *History of the Human Sciences, 22*(1), 5–36.

Volkow, N., & O'Brien, C. (2007). Issues for DSM-V: Should obesity be included as a brain disorder? *American Journal of Psychiatry, 164*(5), 708–710.

Vrecko, S. (2006a). Folk neurology and the remaking of identity. *Molecular Interventions, 6*(6), 300–303.

Vrecko, S. (2006b). Governing desire in the biomolecular era: Addiction and the making of neurochemical subjects. (Doctoral Dissertation, London School of Economics and Political Science).

White, W. (1998). *Slaying the dragon: The history of addiction treatment and recovery in America.* Bloomington, Ill.: Chestnut Health Systems/Lighthouse Institute.

Whitehouse, T. (1999, February 7). Russian addicts cured by surgery. *The Guardian.*

Windholz, G. (1997). The 1950 joint scientific session: Pavlovians as the accusers and the accused. *Journal of the History of the Behavioral Sciences, 33*(1), 61–81.

Wortis, J. (1950). *Soviet psychiatry.* Baltimore: Williams & Wilkins.

Young, A. (2000). History, hystery and psychiatric styles of reasoning. In M. Lock, A. Young A. Cambrosio (Eds.), *Living and working with the new medical technologies: Intersections of inquiry.* Cambridge: Cambridge University Press.

Zenevich, G. (1967). *Vrednaia Privychka ili Bolezn': O vrede alkogolizma.* [*Bad habit or illness: On the harmfulness of alcoholism.*] Leningrad: Meditsina.

Zhislin, S. (1959). [Alcoholic abstinence syndrome.]. *Zhurnal nevropatologii i psikhiatrii imeni S. S. Korsakova, 59*(6), 641–648.

Zhislin, S., & Lukomskii, I. (1963). 30 let uslovnoreflektornoi terapii alkogolizma [30 years of conditioned reflex therapy of alcoholism]. *Zhurnal nevropatologii i psikhiatrii imeni S. S. Korsakova, 63,* 1884.

11

Delirious Brain Chemistry and Controlled Culture

Exploring the Contextual Mediation of Drug Effects[1]

Nicolas Langlitz

Today, there is a growing consensus that the dichotomy of nature and culture does not hold up. Deconstructing this distinction is not only a standard move in humanities scholarship, the emergence of the field of cultural neuroscience indicates that brain researchers are also beginning to explore the hybrid ontology of nature and culture. In psychiatry, biopsychosocial models of mental illnesses have been *en vogue* at least since the 1980s (even if, in clinical practice, pharmacotherapies outdid both psychotherapy and social therapy). Most recently, the proponents of a critical neuroscience (Choudhury, Nagel, & Slaby, 2009) have called for linking neuroscience and society by integrating insights from the social studies of neuroscience into neuroscientific research itself. Often these ontological professions go along with calls for interdisciplinary collaborations between natural and cultural scientists. In practice, however, the development of such interdisciplinary research paradigms has turned out to be difficult.

This chapter explores the potential of critical neuroscience in the context of an ethnographic case study from contemporary neuropsychopharmacology showing how neurochemistry and culture broadly conceived interact. It is based on anthropological fieldwork in two laboratories in Zurich and San Diego, which study the effects of hallucinogenic drugs on humans and animals respectively. Hallucinogen research is particularly suitable to explore the tense relationship between cerebral nature and scientific culture because substances such as LSD (Lysergic Acid Diethylamide) and psilocybin are pharmacologically powerful agents and yet their effects depend on a multitude of non-

---

[1] This chapter is a revised translation of Nicolas Langlitz (2010), "Kultivierte Neurochemie und unkontrollierte Kultur. Über den Umgang mit Gefühlen in der psychopharmakologischen Halluzinogenforschung." *Zeitschrift für Kulturwissenschaften*, no. 2, 61–88. The original German article is followed by a debate between the author and three natural and cultural scientists (Malek Bajbouj, Ludwig Jäger, & Boris Quednow).

Critical Neuroscience: A Handbook of the Social and Cultural Contexts of Neuroscience, First Edition. Edited by Suparna Choudhury and Jan Slaby.
© 2012 John Wiley & Sons, Ltd. Published 2016 by John Wiley & Sons, Ltd.

pharmacological, including cultural, factors. This raises the question whether anthropological second-order observations of how drug researchers observe their scientific objects can be fed back into these neuropsychopharmacological practices of first-order observation. The chapter addresses this issue by critically discussing a stillborn proposal from the 1950s concerning a research paradigm at the intersection of psychopharmacology and anthropology to investigate the cultural determinants of drug action.

Ethnographic Vignette I: Bad Trip

Experiences with hallucinogenic drugs can be emotionally difficult. Therefore, there is a far-reaching consensus within the community of hallucinogen researchers that scientists should familiarize themselves with the effects of these drugs before administering them to test persons. Personal experience is meant to help researchers to treat subjects empathically. Such drug experiences can be acquired legally in the context of so-called pilot studies, which also provide an opportunity to test the experimental setup before the actual trial begins.

During my fieldwork in Franz Vollenweider's laboratory Neuropsychopharmacology and Brain Imaging in Zurich, two scientists were preparing a study involving the drug psilocybin, the pharmacologically active ingredient of magic mushrooms. One of them—let's call her Anna—had never taken this substance. Therefore, Anna and her colleague Patrick decided to conduct a pilot study.[2] When Anna was administered the drug, the experiment worked smoothly. But when her older and more experienced colleague Patrick took the drug he received a nasty surprise. Patrick had already served as a test subject in two psilocybin trials without any difficulties. But this time, his experience was different. The experiment involved an EEG (electroencephalograph) measurement during which the test subject was shown a series of images on a computer screen. These images were part of the International Affective Picture System, which provides photographs of standardized emotional stimuli divided into three categories: pleasant (for example, landscapes, lovers), unpleasant (for example, attack scenes, mutilations), and neutral (for example, furniture, household articles). Even though all images selected for the experiment were meant to be affectively neutral, they scared Patrick. Eventually, he asked for the rest of the measurement to take place without the images.

By losing this attentional anchor his world was thrown completely out of kilter. First, a sweater appeared like a threatening grimace. Then the small EEG chamber grew bigger and bigger. Eventually Patrick saw himself as a midget in a huge white space. He felt like the only human being in the whole universe. In a self-reflexive moment, he began to worry that this onslaught of negative affects might interfere with the measurements. He felt nauseous and wanted to break off the experiment. But this thought scared him even more: didn't it prove that he was indeed in real trouble? As a psychiatric researcher, he conceived of hallucinogen intoxication as a kind of psychosis. Now he experienced how he himself gradually slipped into a schizophrenia-like state and felt threatened. The situation was further complicated by

---

[2]   To protect the researchers' privacy their names have been changed.

the role reversal between Patrick who was responsible for the study and his younger colleague Anna who now had to take care of him without anyone directing her. In retrospect, Patrick said:

> I tried to stay in charge, supervising how Anna was looking after me, checking how I was affected by the stimuli, whether the room would be bearable for the subjects, etc. I tried to evaluate all of this. The problem was that I wanted to keep everything under control, which is simply impossible on psilocybin. That made me fully aware of the fact that I was losing control. So I got all worked up about this. You need to let go.

After this test run the researchers decided to decorate the EEG chamber to make it look friendlier. They also replaced the allegedly neutral images by photographs from the category "pleasant."

Ethnographic Vignette II: "This is it!"

When I entered the EEG laboratory the experiment was already under way. The room was lit only by a computer screen displaying the brain waves of the test person. Through an observation window I could look into a neighboring chamber where the subject was located. At first glance, I could not see anything. But as my eyes got used to the darkness I began to make out the shaved head of a Zen master dimly illuminated by a monitor in front of him. He was sitting upright in a leather armchair. A tangled mass of wires seemed to be growing out of the back of his head only to disappear in the dark. Jan, a Swiss meditation teacher in his 50s, had been administered psilocybin to examine how the drug affected his consciousness. The young brain researcher who had invited me to observe this measurement was very excited. While Jan was meditating his brainwaves were particularly "calm," the scientist explained to me, showing strong activity in the alpha range.

After the measurement, Jan appeared happy and serene. The researcher interviewed him to hear about the experience that had accompanied the peculiar EEG pattern, which had been recorded. Jan reported that, at the beginning, he had seen frightening faces and carnivalesque processions of ghosts. But he remembered the *Tibetan Book of the Dead* and the fact that such visions are a mere projection of the ego. Eventually he turned to a simple mantra and began to focus his awareness on his breath. Thereby, he managed to free himself from this spooky spectacle and moved on to a "higher state of consciousness," as he called it. To his surprise and even disappointment the following experience of cosmic unity was associated with the name of Jesus. This must have been due to his upbringing in a Christian family, the dedicated Buddhist mused. But, finally, he also thought of Buddha and this further deepened his state of ego-dissolution. In comparison with his everyday consciousness, he recounted, he attained a much more profound insight into the fact that all existence was love. "Divine love," he specified, "or even better: being." This realization appeared to him as a perennial truth: "It has always been that way and it will always be that way. When reaching that state," he told us, "I thought: This is it! This is it!"—the state he had sought during three decades of regular meditation exercises.

The Persistence of the Subjective

These two ethnographic vignettes demonstrate that the same pharmaceutical can elicit very different, almost diametrical experiences. As far back as the 1950s, the British writer Aldous Huxley (1954) described that hallucinogenic drugs could take one to heaven or to hell. Accordingly, they were used for both the experimental investigation of mystical states (Griffiths, Richards, McCann, & Jesse, 2006; Langlitz, in press; Pahnke & Richards, 1966) and as pharmacological models of schizophrenic psychoses (Beringer, 1927; Langlitz, 2006; Vollenweider, 1998). However, representatives of these two approaches were often divided by their antagonistic worldviews. One party was indignant at the pathologization of spiritual experience while the other party ridiculed the mystification of a deranged brain metabolism. This conflict is based on the assumption that both camps are talking about the same brain chemistry, which they only interpret differently. As a pharmacologist from the Vollenweider group put it: "Hallucinogens enable you to have limit-experiences. Whether one regards such liminal states as mystical experiences or as psychotic delusions is mostly a matter of interpretation" (Hasler, 2007, p. 39 [my translation]).

However, it was Vollenweider's laboratory, which endowed these antipodal experiences with objectivity by identifying their neural correlates. For this purpose, Vollenweider and colleagues (1997) used positron emission tomography (PET) in order to measure the metabolic activity in the brains of test subjects under the influence of psilocybin. Afterwards, they asked them to fill in questionnaires to record their subjective experiences. Subjects had to rate statements such as "I saw strange things which I now know were not real," "I felt an all-embracing love," or "I felt threatened without realizing by what" on a scale of 1–10. More than 90 items of this sort were supposed to capture and quantify three dimensions of altered states of consciousness: "visionary restructuralization" encompassing hallucinatory phenomena, "oceanic boundlessness" dealing with ecstatic experiences, and "dread of ego-dissolution" covering the more horrifying aspects of their experiences. Such self-rating scales translated inner experiences into numbers, which could then be correlated with PET measurements. Vollenweider's investigation demonstrated that dread of ego-dissolution and the blissful transgression of ego-boundaries went along with the activation of different brain areas (Langlitz, 2008; Vollenweider, Vollenweider-Scherpenhuyzen, Bäbler, Vogel, & Hell, 1998). Consequently, mystical experiences and bad trips are not two interpretations of the same neurophysiological event, but neurophysiologically distinct states.

In scientific practice, however, this objectifying approach to the study of psycho-pharmacologically induced mind–brain states soon reaches an ethico-epistemological limit. Test persons are not objects of investigation that can be observed from a distance. For ethical reasons, researchers cannot passively watch a subject sliding deeper and deeper into a state of horror. The scientists familiarize themselves with the effects of the applied substances precisely to become more empathic, and to be better equipped to take countermeasures if subjects are about to get emotionally unstable. But bad trips are also detrimental to the scientific study as such. As participation in experiments is voluntary, test subjects can break off measurements at any time if they feel too uncomfortable. In that case, the scientists would lose their data. In both their own interest and their subjects' interest, they cannot sit back while their perfectly impartial measuring devices register the neural correlates of exacerbating "dread of ego-dissolution."

Despite all objectivizing procedures (standardized experimental protocols, instrumental recordings, and so forth) the experimental space continues to be pervaded by the subjectivity of both test persons and scientists. The epistemic virtue of objectivity (Daston & Galison, 2007) associates itself with the cultivation of intersubjectivity: the art of taking good care of subjects. In practice, the neuroscientific investigation of (altered states of) consciousness cannot be reduced to the correlation of first-person and third-person perspectives, but crucially involves the second-person perspectives and social interactions of researchers and test persons alike (Roepstorff, 2001). In such experimental settings, neuroscientists cannot adopt the position of detached observers, but must interact with their subjects to obtain data and insights marked by these engagements. Here, brain research—just like anthropology—turns out to be a form of participant observation. Consequently, it is equally entangled in the epistemological problematics of the human sciences (Langlitz, 2010).

Setting Matters: The Limits of Placebo Controls

The ethnographic investigation of the practice of contemporary hallucinogen research shows that scientists are well aware of the impact of environment, interpersonal treatment, and expectations on subjects' experiences and brain states. Nevertheless the predominant study design of pharmacological research continues to be the randomized, double-blind placebo-controlled trial. While all other conditions are supposed to be kept identical, subjects randomly receive a pharmacologically active drug or an inactive placebo. Neither the researcher nor the test person knows whether the former or the latter is administered. The underlying assumption is that all psychosocial and cultural factors are also operative when the placebo is given. Hence, when subtracting the placebo's effects from the effects of the pharmacological agent, the drug's own activity is revealed in its purest form. If the psychotropic effect of the drug should have been affected by the organism's environment or mood, this influence is hereby made to disappear.

But what would happen if the pharmacological activity of a substance also changed the relationship between a living thing and its environment—not in a deterministic and unilinear way, but depending on the quality of the environment? In this case, the particular environment would still be inscribed in the observed pharmacological effect when the placebo effect (measured under identical conditions) has been subtracted. Instead of effectively neutralizing the impact of the environment, placebo controls merely render it invisible.

During my fieldwork in Mark Geyer's animal laboratory in San Diego I discovered a peculiar practice based on such an ecological understanding of psychopharmacology. On the eve of a set of hallucinogen experiments, the rats were brought from their home cages in the basement to the lab facilities to familiarize them with this unknown environment and their handling by the experimenters. This procedure, prescribed by a special protocol, was based on an experiment which had shown that LSD made rats more afraid of new things and open spaces. In an unfamiliar box, in which infrared rays registered the animals' exploratory behavior, the rats moved around less under the influence of LSD. They preferred to stay close to the walls instead of venturing into the center of the box and generally showed less curiosity. Next, the researchers connected the rats' home cages to these unknown motion-tracker boxes, allowing them to go back and forth between the two spaces. In spite of the LSD effects, they moved around normally in the familiar space of the home cage whereas they displayed increased fearfulness in the unfamiliar space (Geyer & Krebs, 1994). Hence, this dread of the new (neophobia) could neither be entirely attributed to the drug nor to the environment, but resulted from a drug-induced change of the animals' attitudes toward these different environments. The custom of familiarizing the rats with the laboratory on the day before the experiment, which had been derived from this finding, was supposed to minimize the impact of the novelty of the lab space on the rodents' behavior. It did not, however, eliminate the ecological conditioning of the animals' minds.

This experiment and the ethnographic observations presented above point to the fact that the realm of the mental cannot be reduced to the brain, but encompasses the organism's surroundings (Clark & Chalmers, 1998; Noë, 2005). To study the psychopharmacological activity of a drug, it is not sufficient to look only at its effects on the mind/brain while turning a blind eye to the environment, as happens in randomized placebo-controlled trials. The physical, atmospheric, social, and—at least in the case of humans—cultural qualities of the setting, in which a drug is taken, also determine how an organism responds to it. In allusion to Margaret Lock's (1995) discovery of "local biologies" (in the sense of biological differences molding and containing subjective experience and cultural interpretations), the decisive role of the circumstances of drug ingestion can be taken as a powerful indicator of the existence of "local pharmacologies."

Controlling for Culture

The fact that the psychopharmacological effects of hallucinogens depend on the complex contexts of drug ingestion was first described in 1959 by anthropologist Anthony Wallace. Wallace was primarily working on Native Americans. At the time, however, he served as research director at an institute of psychiatry where hallucinogen experiments were conducted. Wallace noted that the experience reports of white test persons who had been given mescaline differed significantly from the reports of Native American participants in peyote ceremonies ingesting a cactus also containing mescaline. After administration of the drug, Caucasian experimental subjects experienced extreme mood shifts—from depressive and anxious to euphoric. When eating peyote buttons, indigenous people, on the other hand, displayed an "initial relative stability of mood, followed by religious anxiety and enthusiasm, with tendency to religious reverence and personal satisfaction when vision achieved." The Whites suffered from "unwelcome feelings of loss of contact with reality" whereas the peyotists embraced "feelings of contact with a new, more meaningful, higher order of reality". Wallace attributed these and other differences to two factors: the impact of the setting in which the drug was taken and the different meanings ascribed to the physiological "primary drug effects" (Wallace, 1959 pp. 58–69).

From this observation, Wallace concluded that placebo-controlled studies (which, at the time, were only beginning to get established) had to be supplemented by "cultural controls." He proposed not only to vary the pharmacological activity of the administered substance, but also to test the same drug under different cultural and situational circumstances in order to systematically investigate (and subsequently control) the impact of these conditions on psychotropic effects. In this context, Wallace's notion of culture was quite broad. The suggested culture controls comprised the socio-cultural background of test subjects, their personality and expectations vis-à-vis the experiment, their social treatment by laboratory staff, and the experimental setting as a whole. He speculated that these factors would not only affect the effects of hallucinogens, but of all psychopharmaceuticals.

While placebo-controlled studies soon became the gold standard of pharmacological research, Wallace's culture-controlled trials never really caught on. For scientific, disciplinary, economic, and political reasons, biological psychiatry and psychopharmacology had an interest in attributing the effects of drugs to the drugs alone. This ideology of "pharmacologicalism" helped psychiatry to be acknowledged as part of scientific medicine, enabled pharmaceutical companies to fulfill the Food and Drug Administration's regulatory requirement to demonstrate specificity of drug action, and legitimized the War on Drugs (DeGrandpre, 2006).

At the same time, culturalist approaches gained the upper hand in the field of cultural anthropology, which became increasingly alienated from the biological part of the discipline. In the last quarter of the twentieth century, the disciplinary unity of anthropology broke apart as anthropologists came to reject the association of non-European peoples with early hominids and non-human primates that had been constitutive of US anthropology's holistic agenda, but was enmeshed in the distinction between the West and the rest. The culturalist response to this complicity of anthropological holism and colonialist racism was not to apply a biocultural perspective to humankind overall instead of non-European others alone, but to exclude biological approaches and to focus on the study of cultures—both Western and non-Western (Clifford, 2005; Segal & Yanagisako, 2005). Rather than identifying the "cultural determinants" of psychopharmacological effects as Wallace (1959) had sought to do, culturally oriented studies of drugs focused on the drug as symbol (for example, Myerhoff, 1974) or on historically and culturally different interpretations of identical neurochemical effects (Becker, 1963; Zinberg, 1984). These approaches were based on the implicit ontological assumption that there is one nature and many cultures.

Having fallen between the two stools of cultural anthropology and psychopharmacology, Wallace's "method of cultural and situational controls" led a shadowy existence. However, such marginalized practices can enable a critique that does not come from outside, but from the fringes of psychopharmacology itself (Dreyfus & Rabinow, 1982, pp. 262–263). The question is just whether, 50 years after the publication of Wallace's article, drug researchers are willing to reinvent psychopharmacology as a hybrid of natural and cultural science and whether anthropologists are willing to return to an anthropology that is not split into biological and cultural.

Conclusion

Against the background of ubiquitous calls for interdisciplinary perspectives and for overcoming the nature/culture dichotomy, this question might appear rhetorical. But Wallace's proposal to control for culture presupposes a reification and essentialization of culture, which few cultural anthropologists still subscribe to today. In order to understand the cultural dimension of human life, "thick descriptions" (Geertz, 1973) seem to be more promising than the experimental variation of isolated factors in the laboratory because the effect of each individual factor depends on its role in a whole network of factors (Latour, 1999, pp. 174–215). Such networks might well be too complex to be controlled successfully. This makes it difficult for laboratory scientists to extract statistically significant signals from the cultural noise—even if the entirety of non-pharmacological factors has a powerful impact. Therefore, Wallace's culture-controlled trials do not appear to provide a satisfying answer to the question of how to factor in complex environments. If experience is over-determined by intricate contexts, then field studies appear to be a more suitable approach than controlled experiments, but play only a very marginal role in psychopharmacology. What are still missing in the life sciences are methods which are not—not even for heuristic purposes—based on reduction, but measure up to the complexity of life itself (Mitchell, 2009).

Even though cultural anthropologists often denounce scientific reductionism they will not have much to contribute to overcoming it as long as they reject the dichotomy of nature and culture ontologically while continuing to be committed to culturalist methodologies. All too often they look at the natural sciences exclusively as culture—in other words from the perspective of second-order observation. Second-order observation means to observe how others observe the world while ignoring what they look at. Taking up such a perspective can be important because it reveals the blind spots and contingencies of first-order observations of the world (Luhmann, 1998). For example, it allows us to see, as this chapter has shown, that from the point of view of placebo-controlled trials the contextual mediation of drug effects cannot be recognized. But to the extent that cultural anthropology takes part in the scientific cultures that it observes, it should also contribute to their improvement.

For this purpose, it is not enough to restrict oneself to second-order observations and to uncover contingency after contingency. At some point, second-order observation should inspire the invention of new practices of first-order observation (Langlitz, 2007). Therefore, Wallace's proposal of culturally and situationally controlled trials—however dissatisfying it might be—is well worth a second look. A productive debate between natural and cultural sciences is only possible if observations of the world and observations of such observations are discussed together. This is the project of critical neuroscience. But it is still a long way off for this agenda to translate into non-reductionist research paradigms to study the brain in context.

Acknowledgements

Without the hospitality of Franz Vollenweider, Mark Geyer, and their research groups I would not have been able acquire the insights into neuropsychopharmacological practice informing this chapter. The Volkswagen Foundation's European Platform for the Life Sciences, Mind Sciences and Humanities provided an inspiring and well-funded forum to experience and think through the intricacies of interdisciplinary collaborations. Finally, I would like to thank Donya Ravasani, Suparna Choudhury, Allan Young, Maria-Christina Lutter, and Daniela Hammer-Tugendhat for their helpful responses to earlier drafts of this chapter.

References

Becker, H. (1963). *Outsiders: Studies in the sociology of deviance*. London: The Free Press of Glencoe.

Beringer, K. (1927). *Der Meskalinrausch. Seine Geschichte und Erscheinungsweise*. Berlin: Julius Springer.

Choudhury, S., Nagel, S., & Slaby, J. (2009). Critical neuroscience: Linking neuroscience and society through critical practice. *BioSocieties, 4*(1), 61–77.

Clark, A., & Chalmers, D. (1998). The extended mind. *Analysis, 58*(1), 7–19.

Clifford, J. (2005). Rearticulating anthropology. In D. Segal & S. Yanagisako (Eds.), *Unwrapping the sacred bundle: Reflections on the disciplining of anthropology,* (pp. 25–48). Durham, NC: Duke University Press.

Daston, L., & Galison, P. (2007). *Objectivity*. New York: Zone Books.

DeGrandpre, R. (2006). *The cult of pharmacology: How America became the world's most troubled drug culture*. Durham, NC: Duke University Press.

Dreyfus, H., & Rabinow, P. (1982). *Michel Foucault: Beyond structuralism and hermeneutics*. Chicago: The University of Chicago Press.

Geertz, C. (1973). Thick description: Toward an interpretive theory of culture. In *The interpretation of cultures: Selected essays* (pp. 3–30). New York: Basic Books.

Geyer, M., & Krebs, K. (1994). Serotonin receptor involvement in an animal model of the acute effects of hallucinogens. In G. C. Lin & R. A. Glennon (Eds.), *Hallucinogens: An update. NIDA Research Monograph, 146*, 124–156.

Griffiths, R., Richards, W., McCann, U., & Jesse, R. (2006). Psilocybin can occasion mystical-type experiences having substantial and sustained personal meaning and spiritual significance. *Psychopharmacology, 187*(3), 268–283.

Hasler, F. (2007). LSD macht keinen zum Genie [interview by Thomas Gull & Roger Nick]. *Unimagazin, 1*(2), 39–42.

Huxley, A. (1954) *The doors of perception*. London: Chatto & Windus.

Langlitz, N. (2011). Political neurotheology: Emergence and revival of a psychedelic alternative to cosmetic psychopharmacology. In F. Ortega & F. Vidal (Eds.), *Neurocultures. Glimpses into an expanding universe*. New York: Peter Lang, 141–165.

Langlitz, N. (2010). The persistence of the subjective in neuropsychopharmacology. Observations of contemporary hallucinogen research. *History of the Human Sciences, 23*(1).

Langlitz, N. (2008). Neuroimaging und Visionen. Zur Erforschung des Halluzinogenrauschs seit der 'Dekade des Gehirns'. *Bildwelten des Wissens. Kunsthistorisches Jahrbuch für Bildkritik, 6*(1), 30–42.

Langlitz, N. (2007). What first-order observers can learn from second-order observations. *ARC Concept Note*, http://anthropos-lab.net/wp/publications/2007/08/conceptnoteno3.pdf.

Langlitz, N. (2006). Ceci n'est pas une psychose. Toward a historical epistemology of model psychosis. *BioSocieties, 1*(2), 158–180.

Latour, B. (1999). *Pandora's hope: Essays on the reality of science studies.* Cambridge, MA: Harvard University Press.

Lock, M. (1995). *Encounters with aging: Mythologies of menopause in Japan and North America.* Berkeley: University of California Press.

Luhmann, N. (1998). *Observations on modernity.* Stanford: Stanford University Press.

Mitchell, S. (2009). *Unsimple truths: Science, complexity, and policy.* Chicago: University of Chicago Press.

Myerhoff, B. (1974). *Peyote hunt: The sacred journey of the Huichol Indians.* Ithaca: Cornell University Press.

Noë, A. (2005). *Action in perception.* Cambridge, MA: MIT Press.

Pahnke, W., & Richards, W. (1966). Implications of LSD and experimental mysticism. *Journal of Religion and Health, 5*(3), 175–208.

Roepstorff, A. (2001). Brains in scanners: An Umwelt of cognitive neuroscience. *Semiotica, 134*(1), 747–765.

Segal, D., & Yanagisako, S. (2005). Introduction. In D. Segal & S. Yanagisako (Eds.), *Unwrapping the sacred bundle: Reflections on the disciplining of anthropology* (pp. 1–23). Durham, NC: Duke University Press.

Vollenweider, F. (1998). Recent advances and concepts in the search for biological correlates of hallucinogen-induced altered states of consciousness. *The Heffter Review of Psychedelic Research, 1,* 21–32.

Vollenweider, F., Vollenweider-Scherpenhuyzen, M., Bäbler, A., Vogel, H., & Hell, D. (1998). Psilocybin induces schizophrenia-like psychosis in humans via a serotonin-2 agonist action. *NeuroReport, 9*(17), 3897–3902.

Vollenweider, F., Leenders, K., Scharfetter, C., Maguire, P., Stadelmann, O., & Angst, J. (1997). Positron emission tomography and fluorodeoxyglucose studies of metabolic hyperfrontality and psychopathology in the psilocybin model of psychosis. *Neuropsychopharmacology, 16*(5), 357–372.

Wallace, A. (1959). Cultural determinants of response to hallucinatory experience. *Archives of General Psychiatry, 1*(1), 58–69.

Zinberg, N. (1984). *Drugs, set, and setting: The basis for controlled intoxicant use.* New Haven: Yale University Press.

Part IV
Situating the Brain
From Lifeworld Back to Laboratory?

12

From Neuroimaging to Tea Leaves in the Bottom of a Cup[1]

Amir Raz

Hardly any advance in neuroscience has garnered as much public interest as imaging of the living human brain. The crisp images of brains in action seem to mesmerize the masses, including many a neuroscientist (Dumit, this volume). This trend is especially conspicuous in the cognitive and behavioral sciences, including psychology and psychiatry. Before examining results from any imaging excursion, however, it may be advisable to ruminate about the process and methodology of neuroimaging. After all, it takes a great deal of computer processing and human judgment to get from blood oxygen levels to a snapshot of a higher brain function. Critical neuroscience is an important conceptual call to exercise judicious consideration while the popular media publish stunning pictures, sometimes from the labs of respectable neuroscientists, spanning sexy topics such as political attitudes of voters and commercial ventures such as brain-based lie detection. Such reports capitalize on the scientific cache of brain imaging to increase their clientele, in addition to whatever valid information the imaging findings may suggest. Critical neuroscience should discern good from bad reasons for skepticism about the conclusions of such studies and afford scientific ways to evaluate and validate such claims.

Functional magnetic resonance imaging (fMRI) is one of the main neuroimaging methods in current use. Increasingly ubiquitous, fMRI is a non-invasive technique that permits imaging of the living brain and provides findings that relate neural to cognitive activity by measuring small changes in the magnetic properties of blood (Huettel, Song, & McCarthy, 2004). Given that the density of neurons and synapses in the cerebral cortex is about 12×10^4 and 9×10^8 per mm³, respectively, it becomes evident that fMRI signal is a crude index of the overall activity of many neurons and processes. Most fMRI measurements rely on the Blood Oxygenation Level Dependent (BOLD) signal, which is an indirect measure of neuronal activity. By placing the living

[1] This chapter draws on a Target Article published by the author in *Neuropsychoanalysis*, with kind permission.

Critical Neuroscience: A Handbook of the Social and Cultural Contexts of Neuroscience, First Edition.
Edited by Suparna Choudhury and Jan Slaby.
© 2012 John Wiley & Sons, Ltd. Published 2016 by John Wiley & Sons, Ltd.

brain in a strong magnetic field, we can measure the BOLD signal mainly from the capillaries, venules, and veins (in arteries and arterioles there is little deoxyhemoglobin). Although it is difficult to conclude whether the BOLD signal reflects neurons firing (that is, spiking activity) or synaptic activity, BOLD response directly reflects a local increase in neural activity (for example, as assessed by the mean Extracellular Field Potential signal). Technologies such as fMRI entice researchers to submit higher brain functions, including morality, (Greene, Nystrom, Engell, Darley, & Cohen, 2004) to scientific scrutiny. The images gathered by such efforts, however, may enthrall more than explain (McCabe & Castel, 2008). This type of "neurorealism" speciously leads individuals to believe that images of brain activity make a behavioral observation more scientific (Racine, Bar-Ilan, & Illes, 2006a); consequently media coverage frequently oversimplifies research findings and marginalizes caveats (Racine et al., 2006b). In November 2009, for example, the *New York Times* (*NYT*) published an op-ed column describing fMRI findings from undecided voters who viewed photographs and videos of the major candidates in the 2008 US presidential election (Iacoboni et al., 2007). According to the study's authors as articulated in the *NYT*, the findings revealed "some voter impressions on which this election may well turn." A later editorial in *Nature* lambasted studies that simply place individuals in fMRI scanners and then come up with elaborate stories describing the results ("Mind games", 2007). Consumers of neuroimaging may benefit from a measure of rigor (Kriegeskorte, Simmons, Bellgowan, & Baker, 2009).

While neuroimaging is a relatively young enterprise, it has occupied a prominent place not only within neuroscience but also in popular science, the popular media and contemporary culture. This chapter addresses the pros and cons of using fMRI—one of the many neuroimaging technologies—by taking an overarching approach and touching on some of the important steps followed, from experimental design all the way to the interpretation and dissemination of the results. I aim this exposition at both consumers of neuroimaging (for example, journalists and policymakers) and professional neuroimagers (for example cognitive neuroscientists). As the chapter unfolds, these two distinct viewpoints may become less discernible. Indeed, the perspectives of consumers versus providers of neuroimaging information sometimes intersect and bleed into one another. While social scientists, professionals of the media, and policymakers would benefit from closer engagement with the details and limitations of the methods, neuroscientists would do well to impose a measure of rigor in the communication of their experiments and findings. Using fMRI as a vehicle, I outline a constructive approach and sketch the caveats and merits of this important technique, including how—like a "new phrenology"—neuroimaging can engender useful insights and pave the road to new scientific understandings.

The Perils of Neuroimaging

As seventeen prominent cognitive neuroscientists pointed out in a collective reply to the *NYT* op-ed piece, one of the core shortcomings of a naïve fMRI approach hinges on reverse inferences—inferring a specific mental state from the activation of a particular brain region (Aron et al., 2007). For example, anxiety involves fMRI signal changes in the amygdala, but so do many other things, including intense smells and sexually explicit images. The blunder of "reverse inference" is widespread and many neuroimagers, including signatories to the *NYT* rebuke, have slipped into reverse-inferencing in an attempt to understand how brain mechanisms subserve mental processes (Poldrack & Wagner, 2004). Because cognitive neuroscience is a relatively new field of scientific inquiry, however, some of the same researchers who initially advocated the idea of reverse inferences have grown considerably more skeptical of it in recent times (Poldrack, 2006). Although reverse inferences may still be useful in specific situations, cumulative analyses over the past few years have resulted in marked disillusionment regarding many of the reverse inferences presented in the literature. Thus, past support for reverse inferences has taken a turn against it.

Reverse inferences are particularly common in newer fields, such as social cognitive neuroscience, in which researchers are still trying to identify the cognitive processes underlying the behaviors they investigate. One study, for example, used fMRI to explore the neural underpinnings of individuals who were mulling over moral dilemmas (Greene, Sommerville, Nystrom, Darley, & Cohen, 2001). Brain areas with fMRI signal changes included regions that had been linked to "emotional" and "rational" cognitive processes in previous studies. Researchers therefore concluded that these two types of process are active, in varying degree, in different types of moral judgment. The rigor of such arguments, however, depends on the evidence that a focal brain area instigates a particular mental process. However, at least some of the emotional brain regions in the morality study have also been associated with memory and with language. It is curious that such caveats typically escape mention (Miller, 2008).

Using results from brain imaging as probabilistic markers of brain states may represent a viable approach, but we must scrutinize the probabilities. Testing these odds on real data revealed that while engagement of an individual region did provide some statistical information regarding the engagement of a mental process, the added information was relatively weak (Poldrack, 2006). Cognitive neuroscience may ultimately find ways to predict mental states using neuroimaging data; but even then, rather than surfacing from localized activity in a focal brain region, such predictions will probably result from both subtle activation patterns and coordinated activity across many brain regions.

Using specific reverse inferences (for example, the association of fMRI signal change in the amygdala with anxiety) is a function of previous publications. The distribution of terms in the literature is, however, a function of past theories that have driven publications in particular directions, and which may hardly reflect current perspectives. For example, scientific literature contains many more citations for "amygdala and anxiety" than "amygdala and tranquility." This difference, however, is a reflection of roughly 30 years of research investigating the association between anxiety and amygdala activity whereas only recently have researchers begun to examine the role of the amygdala in positive emotional responses. Thus, to deduce that fMRI signal changes localized to the amygdala are a strong prognostic of negative emotion, may be misleading.

fMRI has transformed neuroscience in fewer than two decades yet many studies, including some of those that garner the most attention in the popular and trade press, shed little light on the neural mechanisms of human cognition, affect, thought, and action. Researchers attempt to confront the limitations of fMRI by conducting experiments that match human fMRI data with analogous fMRI and electro-physiological recordings of neural activity in non-human primates. The general idea is to follow up on the human findings by identifying equivalent regions of the monkey brain using fMRI, and then recording the activity of individual neurons in those locations using microelectrodes. In some cases, single neuron recordings in monkeys have confirmed fMRI findings in humans (Tsao, Freiwald, Tootell, & Livingstone, 2006). Although the parallel human-monkey approach represents an admirable, albeit time-intensive, paradigm, one of its main drawbacks is the difficulty in applying it to study many types of human cognition and social interaction.

Comely fMRI-generated images may seduce the general public, but even neuroscientists seem to fall for them and overlook the limitations of neuroimaging. One constraint is the narrow sliver of the human experience that researchers can capture when a person has to keep still inside a scanner; another limitation has to do with the posture current scanners impose on participants (Raz et al., 2005). Yet another limitation pertains to resolution: using fMRI to measure nuanced neural activity is akin to observing ocean currents to learn about the properties of water droplets. fMRI can only detect large-scale activities; generalizations to subtle local effects are speculative and tenuous at best. In addition, with standard fMRI equipment, even the atomic volume-pixel unit of imaging (that is, the voxel) typically comprises millions of neurons. Neurons can fire hundreds of impulses per second, however, and the fMRI signal—triggered by an increase in oxygenated blood—builds incrementally and peaks after several seconds, not instantaneously. fMRI is therefore an indirect and crude tool for investigating how neuronal ensembles "compute" cognition and behavior. It can be helpful in guiding where something is happening in the brain, but it is considerably more difficult to use this neuroimaging technique to elucidate mechanisms.

fMRI signals are weak and occur amidst much "noise" in the form of false signals. Moreover, the real signals are often so weak that researchers have to stimulate a person's brain time and again to discern an incipient pattern. To study the brain areas that respond to faces, for example, researchers typically present many faces in order to detect an increase in neural activity in a specific brain location. Thereafter, they repeat the experiment on a dozen or more additional individuals to ascertain that the same brain areas consistently light up across people. In many cases, this outcome is unwarranted even though face recognition is a relatively robust process. Critical neuroscientists need therefore to be wary of embracing fMRI findings that purport to index and sometimes even identify higher brain functions.

fMRI studies frequently produce billions of data points—most of them sheer noise—wherein one can find coincidental patterns (Kriegeskorte, Simmons, Bellgowan, & Baker, 2009). Whirl those tea leaves around often enough and recognizable impressions will appear at the bottom of your cup. In addition, many fMRI studies dip into the same data twice: first to pick out which parts of the brain are responding; and second to measure the response strength. This practice is statistically problematic and results in findings that appear stronger than they actually are (Vul, Harris, Winkielman, & Pashler, 2009). Onlookers should know from what messy data these attractive images are formed.

The Promise of Neuroimaging

A very different approach to overcoming some of fMRI's constraints comes from new analysis tools borrowed from machine-learning research. In a standard fMRI study, neuroscientists average together the fMRI activation from neighboring voxels. While averaging makes it easier to detect differences between experimental conditions, this technique follows the assumption that neurons from different voxels all behave the same way, an assumption that is, however, extremely unlikely. Instead, it is possible to use statistical tools, including multivariate pattern classifiers that recover small biases in individual voxels in their responses to different stimuli, to take a finer-grained look at brain activity and consider patterns of activation across many individual voxels without averaging. These methods shift the focus from trying to identify the specific brain regions activated during a particular task, to trying to identify how the brain processes germane information.

An early demonstration of this statistical approach came from a neuroimaging study that presented participants with hundreds of images of faces, cats, houses, and scissors (Haxby et al., 2001). The investigators identified statistically distinct brain activity patterns elicited by each type of object. fMRI activation in the primary visual cortex made it possible to determine the orientation of lines a participant was viewing, a feat previously thought impossible because neurons that share a preference for lines of a particular orientation pack into columns narrower than a voxel (Op de Beeck, Haushofer, & Kanwisher, 2008; Tong, 2003). A variety of new findings illustrate how this new analysis of fMRI data can reveal information processing in the brain that would be overlooked by conventional analyses (Raizada, 2008). Hence, rather than looking at whether a specific brain region is active, researchers are beginning to focus on whether the activity pattern in many different voxels can predict what people are experiencing. In other words, instead of inferring that a spider induces anxiety, researchers could collect patterns of brain activity evoked by known anxiety inducers (images of snakes, accidents about to happen, and presurgical situations, for example) and see whether the spider pattern forms a statistical match. Although it may well be that such classifiers will help rescue fMRI research from the logical perils of reverse inference, even with the promise of these new tools fMRI remains limited to revealing correlations between cognitive processes and activity in the brain.

fMRI may be most effective when people view it as one tool in a toolbox (in other words, by employing converging techniques and evidence). Increasingly, neuroscientists are using fMRI and related methods to investigate the connectivity between different brain regions involved in cognitive functions such as language and memory. One fMRI approach is to identify brain regions showing synchronized activity when subjects perform a given task. In some cases, researchers use diffusion tensor imaging (DTI) to further determine whether physical connections link those areas that fire together. A relatively new MRI method, DTI provides a way to visualize the axon tracts that connect regions. Some researchers are trying to establish causal links between brain and behavior. Having linked a brain region to a particular behavior using fMRI, for example, researchers are following-up with Transcranial Magnetic Stimulation (TMS) experiments; TMS delivers a short burst of a powerful magnetic field to a specific brain area, inducing a temporary brain "lesion" which is reversible in nature and leaves no anatomical traces. If the behavior then changes, the brain region probably plays a role in controlling it.

Conclusion

Most neuroscientists are unlikely to deliberately mislead by manipulating their image processing in immoral ways. Instead, many individuals who are skeptical of the image processing procedures involved in functional neuroimaging often thrive on an overreaction to the realization that functional brain images are hardly as straightforward as a photograph. That neuroscientists can process neuroimaging data to make them show anything is a myth; however, coming up with functional brain images involves large amounts of data processing, which—in the hands of inexperienced or unprincipled researchers—may distort the evidence. On the other hand, functional brain images are probably as prone to fakery as any other kind of scientific evidence.

fMRI is a relatively new method, and its potential for measuring psychological phenomena is still a matter for experimentation and exploration. We should not conclude that imaging simply cannot provide useful information about the mental states of individuals (for example, reactions to specific political candidates), nor indeed that the use of brain imaging for such purposes is, by definition, poor science. Whereas most bread-and-butter applications of fMRI involve extrapolating from many repetitions of tightly controlled experimental tasks over a small number of participants, we should at least entertain the possibility that fMRI may generalize beyond such uses, including perhaps to indexing the kinds of attitudes and feelings that are relevant to political campaigns.

The *NYT* report will most probably receive greater attention because it involved neuroimaging. Furthermore, the general public will probably attribute more credibility to it compared with studies that were to use only behavioral measures such as surveys. Alas, humans fall for the fancy technology of neuroimaging because they erroneously construe it as more scientific and perhaps even more objective. As neuroimaging variants continue to spiral, hyperscanning—a method by which multiple participants, each in a separate MRI scanner, can interact with one another while their brains are simultaneously scanned (Babiloni, Astolfi et al., 2007; Babiloni, Cincotti et al., 2007; Babiloni et al., 2006; Montague et al., 2002)—becomes more prevalent. Hyperscanning technology seems to permit the study of brain responses that underlie important social interactions. It would behoove us, however, to pause and ponder the lessons of critical neuroscience.

Critical neuroscience must offer constructive ways to address—rather than carp about—the inherent shortcomings of neuroimaging research. Fortunately, neuroscientists can distinguish between parsimonious interpretations that are in line with the data and looser explanations that appeal to a story-like narrative. For example, we can spend more time testing neuroimaging methods using questions for which we know the answer. As a case in point, if the neuroscientists who published in the *NYT* were to select a group of individuals whose likely attitudes we can all agree on in advance, they could carry out imaging studies like the ones they reported and then, blind to the identity of personage and participant for each scan, interpret the patterns of activation. Alas, these types of studies are sorely lacking. The *NYT* imaging study may well have extracted some useful information about voter attitudes; but, until further substantiation is available, most cognitive scientists will probably remain unconvinced. The problem has less to do with brain imaging per se and more to do with the human tendency to make up compelling and believable stories. The devil is in the detail and depending on the interpretation of the output from a multi-million dollar brain scanner, the result may be objective and scientific, or of little more value than tea leaves in the bottom of a cup—ambiguous and susceptible to a large number of possible outcomes.

References

Aron, A., Badre, D., Brett, M., Cacioppo, J., Chambers, C., Cools, R., Engel, S., D'Esposito, M., Frith, C., Harmon-Jones, E., Jonides, J., Knutson, B., Phelps, L., Poldrack, R., Wager, T., Wagner, A., Winkielman, P. (2007). Politics and the brain (Letter to the Editor, 2007, November 14). *New York Times.*

Babiloni, F., Astolfi, L., Cincotti, F., Mattia, D., Tocci, A., Tarantino, A., & De Vico Fallani, F. (2007). Proceedings from the IEEE Engineering in Medicine and Biology Society: *Cortical activity and connectivity of human brain during the prisoner's dilemma: An EEG hyperscanning study.*

Babiloni, F., Cincotti, F., Mattia, D., De Vico Fallani, F., Tocci, A., Bianchi, L., & Astolfi, A. (2007). Proceedings from the IEEE Engineering in Medicine and Biology Society: *High resolution EEG hyperscanning during a card game.*

Babiloni, F., Cincotti, F., Mattia, D., Mattiocco, M., De Vico Fallani, F., Tocci, A., & Astolfi, L. (2006). *Hypermethods for EEG hyperscanning.* Paper presented at the Conference Proceedings of the IEEE Engineering in Medicine and Biology Society.

Dumit, J. (this volume). Critically Producing Brain Images of Mind.

Greene, J. D., Nystrom, L. E., Engell, A. D., Darley, J. M., & Cohen, J. D. (2004). The neural bases of cognitive conflict and control in moral judgment. *Neuron, 44*(2), 389–400.

Greene, J. D., Sommerville, R. B., Nystrom, L. E., Darley, J. M., & Cohen, J. D. (2001). An fMRI investigation of emotional engagement in moral judgment. *Science, 293*(5537), 2105–2108.

Haxby, J. V., Gobbini, M. I., Furey, M. L., Ishai, A., Schouten, J. L., & Pietrini, P. (2001). Distributed and overlapping representations of faces and objects in ventral temporal cortex. *Science, 293*(5539), 2425–2430.

Huettel, S., Song, A., & McCarthy, G. (2004). *Functional magnetic resonance imaging.* Sunderland, Massachusetts: Sinauer Associates, Inc.

Iacoboni, M., Freedman, J., Kaplan, J., Jamieson, K. H., Freedman, T., Knapp, B., Fitzgerald, K. (2007, November 11). This is your brain on politics. *New York Times.*

Kriegeskorte, N., Simmons, W. K., Bellgowan, P. S., & Baker, C. I. (2009). Circular analysis in systems neuroscience: The dangers of double dipping. *Nature Neuroscience, 12*(5), 535–540.

McCabe, D. P., & Castel, A. D. (2008). Seeing is believing: The effect of brain images on judgments of scientific reasoning. *Cognition, 107*(1), 343–352.

Miller, G. (2008). Neurobiology: The roots of morality. *Science, 320*(5877), 734–737.

Mind games. How not to mix politics and science (Editorial) (2007), *Nature, 450*, 457 (22 November, 2007), Published online 21 November 2007 http://www.nature.com/nature/journal/v450/n7169/full/450457a.html.

Montague, P. R., Berns, G. S., Cohen, J. D., McClure, S. M., Pagnoni, G., Dhamala, M., Wiest, M. C., Karpov, I., King, R. D., Apple, N., Fisher, R. E. (2002). Hyperscanning: Simultaneous fMRI during linked social interactions. *Neuroimage, 16*(4), 1159–1164.

Op de Beeck, H. P., Haushofer, J., & Kanwisher, N. G. (2008). Interpreting fMRI data: Maps, modules and dimensions. *Nature Reviews Neuroscience, 9*(2), 123–135.

Poldrack, R. (2006). Can cognitive processes be inferred from neuroimaging data? *Trends in Cognitive Sciences, 10*(2), 59–63.

Poldrack, R., & Wagner, A. (2004). What can neuroimaging tell us about the mind? *Current Directions in Psychological Science, 13*(5), 177–181.

Racine, E., Bar-Ilan, O., & Illes, J. (2006a). fMRI in the public eye. *Nature Reviews Neuroscience, 6*(2), 159–164.

Racine, E., Bar-Ilan, O., & Illes, J. (2006b). Brain imaging: A decade of coverage in the print media. *Science Communication, 28*(1), 122–143.

Raizada, R. (2008, April 13). Proceedings from the Cognitive Neuroscience Society Annual Meeting: *Symposium session 2: Pattern-based fMRI analyses as a route to revealing neural representations.*

Raz, A., Lieber, B., Soliman, F., Buhle, J., Posner, J., Peterson, B. S., Posner, M. I. (2005). Ecological nuances in functional magnetic resonance imaging (fMRI): Psychological stressors, posture, and hydrostatics. *Neuroimage, 25*(1), 1–7.

Tong, F. (2003). Primary visual cortex and visual awareness. *Nature Reviews Neuroscience, 4*(3), 219–229.

Tsao, D. Y., Freiwald, W. A., Tootell, R. B. H., & Livingstone, M. S. (2006). A cortical region consisting entirely of face-selective cells. *Science, 311*(5761), 670–674.

Vul, E., Harris, C., Winkielman, P., & Pashler, H. (2009). Puzzlingly high correlations in fMRI studies of emotion, personality, and social cognition. *Perspectives on Psychological Science, 4*(3), 274–290.

13

The Salmon of Doubt
Six Months of Methodological Controversy within Social Neuroscience

Daniel S. Margulies

There was something fishy going on ...
Ed Vul, in an interview with Jonah Lehrer (2009a)

In the final week of 2008 a controversial article began swiftly circulating through the neuroscience community (Vul, Harris, Winkielman, & Pashler, 2009). Although it is generally uncommon for articles in press to slip far beyond personal correspondence, the implications of this publication were such that science blogs immediately broke the story just before the new year (Bell, 2008; joneilortiz, 2008; *Neurocritic*, 2008; Roberts, 2008). Word spread that over two-dozen articles, many in the highest-ranked journals, were openly accused of invalid results. Certainly, for anyone who read the decidedly combative title, "Voodoo Correlations in Social Neuroscience", this urgent manner of dissemination was understandable. Social neuroscience, the most rapidly emerging field of the past decade, entered 2009 in a rather vulnerable position.

Through the month of January researchers within the community, myself included, worried that our methodological foundation was in jeopardy. This chapter will follow the aftermath through the subsequent six months. It was an interval of heated public controversy that ultimately climaxed, I will argue, in a brilliant display of neuroscientific irony. The scientific debate itself focuses on proper statistical practice in functional magnetic resonance imaging (fMRI) research. However, even if the content of the debate may be quite methodologically oriented, it is my hope that the various formulations of criticism, the conventions of debate (and their disregard), as well as the process of ensuing resolution may offer a worthwhile narrative case study to those embarking on analogous campaigns.

Critical Neuroscience: A Handbook of the Social and Cultural Contexts of Neuroscience, First Edition.
Edited by Suparna Choudhury and Jan Slaby.
© 2012 John Wiley & Sons, Ltd. Published 2016 by John Wiley & Sons, Ltd.

PART I: Voodoos and Don'ts

January: The scientific record

For cognitive neuroscientists around the world, the new year in 2009 began with a surprising email attachment. Although it has become common practice for journals to publish articles electronically, sometimes several months before they appear in print, when "Voodoo Correlations in Social Neuroscience" (henceforth "Voodoo Correlations") began circulating, its wide dissemination was reportedly a surprise even to the lead author (Lehrer, 2009a). More damaging for social neuroscientists than even the content of the article—many had not yet read it—were the sensational headlines appearing throughout the blogosphere, such as: "Scan Scandal Hits Social Neuroscience" (*Neurocritic*, 2008) and "Vul on fMRI Abuse in the Cognitive Neuroscience of Social Interaction" (Joneilortiz, 2008). A response, any response, was called for from the social neuroscience community. Making matters more complex, the unfamiliar terrain of the internet-based scientific discourse would render the usual debate strategies ineffective, thus demanding novel approaches for the forum of the 24/7 online community. With the field under attack, responses and rebuttals were published more rapidly than the time usually taken to return page proofs.

The mordant title alone was certainly enough to produce a flurry of reactions. However, at the core of "Voodoo Correlations" lay a nuanced criticism of statistical practice in neuroimaging studies. As Christian Keysers was quoted later in the month in *Nature News*: "We all agree that there is a kernel of truth in what Vul and his colleagues write about some of the literature being shaky ... We can never be reminded often enough of the importance of good statistical practice" (Abbott, 2009). Considering the subsequent impact of media sensationalism, which largely obfuscated the core issues, we should begin by outlining the statistical criticism.

Vul and colleagues introduced their article by giving examples of the "puzzlingly" high correlation values between behavior and fMRI measurements found in numerous social neuroscience studies. There is no doubt that social neuroscience had been remarkably successful in describing the relationship between behavior and brain activity, publishing numerous articles in the most high-ranking journals. Examples cited in the introduction of "Voodoo Correlations" include the 0.88 correlation between anterior cingulate activity during a social rejection game and subsequent self-reports of the amount of distress participants felt (Eisenberger, Lieberman, & Williams, 2003), correlations of 0.52–0.72 between anterior cingulate activity during empathy manipulations with two scales of emotional empathy (Singer et al., 2004), and a massive 0.96 correlation between a scale measuring proneness to anxiety and cuneal activation during angry speech (Sander et al., 2005).

The basis for Vul and colleagues' suspicion, as they explained, is that the strength of a correlation between two variables is not simply a result of their direct relationship, but also a factor of the independent reliability of both measures. For instance, if you want to explore if the intelligence of a carp predicts the intelligence of the fisherman who can catch it, the theoretical maximum correlation value would be based on the reliability of both intelligence measures. As Vul and colleagues applied the calculation, claiming the most optimistic test-retest reliability of psychological scales ranging 0.7–0.8, and fMRI test-retest peaking at approximately 0.7, correlations between the two should have an upper bound of 0.74. The authors thus argued that the less-than-perfect reliability of these two measures rendered many of the reported correlation values, often exceeding 0.8 in social neuroscience studies, to be "impossibly high."

In order to investigate the origins of such correlation values, the authors selected 55 articles for a detailed investigation of the methods. However, upon surveying the publications, they often encountered a lack of clarity in the methods section, and thus conducted a four-question, multiple choice email survey aiming to clarify the analyses underlying the data presentation. An impressive 53 of the 55 authors responded.

Based on the responses, Vul and colleagues concluded that the core statistical error driving the "impossibly high" correlation values was "non-independence error." The basis of such error results from a two-step procedure, in which voxels marked during a first analysis are *selectively* analyzed in a second analysis. This procedure is of significant concern in fMRI analysis due to the large number of variables included. With up to 60,000 voxels in the brain, the likelihood of randomly encountering a significant relationship in any single voxel is rather high. Thus, several solutions to the problem of "multiple comparison correction" have been developed specifically for fMRI data. The non-independence error is committed when selection criteria for a second analysis on a data set are based on results from a first analysis on the same data. In such cases analyses favor areas that have already been demonstrated to be related. Vul and colleagues then compiled a "red-list" of 28 articles, which were deemed guilty of non-independence error, hereby, they claimed, invalidating the results.

When attack is still diffuse and indirect, guilty parties can pretend that the bullets are not intended for them. However, Vul and colleagues had wisely taken a more sniper-like tactic. Their unequivocal critique resounded with the concluding statement: "At present, all studies performed using these methods have large question marks over them. Investigators can erase these question marks by re-analysing their data with appropriate methods." Names had been named; allegations had been made, and drastic terms for exoneration had been laid out. Authors of the cited articles were thus faced with the options of: (1) admitting the wrong and re-analyzing their data; (2) pleading innocence and demonstrating the error in the "Voodoo Correlations" critique; or (3) hoping that nobody had noticed.

Although for weeks "Voodoo Correlations" had been discussed amongst blog-savvy neuro-enthusiasts, on the morning of Friday, January 9, the article finally reached the mainstream. Sharon Begley, author of the *Newsweek* blog Lab Notes, posted "The "Voodoo" Science of Brain Imaging" (Begley, 2008), in which she described the situation for social neuroscientists in terms that leave little wiggle room: "a bombshell has fallen on dozens of such studies: according to a team of well-respected scientists, they amount to little more than voodoo science." Those frustrated with the sensationalism surrounding the rapidly growing field of social neuroscience considered "Voodoo Correlations" a coveted victory ("Editorial: What were the neuroscientists thinking?" 2009; Giles, 2009); those attacked by the article considered it an offense to proper scientific discourse. Regardless of perspective, the article had been noticed, and the indictment of an entire field (not to mention publications in the most reputable scientific journals such as *Nature* and *Science*) made this accusation difficult to shrug off quietly.

As Begley parenthetically suggests: "in fairness, the skewered authors should be given a chance to defend themselves," several accused authors responded by attacking the validity of Vul and colleagues' claim through both conventional and Internet-guerilla tactics. On January 13, Christian Keysers began energetically posting links throughout the blogosphere to a response article he had co-authored with Mbemba Jabbi, Tanya Singer, and Klaas Enno Stephan (Jabbi, Keysers, Singer, & Stephan, n.d.; Lieberman, 2009; *Neurocritic*, 2009). Much of "Response to 'Voodoo Correlations in Social Neuroscience' by Vul et al.—summary information for the press" takes the form of accusations of libel for the indiscriminate and unwarranted criticism made against social neuroscience. As they explain, the field is larger than the details of a specific statistical technique: "statistical arguments that are partially flawed, and misleadingly implies that social neuroscience studies rest entirely on the sort of brain-behaviour correlations that are criticised."

Through eight brief counter-arguments the authors offer broad-spectrum critiques of "Voodoo Correlations," which basically aim to undermine the "outsider" understanding of social neuroscience held by Vul and colleagues. For example, while Vul and colleagues argue that secondary analyses based on regions selected during an initial analysis constitutes non-independence error, Jabbi and colleagues claim that the correlation coefficients and p-values are reported for the purpose of illustrating effect size alone, and thus do not constitute secondary analyses. Furthermore, the response explains that the question underpinning social neuroscience studies is not the strength of correlations between brain and behavior, but rather where in the brain such correlations occur. Finally, there is frustration with the brevity of the questionnaire, and that based on such minimal data Vul and colleagues "flag a set of studies as 'problematic' without discriminating when non-independence errors were committed and when not." The response, aimed at the same lay audience that had embraced the sensationalism of the initial article, established a counter-argument to "Voodoo Correlations," while launching the two-sided debate within the public sphere.

The response, however, came too late for the accelerating media aggression. On January 14, the *Wall Street Journal* quoted the senior author of "Voodoo Correlations," Harold Pashler, in an article about the use of neuroimaging technologies in the courtroom: "In the law, individual differences are the main focus ... and it often could come down to these voodoo statistics" (Hotz, 2009). *New Scientist* had published an editorial on the same day entitled: "What were the neuroscientists thinking?"(2009). The author laments the journal's own involvement in promulgating many of the criticized studies: "We have to eat a little humble pie and resolve that next time a sexy-sounding brain scan result appears we will strive to apply a little more skepticism to our coverage." These same publications that were so quick to embrace social neuroscience's successes were just as quick to propagate its alleged failures—after all, they had been duped as well.

Amidst the popular media hubbub, a more technical dialogue persevered in the blogosphere with a rebuttal by Vul and colleagues. The reply ("Reply to Jabbi et al.") can often be found immediately succeeding Keyser's posts on numerous blogs on January 15. It begins with a disclaimer that Jabbi and colleagues' response has an "evolving rebuttal, but it has changed at least once since we replied to it, so we can't be sure whether our comments below will address the points in this version" (Ed Vul, 2009). They are referring to the removal of two-thirds of the introduction, which in the final version condenses the introduction into a two-sentence lead-in for the summarized points. Vul and colleagues' reply, likewise, contained eight corresponding counterpoints. They argue that multiple comparison correction does not safeguard against the inflation of secondary correlation analyses. And although social neuroscience may not fixate on the amplitude of correlations, Vul and colleagues respond that the scientific literature should nonetheless be free from such statistical errors.

A more thorough critique of "Voodoo Correlations" was posted online beginning on January 27 (Lieberman, 2009). "Correlations in social neuroscience aren't voodoo: A reply to Vul et al." by Matthew Lieberman, Elliot Berkman, and Tor Wager (2009), went as far as to mimic the long list of acknowledgements on the cover page of "Voodoo Correlations." The reply presents a far more exhaustive treatment of "Voodoo Correlations" than the admittedly rushed previous attempt of Jabbi and colleagues. As stated on the cover page, it is an "Invited reply" under submission at the same journal which had accepted "Voodoo Correlations." In general terms Lieberman and colleagues attacked several weaknesses in the methods, including the unexplained absence of 54 correlation values from the meta-analysis conducted in "Voodoo Correlations."

Their primary concern, returning to an outstanding debate, involves the magnitude of the statistical error. While Jabbi and colleagues argued that enhanced correlation values were not of primary concern to social neuroscientists, Vul and colleagues countered that no statistical error belongs in scientific practice. Here, a novel question is raised: if the statistical error only results in minor increases in significance values, does that really merit the designation of "voodoo?" Connoting magical practice and the absence of any genuine scientific support, the term "voodoo" may have been used unfairly.

I have focused the discussion of January on the technical foundation of the debate, along the way highlighting certain players in the story. However, it was not only the accused who felt their fates to be in a tenuous position—cognitive neuroscientists around the world were concerned for the future of their profession. Many had heard the title before working through the paper, and perhaps all were unknowingly guilty of such an error. Nonetheless, a rigorous methodological rebuttal by Lieberman and colleagues brought the story back into perspective by the end of January—and rightly, the optimistic conclusion of their article announces the onward march of social neuroscience:

> There are various ways to balance the concerns of false positive results and sensitivity to true effects, and social neuroscience correlations use widely accepted practices from cognitive neuroscience. These practices will no doubt continue to evolve. In the mean time, we'll keep doing the science of exploring how the brain interacts with the social and emotional worlds we live in.
>
> (Lieberman et al., 2009)

February: The scientific discourse

The heated controversy of January did finally find resolution towards the end of the month. I'll summarize the general sentiment with a personal anecdote. In the first week of February, I received a text message from a colleague in New York: "Went out for drinks with the stats department. Comfortable now that voodoo correlations argument is bullshit." Researchers, even those not directly attacked by "Voodoo Correlations", were relieved that the criticism was not as grave as it initially appeared.

Scientific discourse can take various forms. With the theoretical debate ebbing, the practical implications of the public's involvement became the focus of discussions. Let us examine briefly Tania Singer's assertion in a *Nature News* article on January 15. In what ways was the experience with "Voodoo Correlations," as she claimed, "not the way that scientific discourse should take place" (Abbott, 2009)? The *Nature* article offers a suggestion of the meaning implicit in Singer's comment:

> The swift rebuttal was prompted by scientists' alarm at the speed with which the accusations have spread through the community. The provocative title ... and iconoclastic tone have attracted coverage on many blogs ... Those attacked say they have not had the chance to argue their case in the normal academic channels.
>
> (Abbott, 2009)

Three particular issues emerge here regarding the wider shifts at play in the scientific discourse: (1) the increased speed of dissemination; (2) the sensational title aimed at media coverage; and (3) the role of public debate in discussing complex methodological topics. While these observations had been aimed as accusations against Vul and colleagues, both parties appeared to be equally guilty.

The issue of speed is crucial. "Voodoo Correlations" is often accused of having being released before journal publication, but whilst such statements may be subtly underplayed, and although the article had in fact been accepted by the time of release, the suggestion is that "Voodoo Correlations" leaked in a "scientifically inappropriate" manner. While there was certainly much surprise at the speed of dissemination, it is important to note that there was nothing unusual about the release protocol of the article. Oddly, the most classically "improper" scientific proceeding during the month of January was the release of Jabbi and colleagues' rebuttal, which contained a seemingly self-aware admission of its own prematurity: "A detailed analysis will be submitted to a peer reviewed scientific journal shortly" (Jabbi et al., 2009). Strangely, the eventual details of their analysis do not extend beyond the original online publication, although a second online version was released shortly thereafter with a revised introduction.

With respect to sensationalism in the article title, Ed Vul was candid about the choice: "We wanted to make the paper entertaining and to increase its readership. We wanted our paper to have some impact. If people don't know about these statistical problems, nothing will be done to fix them." The accusation there is certainly justified. However, the necessity of such sensationalism in order to elicit a response from the research community raises a question with respect to the role of the public voice in motivating innovation in science.

The public's role in the debates surrounding "Voodoo Correlations" was a contentious topic for those accused by the article. The authors of the reply rejected an invitation to openly debate the issues on a public blog, claiming instead that "the critique will be dealt with in peer-reviewed literature in forthcoming papers by the scientific community" (Klincewicz, 2009). They went on to state that they would postpone further discussion until "a proper scientific dialogue occurs; not a dialogue by press and anonymous blogs who cannot evaluate the statistical claims made by Vul et al. Popular opinion asserts that the way in which the paper is discussed does not support fair and suitable scientific manner" (Klincewicz, 2009). At the heart of such debate tactics is the trope of "proper science" and the discursive weight of the objective scientist. "Science" in this context implies a private, "expert," insular discourse, that is, not for laymen—but with "valid scientists" on both sides of the debate, such expertise cancels itself out. For instance, in her *Newsweek* blog, while buttressing the criticism by Vul and colleagues by describing the authors as "well-respected scientists," Begley also makes claims that seemingly contradict the trope of objectivity in scientific practice:

> If you were wondering how, exactly, problematic studies got past the peer review at these top journals, that's a clue: scientists no less than other mortals love to have their hunches, prejudices and stereotypes validated by empirical evidence. Maybe they didn't look too critically at studies that did exactly that.
>
> (Begley, 2008)

Contradictions such as this appear throughout the debates surrounding "Voodoo Correlations." After all, both the accused and accusing parties are equally well-respected scientists, and both the accused and accusing articles are published in peer-reviewed scientific journals. The traditional strategies used in public scientific debate—namely, attempting to invalidate the scientific legitimacy of your opponent—are at times attempted, but are largely ineffective. In the case of the "Voodoo Correlations" debate, the non-scientific public was given the role of the invalid scientist, unequipped to engage in thoughtful criticism.

Behind all the well-worn polemics, what was at the core of the frantic opposition to the publicity received by "Voodoo Correlations"? Again, one indication can be found towards the conclusion of the *Nature News* article, where Chris Frith, author on several of the red-listed articles, strips away the rhetoric of "proper scientific discourse," revealing the issues in more realistic terms: "We are not worried about our close colleagues, who will understand the arguments. We are worried that the whole enterprise of social neuroscience falls into disrepute" (Abbott, 2009). Disrepute can indicate many things, but the sentiment takes another form during interviews with Tor Wager, who was also one of the authors of the reply article with Lieberman. Wager makes clear that "Voodoo Correlations" negatively (and unwarrantedly) biases funding and top journals (Lehrer, 2009b). Even if at times researchers may want the public to remain out of specialized debates until resolution within the community can be achieved, public opinion still contributes to the practical outcome for the field as a whole. Interestingly, although scientists are usually also the advisors or leaders of funding agencies and top journals, it is made clear here how biased they too become through the public discourse.

Perhaps *Seed Magazine* contributor, Jon Bardin, summarized the events of February best: "When findings are debated online, as with a yet-to-be-released paper that calls out the field of social neuroscience, who wins?" (Bardin, 2009).

March: The scientific agenda

With little changes in the "Voodoo Correlations" story during the month of March 2009, perhaps this presents an ample moment to reflect on the distance that criticism travelled within cognitive neuroscience over that year. This was not the first time social neuroscience had been under attack—the statistical debate of 2009 was, in many ways, a reincarnation of former critiques of overvaluation and over-interpretation of social neuroscience imaging results. If in 2009 the critical approach to cleaning up statistics was challenged, so in 2008 the over-interpretation of fMRI results was contested. Articles with self-reflective titles were being published that year, such as "What we can do and what we cannot do with fMRI" (Logothetis, 2008) and "The role of fMRI in cognitive neuroscience: Where do we stand?" (Poldrack, 2008). Where in 2009 the propriety of using Internet media as a forum of scientific debate was disputed, in 2008 the focus was on printed, but no less public forms.

Much was spurred by the ground-breaking *New York Times* op-ed piece "This is your Brain on Politics" of November 11, 2007 (Iacoboni et al., 2007), which set a precedent for self-publishing results in the popular media in advance of journal acceptance. Marco Iacoboni and colleagues presented an fMRI study of 20 swing votes viewing the political candidates (almost replicating a study they had published three years previously; see Kaplan, Freedman, & Iacoboni, 2007). The piece was written in the standard journalistic style and reflected extensively how brain data impacted on the perceived status of the various presidential candidates. It was sensational in the extreme and, needless to say, provoked a commensurate retaliation from the neuroimaging community in *The New York Times* a few days later, signed by no fewer than 17 leaders in the field (Aron et al., 2007).

A poignant generational gap is evident in the comfort with which the individuals involved in the "Voodoo Correlations" debate navigated the social community on the Internet. One telling example is the alacrity of Vul and colleagues' reply to the Jabbi and colleagues' response article. Within 24 hours they had published a response online.

Unlike "Voodoo Correlations," the controversy in the *New York Times* piece was spurred by the true lack of peer-review in the article's publication. Perhaps some residual strategies had made their way into the Vul debate without recognition of the numerous differences. While a shift in critical priorities occurred from 2008 to 2009, the underlying motivation may have persisted: a general concern over the increasing popularity and ubiquity of social neuroscience.

PART II: Critical Tactics in Action

April: Naming names

With the controversy of "Voodoo Correlations" as yesterday's news, Nikolaus Kriegskorte and colleagues at the National Institute of Mental Health published a second, albeit gentler, reprimand of the neuroimaging community in the high-impact journal *Nature Neuroscience* (Kriegeskorte, Simmons, Bellgowan, & Baker, 2009:

online publication April 26, 2009). Much theoretically aligned with "Voodoo Correlations," the article also carried the requisitely catchy title "Circular analysis in systems neuroscience: The dangers of double dipping." The content of the critique centers on selection bias in analyses—almost identical to the critique of non-independence error. Thus, *voodoo correlation* was rechristened the less exotic, though equally condescending *double dipping*. Nonetheless, the determinedly gentler title, invoking ice-cream or Seinfeldesque taboos, did not usurp voodoo correlation as the descriptive term within the field—nor did any comparable media coverage follow.

The lack of reaction is probably attributable to the lack of direct finger pointing, characteristic of "Voodoo Correlations." With nobody feeling the need to defend his or her careers, nobody really did. When all were arguing about the proper *place* to debate science in the early part of the year, they may rather have meant to say that proper science simply should not get personal.

May: Name dropping

By May, the anxiety had subsided—scanners were still running, funding had not been revoked, and social neuroscience had not been disbanded—just in time for the relatively quiet publication of volume 4, issue 3 of *Perspective on Psychological Science*. On the surface the final version of Vul and colleagues' article is unrecognizable with reparations apparent in the new title, "Puzzlingly High Correlations in fMRI Studies of Emotion, Personality, and Social Cognition" (Ed Vul, Harris, Winkielman, & Pashler, 2009). An awkward title footnote marks the emasculation: "This article was formerly known as 'Voodoo Correlations in Social Neuroscience.'" The text remains largely unchanged, however "social neuroscience" is nowhere to be found, and in its place is a thorough substitution of "fMRI studies of emotion, personality, and social cognition." The issue also contained six response articles, a "reply to comments" from Ed Vul and colleagues, and an editor's introduction.

In addition to referring to the rechristening in final publication of "Voodoo Correlations," the title of this section also refers to another theme raised in May 2010: the suggestion that other issues should also be marked for urgent critical discourse within the neuroimaging community. While the question of statistics was of crucial importance, its relative resolution gave the illusion that social neuroscience had been packaged up neat and tidy again.

Of all the contributions, only that of the editor Ed Diener, commented on the peripheral issues raised by the community's handling of the article, suggesting "that the debate can itself stimulate useful discussions about scientific practices and communication" (Diener, 2009). He then suggests that his journal is not an appropriate forum to continue the debate, instead asking that "further discussion of the issues should now take place in journals that are focused on imaging and neuroscience." However, just before distancing himself, he manages to slip a brief editorial comment into the penultimate paragraph: "In addition, there are questions related to what relative blood-oxygen levels actually signify about the mind when they are uncovered." The suggestion here is that amidst the hubbub surrounding "Voodoo Correlations," there are other fundamental issues of cognitive neuroimaging with fMRI that may reflect equally unfavorably if treated with similiar attention.

Diener significantly calls attention to the fact that just because the crisis caused by the "Voodoo Correlations" article has been resolved does not mean that all is well and good in fMRI studies in social neuroscience research. Many more important questions and assumptions remain untested and unanswered. The question raised by Diener, though tactfully underplayed, may be of too grand a scale. Greater impetus, on the scale of "Voodoo Correlations," may be truly necessary to engage a community in earnest self-reflection and productive criticism.

June: Swimming upstream

Perhaps the most globally appreciated prank to ever make use of an fMRI scanner was brought to the attention of the neuroimaging community at the Organization for Human Brain Mapping's annual conference. During the final lecture on June 22, the past chair, Rainer Goebel, delivered his "closing comments and meeting highlights" to a full auditorium in San Francisco. After reviewing many of the emerging directions in the field, he displayed what he described as one of his favorite posters from the conference: "Neural Correlates of Interspecies Perspective Taking in the Post-Mortem Atlantic Salmon: An Argument for Multiple Comparisons Correction." It was greeted with a cathartic laughter of recognition.

The abstract, by Craig Bennett, Michael Miller and George Wolford (Bennett, Miller, & Wolford, 2009: later to include Abigail Baird on the poster), described a study of social cognition in "one mature Atlantic Salmon (*Salmo salar*)." In keeping with scientific punctiliousness, and no doubt to thwart any appropriation of their study by the overzealous, the authors then noted: "The salmon was approximately 18 inches long, weighed 3.8 lbs, and was not alive at the time of scanning." The task paradigm was delivered with the familiar laconic methods section:

> The task administered to the salmon involved completing an open-ended mentalising task. The salmon was shown a series of photographs depicting individuals in social situations with a specified emotional valence. The salmon was asked to determine what emotion the individual in the photo must have been experiencing. Stimuli were presented in a block design
> (Bennett et al., 2009)

Before proper correction for multiple comparisons, a cluster 27 mm³ was found to be significant within the brain cavity; however, the authors dutifully noted that "due to the coarse resolution of the echo-planar image acquisition and the relatively small size of the salmon brain further discrimination between brain regions could not be completed" (Bennett et al., 2009). Of course (and thankfully), after proper statistical correction, no active voxels were detected.

To those unfamiliar with the techniques, this appeared to be another successful attack against social neuroscience—it conclusively demonstrated the virtually limitless potential of opaque fMRI statistics. With the same sense of vindication with which "Voodoo Correlations" had previously been disseminated, the "Atlantic Salmon" poster filled inboxes and blogs across the community in the following months. However, those within the community understood that the obvious tongue-in-cheek presentation was far from being an attempt to invalidate fMRI approaches to questions of social cognition. Rather, it was an example of statistical criticism, which reinforced the validity of correction techniques that have long been argued as essential. In fact, the common statistical error cajoled by Bennett and colleagues was also critically addressed as a subpoint by "Voodoo Correlations" in the results and discussion section.

Before 2009, the nuanced debate over proper multiple comparison correction had rarely leaked beyond the fMRI methods and statistics community. By integrating the criticism into a tongue-in-cheek experimental context, the point was made while garnering public attention. Reiterated here in a more accessible form, perhaps this last rendition provided precisely the emotional closure the fMRI research community needed. The problematic was genuine, but there was an effective solution. The ominous implications of voodoo correlations were finally transformed into a unifying mascot: *the salmon of doubt*. Of course, the content of the salmon poster was not the same precise criticism at the centre of "Voodoo Correlations." The former addressed the problem of multiple comparisons, while the latter dealt primarily with the non-independence error (although it also addressed problems with certain forms of multiple comparison correction). Nonetheless, a similar approach could well be used to represent the dangers of the biased selection of regions for secondary analysis.

Rooted in an intimate knowledge of statistics, the methodological critique could only come from those who were, at least to some extent, within the field. For instance, Ed Vul explained in a interview with Jonah Lehrer that "Voodoo Correlations" began the year prior with a sense he had that "there was something fishy going on … despite our suspicions, we didn't know exactly what that fishy thing was, so we put the topic aside" (Lehrer, 2009a). After joining Nancy Kanwisher's lab at MIT, he began "working directly with fMRI data" and "learned the relevant jargon and statistics." Certainly, to understand the criticism, one needs the insider knowledge described by Vul. Thus, Vul also recognized that such knowledge would not be easily accessible to those outside the community.

Vul and colleagues' criticism was, however, also positioned from outside the field it attacked. While it certainly had impact, that effect may have been limited as researchers quickly scrambled to resolve the methodological dispute (which was statistical, and could eventually be answered). Rather, by sketching the critique within neuroscience, as Bennett and colleagues did with such humor, the argument may have been received more productively. The salmon study does not encourage a rebuttal; simply recognizing its irony is a form of corroboration.

As the 1990s was the "decade of the brain," the 2000s are already being labeled the "decade of social neuroscience." Perhaps a bit of unrealistic optimism urges the question: could the looming discomfort in recent years with neuroimaging studies signal the start of the "decade of ironic neuroscience"? The critical strength of irony may be its potential to unify while still making its argument understood within the target community—a valuable tool for any critical neuroscience endeavor.

Acknowledgements

The author is deeply indebted to Jan Slaby for suggesting the title of this chapter, and to Douglas Adams who coined the original. The author also wishes to thank the editors, as well as Felicity Callard, Alan Fishbone and The Neuro Bureau for their helpful suggestions.

References

Abbott, A. (2009). Brain imaging studies under fire. *Nature News, 457*(245). doi:10.1038/457245a.

Aron, A., Badre, D., Brett, M., Cacioppo, J., Chambers, C., Cools, R., & Winkielman, P. (2007, November 14). LETTER; Politics and the brain. *The New York Times.* Retrieved from http://query.nytimes.com/gst/fullpage.html?res=9907E1D91E3CF937A25752C1A9619C8B63.

Bardin, J. (2009, February 24). Voodoo that scientists do. *Seed Magazine.* Retrieved from http://seedmagazine.com/content/article/that_voodoo_that_scientists_do/.

Begley, S. (2008, January 9). The "voodoo" science of brain imaging. *Lab Notes.* Retrieved January 11, 2010, from http://blog.newsweek.com/blogs/labnotes/archive/_2009/01/09/the-voodoo-science-of-brain-imaging.aspx.

Bell, V. (2008, December 29). Voodoo correlations in social brain studies: *Mind Hacks.* Retrieved December 23, 2009, from http://www.mindhacks.com/blog/2008/12/voodoo_correlations_.html.

Bennett, C., Miller, M., & Wolford, G. (2009). Neural correlates of interspecies perspective taking in the post-mortem Atlantic salmon: An argument for multiple comparisons correction. *NeuroImage, 47*(Supplement 1), S125. doi:10.1016/S1053-8119(09)71202-9.

Diener, E. (2009). Editor's introduction to Vul et al. (2009) and comments. *Perspectives on Psychological Science, 4,* 272. doi:10.1111/j.1745-6924.2009.01124.x

Editorial: What were the neuroscientists thinking? (2009, January 14). *New Scientist,* (2691). Retrieved from http://www.newscientist.com/article/mg20126912.800-editorial-what-were-the-neuroscientists-thinking.html.

Eisenberger, N. I., Lieberman, M. D., & Williams, K. D. (2003). Does rejection hurt? An fMRI study of social exclusion. *Science, 302*(5643), 290–292. doi:10.1126/science.1089134.

Giles, J. (2009, January 14). Doubts raised over brain scan findings. *New Scientist,* (2691). Retrieved from http://www.newscientist.com/article/mg20126914.700-doubts-raised-over-brain-scan-findings.html.

Hotz, R. L. (2009, January 15). The brain will take the stand. *The Wall Street Journal,* A7. Retrieved from http://online.wsj.com/article/SB123205921925787437.html.

Iacoboni, M., Freedman, J., Kaplan, J., Jamieson, K., Knapp, B., & Fitzgerald, K. (2007, November 11). This is your brain on politics. *The New York Times.* Retrieved from http://www.nytimes.com/2007/11/11/opinion/11freedman.html?_r=1.

Jabbi, M., Keysers, C., Singer, T., & Stephan, K. E. (2009). "Voodoo correlations in social neuroscience" by Vul et al. – summary information for the press. Retrieved from www.bcn-nic.nl/replyVul.pdf.

Joneilortiz. (2008, December 30). Vul on fMRI abuse in the cognitive neuroscience of social interaction. *Mutually occluded media & film, design, philosophy, politics.* Retrieved December 23, 2009, from http://www.mutuallyoccluded.com/2008/12/vul_fmri_cognitive_neuroscience_social_interaction/.

Kaplan, J. T., Freedman, J., & Iacoboni, M. (2007). Us versus them: Political attitudes and party affiliation influence neural response to faces of presidential candidates. *Neuropsychologia, 45*(1), 55–64. doi:10.1016/j.neuropsychologia.2006.04.024.

Klincewicz, M. (2009, January 23). Does fMRI data show implausibly high correlation? *MedicExchange.com.* Retrieved December 27, 2009, from http://www.medicexchange.com/MRI-Clinical/does-fmri-data-show-implausibly-high-correlation.html.

Kriegeskorte, N., Simmons, W. K., Bellgowan, P. S. F., & Baker, C. I. (2009). Circular analysis in systems neuroscience: The dangers of double dipping. *Nature Neuroscience, 12*(5), 535–540. doi:10.1038/nn.2303.

Lehrer, J. (2009a, January 29). Voodoo correlations: Have the results of some brain scanning experiments been overstated? *Mind Matters.* Retrieved December 24, 2009, from http://www.scientificamerican.com/article.cfm?id=brain-scan-results-overstated.

Lehrer, J. (2009b, February 17). In defense of the value of social neuroscience. *Mind Matters.* Retrieved December 28, 2009, from http://www.scientificamerican.com/article.cfm?id=defense-social-neuroscience.

Lieberman, M. (2009, January 27). Seth's blog "Blog archive" voodoo correlations in social neuroscience. *Seth's blog: Self-experimentation, scientific method, the Shangri-La diet, etc.* Retrieved December 27, 2009, from http://www.blog.sethroberts.net/2008/12/28/voodoo-correlations-in-social-neuroscience/#comment-258657.

Lieberman, M. D., Berkman, E. T., & Wager, T. D. (2009). *Correlations in social neuroscience aren't voodoo: A reply to Vul et al.* http://www.scn.ucla.edu/pdf/LiebermanBerkman-Wager(invitedreply).pdf.

Logothetis, N. K. (2008). What we can do and what we cannot do with fMRI. *Nature, 453*(7197), 869–878. doi:10.1038/nature06976.

Neurocritic. (2008, December 31). Scan scandal hits social neuroscience. *The Neurocritic: Deconstructing the most sensationalistic recent findings in Human Brain Imaging, Cognitive Neuroscience, and Psychopharmacology.* Retrieved December 23, 2009, from http://neurocritic.blogspot.com/2008/12/scan-scandal-hits-social-neuroscience.html.

Neurocritic. (2009, January 13). Voodoo counterpoint. *The Neurocritic: Deconstructing the most sensationalistic recent findings in Human Brain Imaging, Cognitive Neuroscience, and Psychopharmacology.* Retrieved December 29, 2009, from http://neurocritic.blogspot.com/2009/01/voodoo-counterpoint.html.

Poldrack, R. A. (2008). The role of fMRI in cognitive neuroscience: Where do we stand? *Current Opinion in Neurobiology, 18*(2), 223–227. doi:10.1016/j.conb.2008.07.006.

Roberts, S. (2008, December 28). Voodoo correlations in social neuroscience. *Seth's blog: Self-Eperimentation, Scientific Method, the Shangri-La Diet, etc.* Retrieved December 20, 2009, from http://www.blog.sethroberts.net/2008/12/28/voodoo-correlations-in-social-neuroscience/.

Sander, D., Grandjean, D., Pourtois, G., Schwartz, S., Seghier, M. L., Scherer, K. R., & Vuillemier, P. (2005). Emotion and attention interactions in social cognition: Brain regions involved in processing anger prosody. *NeuroImage, 28*(4), 848–858. doi:10.1016/j.neuroimage.2005.06.023.

Singer, T., Seymour, B., O'Doherty, J., Kaube, H., Dolan, R. J., & Frith, C. D. (2004). Empathy for pain involves the affective but not sensory components of pain. *Science, 303*(5661), 1157–1162. doi:10.1126/science.1093535.

Vul, E. (2009). Ed Vul – Voodoo rebuttal. Retrieved January 11, 2010, from http://www.edvul.com/voodoorebuttal.php.

Vul, E., Harris, C., Winkielman, P., & Pashler, H. (2009). Puzzlingly high correlations in fMRI studies of emotion, personality, and social cognition. *Perspectives on Psychological Science, 4*(3), 274–290. doi:10.1111/j.1745-6924.2009.01125.x.

14

Cultural Neuroscience as Critical Neuroscience in Practice

Joan Y. Chiao and Bobby K. Cheon

Throughout the history of science, philosophers from Kant to Kuhn have inquired about the nature of scientific progress and the conditions under which scientific progress occurs. Over recent decades, rapid advances in the behavioral and brain sciences have led to the emergence of a new line of critical inquiry. The contemporary field of critical neuroscience aims to investigate the social, cultural, economic, and political contexts that subtly and directly shape the way that researchers in the behavioral and brain sciences conduct research, interpret, and communicate their findings to society as a whole. It also raises the question of whether this contextualization can contribute in any way to scientific practice. Here we illustrate how the study of cultural neuroscience, with its emphasis on examining neurobiological phenomena across cultural contexts and time scales, addresses important challenges posed by critical neuroscience. We review the framework of cultural neuroscience and the recent empirical evidence of cultural variation in the neurobiological mechanisms underlying the self, empathy, and mental health. Finally, we discuss the implications of these cultural neuroscientific findings in relation to achieving the goals of critical neuroscience.

It is particularly in periods of acknowledged crisis that scientists have turned to philosophical analysis as a device for unlocking the riddles of their field. Scientists have not generally needed or wanted to be philosophers. Indeed, normal science usually holds creative philosophy at arm's length, and probably for good reason. But that is not to say that the search for assumptions cannot be an effective way to weaken the grip of a tradition upon the mind and to suggest the basis for a new one (Kuhn, 1970).

Critical Neuroscience: A Timely Challenge for Behavioral and Brain Scientists

In a recent article, Choudhury, Nagel, and Slaby (2009) provide a compelling argument for why a "critical neuroscience" is needed, now more than ever, in the

Critical Neuroscience: A Handbook of the Social and Cultural Contexts of Neuroscience, First Edition.
Edited by Suparna Choudhury and Jan Slaby.

behavioral and brain sciences. Given the immense influence of scientific knowledge on our cultures and the reciprocal influence of cultural contexts on shaping the process of scientific discovery, it is critical that we pay attention to the social, economic, and cultural climates that produce neuroscientific knowledge and to the influence of our growing neuroscientific understanding on our daily lives. Is what we know about human neuroscience changing how we think about our health, our society, even ourselves? What is at stake when a body of scientific knowledge is vulnerable to social, cultural, economic, or political pressures? Would we even be able to recognize such influences when creating scientific knowledge in the behavioral and brain sciences?

The arrival of critical neuroscience alerts us to a shift in the behavioral and brain sciences, perhaps even in a Kuhnian sense, whereby a richer awareness of the social, cultural, economic, or political contexts surrounding normal scientific practices in the behavioral and brain sciences may be the key to developing a deeper and more complete understanding of the human mind and brain. A growing body of research in this field supports the notion that cultural context affects how the mind and brain work (Chiao, 2009; Kitayama & Cohen, 2007) and both behavioral and brain scientists cannot readily assume minimal variability across human populations (Chiao & Cheon, in press; Henrich, Heine, & Norenzayan, in press). For instance, behavioral and brain scientists typically sample from a thin slice of the species. Within the field of psychology, 95% of psychological samples come from countries with only 12% of the world's population (Arnett, 2008). Similarly, within the field of human neuroimaging, 90% of peer-reviewed neuroimaging studies come from Western countries (Chiao, 2009). What kinds of social, cultural, and economic pressures have led to these research biases in the behavioral and brain sciences? Choudhury, Nagel, and Slaby (2009) provide a number of creative ways for addressing these kinds of questions using a critical neuroscience perspective, from ethnographic analysis of how neuroscience is practiced to integrating the knowledge of social and cultural contexts into experimental research. Here we discuss the role that the emerging field of cultural neuroscience can play in fulfilling the goals of critical neuroscience, particularly that of achieving critical neuroscience in the laboratory setting. In this chapter, we describe the aims of cultural neuroscience and then provide examples of how studying different facets of culture and mental illness from a reflexive cultural neuroscientific perspective may serve as critical neuroscience achieved in the laboratory setting.

Cultural Neuroscience: Bridging Cultural and Biological Sciences

Cultural neuroscience is an emerging research discipline that investigates cultural variation in psychological, neural, and genomic processes as a means of articulating the bidirectional relationship of these processes and their emergent properties (Chiao & Ambady, 2007, see Figure 14.1). Research in cultural neuroscience is motivated by two intriguing questions of human nature: how do cultural traits (such as values, beliefs, and practices) shape neurobiology (for example, genetic and neural processes) and behavior? And how do neurobiological mechanisms (for example, genetic and neural processes) facilitate the emergence and transmission of cultural traits?

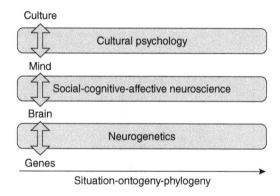

Figure 14.1 Example of the cultural neuroscience framework (from Chiao & Ambady, 2007)

The idea that complex behavior results from the dynamic interaction of genes and cultural environment is not new (Caspi & Moffitt, 2006; Johnson, 1997; Li, 2003); however, cultural neuroscience represents a novel empirical approach to demonstrating bidirectional interactions between culture and biology by integrating theory and methods from cultural psychology (Kitayama & Cohen, 2007), neuroscience (Gazzaniga, Ivry, & Mangun, 2002), and neurogenetics (Canli & Lesch, 2007; Green et al., 2008, Hariri, Drabant, & Weinberger, 2006). Similar to other interdisciplinary fields such as social neuroscience (Cacioppo, Berntson, Sheridan, & McClintock, 2000) or social cognitive neuroscience (Ochsner & Lieberman, 2001), affective neuroscience (Davidson & Sutton, 1995), and neuroeconomics (Glimcher, Camerer, Poldrack, & Fehr, 2009), cultural neuroscience aims to explain a given mental phenomenon in terms of a synergistic product of mental, neural, and genetic events. Cultural neuroscience shares overlapping research goals with social neuroscience, in particular, in that both understand that the way neurobiological mechanisms facilitate cultural transmission, involves investigating primary social processes that enable humans to learn from one another, such as imitative learning. However, cultural neuroscience is also unique from related disciplines in that it focuses explicitly on ways that mental and neural events vary as a function of cultural traits (for example values, practices, and beliefs) in some meaningful way. Additionally, cultural neuroscience illustrates how cultural traits may alter neurobiological and psychological processes beyond those that facilitate social experience and behavior, such as perception and cognition.

There are at least three reasons why understanding cultural and genetic influences on brain function likely holds the key to articulating better psychological theory. First, a plethora of evidence from cultural psychology demonstrates that culture influences psychological processes and behavior (Kitayama & Cohen, 2007). Insofar as human behavior results from neural activity, cultural variation in behavior very probably emerges from cultural variation in neural mechanisms underlying these behaviors. Second, cultural variation in neural mechanisms may exist even in the absence of cultural variation at the behavioral or genetic level. That is, people living in different cultural environments may develop distinct neural mechanisms that underlie the same

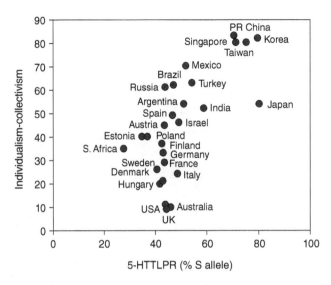

Figure 14.2   Collectivistic nations show greater prevalence of S allele carriers of the serotonin transporter gene (5-HTTLPR) (from Chiao & Blizinsky, 2009)

observable behavior or recruit the same neural mechanism to varying extents during a given task (Chiao et al., 2008; Goh et al., 2007; Gutchess, Welsh, Boduroglu, & Park, 2006; Hedden, Ketay, Aron, Markus, & Gabrieli, 2008). Third, population variation in the genome exists, albeit on a much smaller scale relative to individual variation, and 70% of genes express themselves in the brain (Hariri, Drabant, & Weinberger, 2006). For instance, collectivistic cultures are significantly more likely to be comprised of individuals carrying the short (S) allele of the serotonin transporter gene (5-HTTLPR), a functional polymorphism associated with emotion and social cognition (Chiao & Blizinsky, 2009, see Figure 14.2). This population variation in allelic frequency in functional polymorphisms—such as those that regulate neural activity—may exert influence on subsequent mental processes and behavior. Insofar as behavior arises from neural events and both cultural and genetic factors influence neural events, a comprehensive understanding of the nature of the human mind and behavior is impoverished without a theoretical and empirical approach that incorporates these multiple levels of analyses.

Indeed, early efforts by cultural neuroscientists to address the question of how culture influences brain function have proven fruitful, particularly for understanding differences in neurobiological processing between Westerners and East Asians (Chiao & Ambady, 2007; Han & Northoff, 2008; Park & Gutchess, 2006). Westerners engage brain regions associated with object processing to a greater extent compared to East Asians, who are less likely to focus exclusively on objects within a complex visual scene (Goh et al., 2007; Gutchess et al., 2006). Westerners show differences in medial prefrontal activity when thinking about themselves, relative to close others, but East Asians do not (Zhu, Zhang, Fan, & Han, 2007). Individual differences in endorsement of cultural values of individualism and collectivism predict variability in medial prefrontal activity during self judgments (Chiao et al., 2009a, 2010).

Activation in frontal and parietal regions associated with attentional control show greater response when Westerners and East Asians are engaged in culturally preferred judgments (Hedden et al., 2008). Even evolutionarily ancient limbic regions, such as the human amygdala, respond preferentially to fearful faces of one's own cultural group (Chiao et al., 2008). Taken together, these findings reveal cultural differences in brain functioning across a wide variety of psychological domains and provide concrete examples for how cultural contexts can be successfully incorporated in neuroscience investigations in the laboratory setting.

As the study of cultural influences on brain function is still in its nascent stages, it is important that researchers apply some caution to avoid drawing premature conclusions from initial findings and be aware of methodological and conceptual limitations. Critical neuroscience aims to sensitize practitioners towards conceptual and methodological difficulties wherever these arise. It highlights the need for researchers to maintain reflexivity in the experimental process by carefully considering aspects of design and interpretation in such a way that is open to debates in anthropology—for example in how to conceptualize culture. With regard to conceptualizations so far employed in the field, one focus certainly should be on attempts to supply more fine-grained, context-sensitive notions of culture that go beyond comparisons of country of origin and help capture the rich detail of cultural practices. The cross-fertilization of neuroscience and anthropology has the potential to offer helpful insights to both disciplines and to enrich findings through multiple methodologies (Seligman & Brown, 2010).

Culture and Mental Illness: An Example of Cultural Neuroscience as Critical Neuroscience in Practice

Cultural neuroscience can serve as a conceptual and empirical framework for providing a critical approach to the study of the brain in other domains of application as well. One especially promising application of cultural neuroscience to critical neuroscience is the examination of the significance of culturally-situated beliefs, values, preferences, and meanings on the social consequences and experience of mental illnesses. Psychological and emotional distress may manifest at the neural level but, as noted above, patterns of neural activity can also vary as a function of one's culture. Moreover, local norms and systems of beliefs regarding mental illness may also impact on one's cognitions and attitudes towards psychopathology, which can shape manifestations of mental illness at the neural level. Reductionistic approaches to examining the neural bases of psychopathology risk overlooking and minimizing the crucial role that the social and cultural context may play on observable neural responses (Choudhury et al., 2009; Kirmayer, 2006). For this reason, it is especially important that approaches in biological psychiatry and neuropsychiatry consider the dynamic interactions between potential variations at the cultural level of analysis—such as attitudes, beliefs, norms, and values—and potential variations at the neural level of analysis when exploring the relationship between the brain and psychopathology (Chiao & Ambady, 2007). Rather than pinning the cause of cultural differences in the experience of mental illnesses to biological differences in the brain or genes, cultural neuroscience seeks to demonstrate the interplay of cultural phenomonena and biological processes in shaping cognition, attitudes, behaviors, and human experience.

However, in order to achieve this important enrichment of psychiatry, it is essential that cultural neuroscience is linked productively to existing bodies of work in cultural anthropology and the social sciences more broadly. Anthropology in particular has the potential to become a nuanced contributor to cultural neuroscience through its views on how culture shapes social environments and partakes in the structuring of individual minds (Seligman & Brown, 2010). The part of the agenda of critical neuroscience that aims to promote and facilitate genuinely interdisciplinary approaches would be helpful in introducing the concepts, vocabularies, and methodologies of anthropology to cultural neuroscience. Given the history of the study of culture in the brain sciences, it is crucial to apply particular attention to the conceptualization of culture in these studies. Narrowly-defined or "broad brush stroke" labels can be informed by rigorous debates and fieldwork from anthropology, to strive to use groupings that are meaningful and get closer to the complexity of culture in the real world, where forces of globalization, for instance, make us reconsider culture as fixed, bounded entities (Choudhury & Kirmayer, 2009; Langlitz, Chapter 11, this volume). However, the sometimes narrow renderings of culture currently in the neuroscience literature, result from very real constraints imposed by the methodologies and technologies involved in neuroscience experiments. Clearly, these differences between disciplines must be acknowledged—and worked with. Critical neuroscience reminds us of the historical contingencies and cultural variation of the categories and phenomena that we study in the lab, an awareness of which might, in turn, open up alternative variables to study in relation to brain activity (Choudhury, 2010). These alternatives may be equally operationalizable in the lab setting but might open up the space of conceptual possibilities for interpretation.

Below, we highlight examples of three domains of inquiry relevant to the study of mental illness and emotional distress in which the cultural neuroscientific framework could provide deeper critical insight into the relationship between culture, biology, and mental illness. All of these domains have received little, if any, attention through neuroscientific investigation to date, but hold exceptional promise for providing a critical understanding of the role of the brain in mental illnesses and their treatment.

The stigma of mental illness

One research domain relevant to mental illness for which cultural neuroscience may be particularly promising is the study of the stigma of mental illness. Social stigma towards mental illness is universal and ubiquitous (Guimon, Fischer, & Satorius, 1999; Lauber & Rossler, 2007), posing one of the most significant barriers to improved quality of life for people with mental illnesses and their families (US Department of Health and Human Services, 1999; World Health Organization, 2001). Whilst the stigma and the marginalization of people with mental illness are universal, the immediate social and cultural contexts serve as profound determinants of which social identities and attributes will be stigmatized and which will be accepted (Crocker, Major, & Steele, 1998; Jones et al., 1984). Ultimately, the perceived boundary that distinguishes what cognitions and behaviors are normal or pathological fluctuates widely based on contextual factors. This is evidenced by culture-bound syndromes, such as *taijin kyoufushou, malgri, amok,* or even eating disorders, which are largely localized within specific cultural environments and represent maladaptive levels of concerns, anxieties, and fears grounded in the culture's system of beliefs and values. Similarly, for the stigma of mental illness, culturally-situated values and beliefs regarding the causes and meanings associated with mental illness can shape people's attitudes and behaviors towards mental illnesses and treatment seeking.

Supporting this view, prior cross-cultural research on mental illness stigma has demonstrated substantial cultural variations in the stigma towards mental illness. For instance, numerous studies have consistently revealed that members of East-Asian cultures maintain stronger stigmatizing attitudes towards mental illness compared to members of Western cultures. East-Asians and Asian-Americans tend to perceive people with mental illness as more dangerous, more unpredictable, and more abnormal than Western Europeans and European-Americans do (Furnham & Chan, 2004; Furnham & Murao, 2000; Furnham & Wong, 2007; Griffiths et al., 2006; Rao, Feinglass, & Corrigan, 2007); and Asian-Americans typically report experiencing greater levels of discrimination from friends, employers, and family members than European-Americans (Fogel & Ford, 2005). Furthermore, cultural differences in mental illness stigma may also affect the extent to which individuals seek and utilize treatment for mental illnesses. In the United States, Asian-Americans consistently display the lowest rates of mental health service use compared to all other major ethnic groups in the country (US Department of Health and Human Services, 2001; Yang & Wonpat-Borja, 2006) and exhibit longer delays in seeking treatment after illness onset compared to Caucasian-Americans (Hsu & Alden, 2008; Lin, Inui, Kleinman, & Womack, 1982; Lin, Tardiff, Donetz, & Goresky, 1978). Though many factors may contribute to these disparities, such as discrepancies in the availability of mental health services across communities or language barriers between patients and service providers, shame and disgrace associated with utilizing mental health services are also a major barrier to service use for Asians living in the United States (Leong & Lau, 2001; Yang, Phelan, & Link, 2008).

These profound cultural discrepancies in mental illness stigma are believed to arise from variations in social, moral, or spiritual values and beliefs associated with mental disorders. For example, within cultures oriented towards collectivism, such as in East-Asia, the stigma of mental illness not only affects the individual, but may have greater transmissibility across one's family and social network compared to in less collectivistic cultures (Kirmayer, 1989; Yang, 2007). As a result, the negative social repercussions of having a mental illness are greater within some cultures relative to others. Cultural concepts regarding social and moral status, such as face, also contribute to cultural discrepancies in stigma. In Chinese society, face represents one's social capital and moral standing within the local society, and concerns regarding face permeate social behaviors and relationships. The shame associated with mental illness is considered to severely undermine one's ability to accrue and maintain face, which is associated with overwhelming feelings of anxiety, humiliation, and dread for individuals in Chinese society (Yang et al., 2007; Yang & Kleinman, 2008). Cultures may vary in their beliefs about the role of biological, social, or superstitious factors as the cause of mental illnesses (Furnham & Chan, 2004; Furnham & Wong, 2007). Such beliefs regarding lay etiologies of mental illness have also been noted to influence stigmatizing attitudes and behaviors towards people with mental illness. These causal attributions can guide subsequent emotional reaction and behaviors towards people with mental illnesses. For example, beliefs about mental illness arising from uncontrollable factors (that is, neural and biological causes), typically lead to less blame and greater feelings of pity and sympathy, as well as helping behaviors, compared to beliefs about mental illness arising from controllable factors (such as weakness of will, punishment for sinful acts), which typically lead to greater blame, anger, and rejection (Corrigan, 2000; Weiner, 1993).

Cultural neuroscience offers exceptional promise for developing a critical understanding of why these cultural differences in stigmatizing beliefs, attitudes, and behaviors towards mental illness may arise and how they may contribute to a culturally-situated understanding of the neurobiology of mental illnesses and their interventions. Nearly all of the cross-cultural research to date comparing stigmatizing attitudes and behaviors associated with mental illness and the use of mental health services has been based on self-report, such as interviews or questionnaires. Given advances in understanding of how attitudes, evaluations, and emotions are represented in the brain, the integration of neuroscience to examining cultural variations in mental illness stigma would provide a powerful methodological tool for clarifying how cultural values and beliefs influence the stigma of mental illness.

One promising application of cultural neuroscience to examining cultural variations in the neural correlates of mental illness stigma would be to examine how neural activity associated with stigma may be influenced by adherence to culturally-situated values, such as face-related concerns. fMRI investigations of prejudice and discrimination have implicated the amygdala and insula in evaluations of stigmatized social groups that elicit negative affect and arousal, such as anxiety, fear and disgust (Amodio, 2008; Harris & Fiske, 2006; Krendl, Macrae, Kelley, Fugelsang, & Heatherton, 2006). Future research could examine whether activity in these regions is moderated by concerns with face or maintaining social standing when one is evaluating others who are believed to have mental illnesses or have utilized mental health services.

The impact of lay etiologies or causal attributions of mental illness on stigma could be studied in a similar manner. Given that biological attributions of mental illness are associated with greater feelings of pity than of blame, the impact of culturally-prevalent beliefs about the causes of mental illness on behavior could be studied by examining reactivity in networks associated with empathy, such as the anterior insula and anterior cingulate cortex (ACC), when observing the suffering of a person with mental illness. Consistent with this notion, a recent study by Decety, Echols, and Correll (2009) demonstrated that greater activity in regions associated with empathy were recruited when participants viewed the suffering of an AIDS patient who wasn't responsible for contracting the disease compared to an AIDS patient who was perceived as blameworthy. Similar paradigms could be employed to examine how culturally varying attributions and causal beliefs about mental illness may modulate activity in brain regions associated with empathic and pro-social processing.

Overall, examining the neural correlates for cultural variations in stigma towards mental illness may provide researchers and practitioners with insights into not only how mental disorders are represented at the level of the brain, but also how attitudes and cognitions about mental illness may be expressed within the brain. Exploring how culture influences stigma may be especially useful for future endeavors both in the identification of the universal and culturally-specific neural correlates of stigma towards mental illness.

Emotional experience and regulation

Another research domain relevant to mental illness and emotional distress into which cultural neuroscience could provide critical insight is the study of the neural underpinnings of emotional experience and regulation. As culturally-situated beliefs and values may impact attitudes towards mental illness, they may also influence how individuals experience, interpret, express, and control their emotions. While social neuroscience has greatly advanced our understanding of the neural systems involved in emotion and emotion regulation, there has been limited neuroscience research that has situated healthy and distressing emotional experiences within a cultural context. Given that the neural systems underlying emotion can vary between populations and cultures (Chiao et al., 2008; Hot et al., 2006; Moriguchi et al., 2005) and that cultural values and norms can shape the regulation of emotions (Butler, Lee, & Gross, 2007), the complex interplay between biological and socio-cultural influences may not be represented in our current understanding of the neural bases of emotional experience and regulation.

A promising application for cultural neuroscience to provide a more comprehensive and critical approach to the study of the neural bases of emotion would be to examine the impact of cultural display rules on neural activity during the experience of normal, as well as pathological, emotional states. Display rules refer to culture-specific rules that dictate when specific facial expressions of emotion are appropriate, as well as the appropriateness of the intensity of expression. While individualistic cultures value personal expression and autonomy and the freedom to express oneself is highly encouraged, in collectivistic cultures, maintaining harmonious social relationships are valued over individual expression (Markus & Kitayama, 1991; Triandis, 1989). As a result, cultures may differ in the extent to which they promote norms that encourage restraint of intense emotional expression during interpersonal interactions.

Consistent with these ideas, early studies on cultural variations in display rules demonstrated that while American participants expressed negative emotions to a similar extent when alone or in the presence of a higher-status experimenter, Japanese participants were more likely to control or mask the expression of negative emotions in the presence of the experimenter compared to being alone (Ekman, 1972; Friesen, 1972). Since then, a number of studies have demonstrated the important role of display rules on influencing cultural variation in emotion expression (Matsumoto, 1993; Matsumoto, Takeuchi, Andayani, Kouznetsova, & Krupp, 1998). Particularly noteworthy is a recent large-scale study by Matsumoto and colleagues (2008) spanning 32 countries, which demonstrated that greater levels of individualism are associated with higher levels of emotional expressivity.

Based on these findings on cultural display rules, a critical application of neuroscience to the study of emotion could consist of examining how one's adherence to cultural display rules influences emotional reactivity at the neural level. Cultures that value greater emotional restraint may differentially activate regions associated with monitoring and regulating emotions, such as the dorsal ACC and lateral prefrontal cortex (PFC), relative to cultures that value greater emotional expressiveness.

Methodologically, concerns about adhering to display rules could be experimentally induced before or during neuroimaging by cultural priming of representations of social others who are of equal status or superior status.

Moreover, cultural differences in the regulation of emotional expressions may lead to variations in the subjective experience of emotions themselves. Though facial expressions of emotions are usually assumed to be a consequence of subjective emotional experiences, the relationship between emotional experience and expression is bidirectional. Network theories of emotion postulate that emotional experiences—and their associated stimuli and action tendencies—are organized into a network in memory that comprises an emotion schema (Lang, 1985; Leventhal, 1980; Moors, 2009). While specific external stimuli may activate an emotion and its behavioral responses, facial expressions of emotions can also activate their respectively associated emotions (Ekman, Levenson, & Friesen, 1983; Strack, Martin, & Stepper, 1988). Given that facial expressions can serve as feedback for emotional experience, cultural display rules may moderate response in brain regions associated with emotional experience. As a result, when viewing stimuli that may induce fear or disgust within a social or interpersonal context, members of cultures with adherence to strict display rules may exhibit reduced reactivity in regions which process aversive emotional stimuli. These types of studies may be especially relevant when examining individuals with emotional disorders at the neural level of analysis, such that cultural display rules may influence how abnormal emotional states are expressed in the brain.

Cultural display rules may, furthermore, also impact the neural processes involved in the interpretation of the emotional expressions of others. Members of cultures emphasizing high restraint of emotional expression may have to rely on cognition to a greater extent to read the internal mental and emotional states of others who may be adhering to the same strict display rules. Thus, individuals may recruit regions—such as the medial PFC and temporal-parietal junction—that are associated with mentalizing, perspective taking and theory of mind processes, differentially during the interpretations of cultural ingroup members' facial expressions as a function of culturally-situated display rules.

In sum, much of our understanding of the neural correlates of emotion expression and regulation have been developed without adopting a critical approach that accounts for the role of cultural meanings, values, and norms on shaping how emotions are understood, experienced, and regulated. Display rules are only one factor that may produce cultural variations in the neural correlates of emotions across cultures; but as illustrated above, adopting a cultural neuroscientific approach to the study of emotion by considering such cultural phenomena would provide more critical insights into the neural underpinnings of healthy and maladaptive emotional states.

Somatization of mental illness

The role of the cultural context on patterns of symptom presentation is yet another avenue for applying cultural neuroscience to framing the neurobiological bases of mental illness within the context of cultural systems of meanings and beliefs. Exploring the relationship between the neural correlates of psychological distress and the somatization of depressive symptoms may be a promising future direction in regard to this endeavor. Relative to members of Western cultures, members of East-Asian cultures suffering from depression and emotional distress are less likely to verbally express psychological symptoms, and instead emphasize physical and bodily symptoms (Chang, 1985; Kleinman, 1982; Ryder et al., 2008). Even after controlling for variations in language-specific variations in differentiating emotional and somatic words between Chinese and English, through comparing Chinese-Americans and European-Americans, Chinese-Americans are more likely to utilize somatic terms than European-Americans to describe emotional events (Tsai, Simeonova, & Watanabe, 2004).

Though cultural variations in the somatization or psychologization of symptoms have been well documented, a clear explanation for the reason for these differences remains elusive. Given the greater severity of stigma and shame that would be incurred by individuals with mental illnesses in some cultures, emphasizing somatic rather than psychological symptoms may be a more socially acceptable manner for expressing psychological distress. Despite the appeal of this theory, perceived stigma was not identified as a mediator of culture and patterns of symptom expression, suggesting that other factors may be contributing to cultural variations in somatization besides stigma of mental illness (Ryder et al., 2008). Another explanation is based on the notion that cultures also differ in the degree to which emotions are conceptualized as being intertwined or independent from physical and bodily states. Following the philosophical tradition of Descartes' mind–body distinction, Western culture has largely conceptualized emotional and psychological experiences as distinct from physical and somatic ones. On the other hand, the holistic approach to medicine adopted by Chinese culture conceptualized the integration and balance of the mind and body (Heine, 2008; Nisbett, 2003). Accordingly, variations in somatization may arise because emotions are perceived to be inherently more related to bodily states in some cultures compared to others. This leads to the possibility that some cultures may actually experience greater degrees of somatic symptoms (such as exhaustion, aching body, headaches, dizziness, poor sleep, and appetite) associated with psychological distress relative to emotional and psychological symptoms (such as anxiety, sadness, and troubling thoughts).

A third explanation posits the role of cultural differences in the processing and expression of affect as a source of variations in somatization. For instance, Ryder and colleagues (2008) explored the relationship between the trait alexithymia, in which individuals have difficulty experiencing and articulating emotional states, and somatization of depressive symptoms. The authors discovered that a component of alexithymia—preference for an externally-oriented style of thinking—mediated the relationship between culture and somatization of depressive symptoms. An externally focused style of cognition is consistent with preferences and values prevalent in collectivistic cultures, which emphasize the importance of attending to relational needs, social obligations, and maintenance of social harmony (Lehman et al., 2004; Markus & Kitayama, 1991). This suggests that differences in culturally-situated cognitive styles may contribute to variations in somatization across cultures.

Cultural neuroscience may be able to provide a critical approach to elucidating the underlying causes for the observed cultural variations in somatization. Theories that seek to explain the relationship between culture and somatization may be clarified and tested by utilizing neuroimaging approaches. Whether variations in tendencies for somatization represent differences in verbal expression or actual subjective experience of symptoms could be examined by investigating brain structures that dissociably represent objective and subjective experience. For instance, structures involved in representing pain and discomfort (such as ACC and insula) show differential patterns of activation that correlate with the subjective affective experience and the objective sensory properties of the experience (that is, the intensity of the pain-inducing stimulus). While the posterior portions of the ACC and insula represent objective aspects of pain, such as the quality and intensity of the sensory input, the anterior portions of these regions typically represent the subjective experience and awareness of pain (Amodio & Frith, 2006; Singer et al., 2004; Jackson, Rainville, & Decety, 2006; Craig, 2009). These dissociable regions for pain may provide an initial basis for examining whether individuals who somaticize symptoms objectively experience bodily and physiological states of pain and discomfort associated with their symptoms, or whether they are experienced subjectively in the absence of direct sensory input.

Neural correlates of alexithymia have also been identified, which include differences in reactivity in self-monitoring and emotion regulation regions, such as the anterior and posterior cingulate cortices and the dorsal-lateral PFC, between alexithymic and non-alexithymic individuals (Aleman, 2005; Berthoz et al., 2002; Moriguchi et al., 2007). This may help clarify whether individuals who exhibit greater somatization of symptoms demonstrate patterns of neural activity that resemble alexithymic individuals, such as high preferences for externally-focused cognition and attention. Furthermore, moderation of activity in these regions by cultural values that endorse the importance of maintaining social harmony and attending to the needs of others would also lend support to the theory that somatization arises from greater levels of other-oriented rather than self-oriented cognitions.

Conclusion

Critical neuroscience presents a host of important and timely challenges to behavioral and brain scientists such as the need to critically examine how research questions are chosen and addressed. Here we propose at least two ways in which cultural neuroscience serves as critical neuroscience in practice. First, by studying cultural values, practices, and beliefs at a neural level, we gain leverage on understanding how cultural context affects normal brain functioning in the laboratory setting. Second, by using a cultural neuroscientific approach to examine complex, atypical human behaviors, such as mental illness, we may be able to gain traction on a phenomenon where critical examination is especially needed. Cultural variation in how symptoms of the same disorder are expressed or even experienced has significant implications for clinical diagnoses, as well as for the classification of mental disorders.

Given that proper classification, diagnosis, and treatment of disorders is a critical determinant for outcomes of those suffering from mental illness, a critical examination of the underlying causes of somatization is imperative. By integrating cultural neuroscience with clinical practice, researchers and practitioners may benefit from a deeper and more critical understanding of the intricate relationship between cultural context and mental illness. While pursuing these questions, it is important that cultural neuroscience in its current form maintains critical reflection directed at its concepts, operationalizations, and theoretical background. As a young field of research currently taking shape—and in view of the methodological constraints imposed by the nature of cognitive neuroscience experimental set-ups—it is vital to remain open to influences from a multitude of relevant domains of inquiry and to constantly problematize concepts, methods, and interpretations of results. In line with the interdisciplinary orientation of critical neuroscience, it is our hope that a sustained dialogue be established between cultural neuroscience and other disciplines that investigate cultural traits and cultural differences—such as anthropology, cultural studies, sociology, and transcultural psychiatry. Such a mutual exchange would without doubt benefit all fields involved.

References

Aleman, A. (2005). Feelings you can't imagine: Towards a cognitive neuroscience of alexithymia. *Trends in Cognitive Sciences, 9*, 553–555.

Amodio, D. M. (2008). The social neuroscience of intergroup relations. *European Review of Social Psychology, 19*, 1–54.

Amodio, D. M., & Frith, C. D. (2006). Meeting of minds: The medial frontal cortex and social cognition. *Nature Reviews Neuroscience, 7*, 268–277.

Arnett, J. J. (2008). The neglected 95%: Why American psychology needs to become less American. *American Psychologist, 63*(7), 602–614.

Berthoz, S., Artiges, E., Van de Moortele, P., Poline, J., Rouquette, S., Consoli, S. M., & Martinot, J. (2002). Effect of impaired recognition and expression of emotions on frontocingulate cortices: An fMRI study of men with alexithymia. *American Journal of Psychiatry, 159*, 961–967.

Butler, E. A., Lee, T. L., & Gross, J. J. (2007). Emotion regulation and culture: Are the social consequences of emotion suppression culture-specific? *Emotion, 7*, 30–48.

Cacioppo, J. T., Berntson, G. G., Sheridan, J. F., & McClintock, M. K. (2000). Multi-level integrative analyses of human behavior: Social neuroscience and the complementing nature of social and biological approaches. *Psychological Bulletin, 126*, 829–843.

Canli T., & Lesch, K. P. (2007). Long story short: The serotonin transporter in emotion regulation and social cognition. *Nature Neuroscience, 10*, 1103–1109.

Caspi, A., & Moffitt, T. E. (2006). Gene-environment interactions in psychiatry: Joining forces with neuroscience? *Nature Reviews Neuroscience, 7*, 583–590.

Chang, W. C. (1985). A cross-cultural study of depressive symptomatology. *Culture, Medicine, and Psychiatry, 9*, 295–317.

Chiao, J. Y. (2009). Cultural neuroscience: A once and future discipline. *Progress in Brain Research, 178*, 287–304.

Chiao, J. Y., & Ambady, N. (2007). Cultural neuroscience: Parsing universality and diversity across levels of analysis. In S. Kitayama & D. Cohen (Eds.), *Handbook of Cultural Psychology* (pp. 237–254). New York: Guilford Press.

Chiao, J. Y., & Blizinsky, K. D. (2009). Culture-gene coevolution of individualism-collectivism and the serotonin transporter gene (5-HTTLPR). *Proceedings of the Royal Society B: Biological Sciences.*

Chiao, J. Y., & Cheon, B. K. (in press). The weirdest brains in the world. *Behavioral and Brain Sciences.*

Chiao, J. Y., Harada, T., Komeda, H., Li, Z., Mano, Y., Saito, D. N., & Iidaka, T. (2009a). Neural basis of individualistic and collectivistic views of self. *Human Brain Mapping, 30*(9), 2813–2820.

Chiao, J. Y., Harada, T., Komeda, H., Li, Z., Mano, Y., Saito, D. N., Iidaka, T. (2010). Dynamic cultural influences on neural representations of the self. *Journal of Cognitive Neuroscience, 22*(1), 1–11.

Chiao, J. Y., Iidaka, T., Gordon, H. L., Nogawa, J., Bar, M., Aminoff, E., & Ambady, N. (2008). Cultural specificity in amygdala response to fear faces. *Journal of Cognitive Neuroscience, 20*(12), 2167–2174.

Choudhury, S. (2010). Culturing the adolescent brain: What can neuroscience learn from anthropology? *Social Cognitive and Affective Neuroscience, 5*, 159–167.

Choudhury, S., & Kirmayer, L. J. (2009). Cultural neuroscience and psychopathology: Prospects for cultural psychiatry. *Progress in Brain Research, 178*, 263–283.

Choudhury, S., Nagel, S. K., & Slaby, J. (2009). Critical neuroscience: Linking neuroscience and society through critical practice. *BioSocieties, 4*, 61–77.

Corrigan, P. W. (2000). Mental health stigma as social attribution: Implications for research methods and attitude change. *Clinical Psychology: Science and Practice, 7*, 48–67.

Craig, A. D. (2009). How do you feel now? The anterior insula and human awareness. *Nature Reviews Neuroscience, 10*, 59–70.

Crocker, J., Major, B., & Steele, C. (1998). Social stigma. In D. T. Gilbert & S. T. Fiske (Eds.), *The handbook of social psychology*. Boston, MA: McGraw-Hill.

Davidson, R. J., & Sutton, S. K. (1995). Affective neuroscience: The emergence of a discipline. *Current Opinion in Neurobiology, 5*, 217–224.

Decety, J., Echols, S., & Correll, J. (2009). The blame game: The effect of responsibility and social stigma on empathy for pain. *Journal of Cognitive Neuroscience, 22*(5), 985–997.

Ekman, P. (1972). Universals and cultural differences in facial expressions of emotion. In J. Cole (Ed.), *Nebraska symposium on motivation 1971* (Vol. 19, pp. 207–283). Lincoln, NE: University of Nebraska Press.

Ekman, P., Levenson, R. W., & Friesen, W. V. (1983). Autonomic nervous system activity distinguishes among emotions. *Science, 221*, 1208–1210.

Fogel, J., & Ford, D. E. (2005). Stigma beliefs of Asian Americans with depression in an internet sample. *Canadian Journal of Psychiatry, 50*, 470–478.

Friesen, W. V. (1972). *Cultural differences in facial expressions in a social situation: An experimental test of the concept of display rules.* (Unpublished doctoral dissertation, University of California, San Francisco, 1972).

Furnham, A., & Chan, E. (2004). Lay theories of schizophrenia: A cross-cultural comparison of British and Hong Kong Chinese attitudes, attributions and beliefs. *Social Psychiatry and Psychiatric Epidemiology, 39*, 543–552.

Furnham, A., & Murao, M. (2000). A cross-cultural comparison of British and Japanese lay theories of schizophrenia. *International Journal of Social Psychiatry, 46*, 4–20.

Furnham, A., & Wong, L. (2007). A cross-cultural comparison of British and Chinese beliefs about the causes, behavior manifestations and treatment of schizophrenia. *Psychiatry Research, 151*, 123–138.

Gazzaniga, M. S., Ivry, R., & Mangun, G. R. (2002). *Cognitive neuroscience: The biology of the mind* (2nd ed.). New York: W.W. Norton.

Glimcher, P. W., Camerer, C. F., Fehr, E., & Poldrack, R. A. (Eds.), (2009). *Neuroeconomics: Decision-making and the brain*. New York: Academic Press.

Goh, J. O., Chee, M. W., Tan, J. C., Venkatraman, V., Hebrank, A., Leshikar, E. D., & Park, D. C. (2007). Age and culture modulate object processing and object-scene binding in the ventral visual area. *Cognitive, Affective, Behavioral Neuroscience, 7*, 44–52.

Green, A. E., Munafò, M., DeYoung, C. G., Fossella, J., Fan, J., & Gray, J. R. (2008). Using genetic data in cognitive neuroscience: From growing pains to genuine insights. *Nature Reviews Neuroscience, 9,* 710–720.

Griffiths, K. M., Nakane, Y., Christensen, H., Yoshioka, K., Jorm, A. F., & Nakane, H. (2006). Stigma in response to mental disorders: A comparison of Australia and Japan. *BMC Psychiatry, 6,* 21.

Guimon, J., Fischer, W., & Satorius, N. (Eds.). (1999). *The image of madness: The public facing mental illness and psychiatric treatment.* Basel, Switzerland: Karger.

Gutchess, A. H., Welsh, R. C., Boduroglu, A., & Park, D. C. (2006). Cultural differences in neural function associated with object processing. *Cognitive, Affective, and Behavioral Neuroscience, 6*(2), 102–109.

Han, S., & Northoff, G. (2008). Culture-sensitive neural substrates of human cognition: A transcultural neuroimaging approach. *Nature Review Neuroscience, 9,* 646–654.

Hariri, A. R., Drabant, E. M., & Weinberger, D. R. (2006). Imaging genetics: Perspectives from studies of genetically driven variation in serotonin function and corticolimbic affective processing. *Biological Psychiatry, 59*(10), 888–897.

Harris, L. T., & Fiske, S. T. (2006). Dehumanizing the lowest of the low: Neuroimaging responses to extreme out-groups. *Psychological Science, 17,* 847–853.

Hedden, T., Ketay, S., Aron, A., Markus H. R., & Gabrieli, J. D. (2008). Cultural influences on neural substrates of attentional control. *Psychological Science, 19*(1), 12–16.

Heine, S. J. (2008). *Cultural Psychology.* New York: W. W. Norton & Company, Inc.

Henrich, J., Heine, S., & Norenzayan, A. (in press). The weirdest people in the world? *Behavioral and Brain Sciences.*

Hot, P., Saito, Y., Mandai, O., Kobayashi, T., & Sequeira, H. (2006). An ERP investigation of emotional processing in European and Japanese individuals. *Brain Research, 1122*(1), 171–178.

Hsu, L., & Alden, L. E. (2008). Cultural influences on willingness to seek treatment for social anxiety in Chinese- and European-heritage students. *Cultural Diversity and Ethnic Minority Psychology, 14*(3), 215–223.

Jackson, P. L., Rainville, P., & Decety, J. (2006). To what extent do we share the pain of others? Insight from the neural bases of pain empathy. *Pain, 125,* 5–9.

Johnson, M. (1997). *Developmental cognitive neuroscience.* Oxford: Blackwell.

Jones E., Farina, A., Hastorf, A., Markus, H., Miller, D. T., & Scott, R. (1984). *Social stigma: The psychology of marked relationships.* New York: Freeman.

Kirmayer, L. J. (2006). Beyond the "new cross-cultural psychiatry": Cultural biology, discursive psychology and the ironies of globalization. *Transcultural Psychiatry, 43,* 126–144.

Kirmayer, L. (1989). Cultural variations in the response to psychiatric disorder and emotional distress. *Social Science and Medicine, 29,* 327–329.

Kitayama, S., & Cohen, D. (2007). *Handbook of cultural psychology.* New York: Guilford Press.

Kleinman, A. (1982). Neurasthenia and depression: A study of somatization and culture in China. *Culture, Medicine and Psychiatry, 6,* 117–190.

Krendl, A. C., Macrae, C. N., Kelley, W. M., Fugelsang, J. F., & Heatherton, T. F. (2006). The good, the bad and the ugly: An fMRI investigation of the functional anatomic correlates of stigma. *Social Neuroscience, 1,* 5–15.

Kuhn, T. S. (1970). *The structure of scientific revolutions* (2nd ed.). Chicago: University of Chicago Press.

Lang, P. J. (1985). The cognitive psychophysiology of fear and anxiety. In A. H. Tuma & J. D. Maser (Eds.). *Anxiety and the anxiety disorders* (pp. 131–170). Hillsdale, NJ: Lawrence Erlbaum Associates, Inc.

Lauber, C., & Rossler, W. (2007). Stigma towards people with mental illness in developing countries in Asia. *International Review of Psychiatry, 19*, 157–178.

Lehman, D. R., Chiu, C., & Schaller, M. (2004). Psychology and culture. *Annual Reviews Psychology, 55*, 689–714.

Leong, F. T. L., & Lau, A. S. (2001). Barriers to providing effective mental health services to Asian Americans. *Mental Health Services Research, 3*, 201–214.

Leventhal, H. (1980). Toward a comprehensive theory of emotion. In L. Berkowitz (Ed.), *Advances in experimental social psychology* (Vol. 13, pp. 139–197). New York: Academic Press.

Li, S. C. (2003). Biocultural orchestration of developmental plasticity across levels: The interplay of biology and culture in shaping the mind and behavior across the life span. *Psychological Bulletin, 129*(2), 171–194.

Lin, K., Inui, T. S., Kleinman, A., & Womack, W. M. (1982). Sociocultural determinants of the help-seeking behavior of patients with mental illness. *Journal of Nervous and Mental Disease, 170*, 78–85.

Lin, T., Tardiff, K., Donetz, G., & Goresky, W. (1978). Ethnicity and patterns of help-seeking. *Culture, Medicine & Psychiatry, 2*, 3–13.

Markus, H. R., & Kitayama, S. (1991). Culture and the self: Implications for cognition, emotion and motivation. *Psychological Review, 98*, 224–253.

Matsumoto, D. (1993). Ethnic differences in affect intensity, emotion judgments, display rule attitudes and self-reported emotional expression in an American sample. *Motivation and Emotion, 17*, 107–123.

Matsumoto, D., Takeuchi, S., Andayani, S., Kouznetsova, N., & Krupp, D. (1998). The contribution of individualism-collectivism to cross-national differences in display rules. *Asian Journal of Social Psychology, 1*, 147–165.

Matsumoto, D., Yoo, S. H., Fontaine, J., Anguas-Wong, A. M., Arriola M., Ataca, B., & Granskaya, J. V. (2008). Mapping expressive differences around the world: The relationship between emotional display rules and individualism versus collectivism. *Journal of Cross-Cultural Psychology, 39*, 55–74.

Moors, A. (2009). Theories of emotion causation: A review. *Cognition and Emotion, 23*, 625–662.

Moriguchi, Y., Decety, J., Ohnishi, T., Maeda, M., Mori, T., Nemoto, K., & Komaki, G. (2007). Empathy and judging other's pain: An fMRI study of alexithymia. *Cerebral Cortex, 17*, 2223–2234.

Moriguchi, Y., Takashi, O., Takashi, K., Takeyuki, M., Makiko, H., Minoru, Y., Hiroshi, M., & Gen, K. (2005). Specific brain activation in Japanese and Caucasian people to fearful faces. *Neuroreport, 16*(2), 133–136.

Nisbett, R. E. (2003). The geography of thought: How Asians and Westerners think differently … and why. New York: The Free Press.

Ochsner, K. N., & Lieberman, M. D. (2001). The emergence of social cognitive neuroscience. *American Psychologist, 56*, 717–734.

Park, D. C., & Gutchess, A. H. (2006). The cognitive neuroscience of aging and culture. *Current Directions in Psychological Science, 15*(3), 105–108.

Rao, D., Feinglass, J., & Corrigan, P. (2007). Racial and ethnic disparities in mental illness stigma. *Journal of Nervous and Mental Disease, 195*(12), 1020–1023.

Ryder, A. G., Yang, J., Zhu, X., Yao, S., Yi, J., Heine, S. J., & Bagby, R. M. (2008). The cultural shaping of depression: Somatic symptoms in China, psychological symptoms in North America? *Journal of Abnormal Psychology, 117*, 300–313.

Seligman, R., & Brown, R. A. (2010). Theory and method at the intersection of anthropology and cultural neuroscience. *Social Cognitive & Affective Neuroscience, 5*, 130–137.

Singer, T., Seymour, B., O'Doherty, J. P., Klaube, H., Dolan, R. J., & Frith, C. D. (2004). Empathy for pain involves the affective but not sensory components of pain. *Science, 303,* 1157–1162.

Strack, F., Martin, L. L., & Stepper, S. (1988). Inhibiting and facilitating conditions of the human smile: A non-obtrusive test of the facial feedback hypothesis. *Journal of Personality and Social Psychology, 53,* 768–777.

Triandis, H. C. (1989). The self and social behavior in differing cultural contexts. *Psychological Review, 96,* 506–520.

Tsai, J. L., Simeonova, D. I., & Watanabe, J. T. (2004). Somatic and social: Chinese Americans talk about emotion. *Personality and Social Psychology Bulletin, 30,* 1226–1238.

US Department of Health and Human Services. (1999). *Mental health: A report of the surgeon general.* Rockville, MD: US Department of Health and Human Services.

US Department of Health and Human Services (2001). *Mental health: Culture, race, and ethnicity—A supplement to mental health: A report of the surgeon general.* Rockville, MD: Public Health Service, Office of the Surgeon General.

Weiner, B. (1993). On sin versus sickness: A theory of perceived responsibility and social motivation. *American Psychologist, 48,* 957–965.

World Health Organization. (2001). *Stigma and global health: An international perspective.* New York: World Health Organization.

Yang, L. H. (2007). Applications of mental illness stigma theory to Chinese societies: Synthesis and new directions. *Singapore Medical Journal, 48,* 977–985.

Yang, L. H., & Kleinman, A. (2008). "Face" and the embodiment of stigma in China: The cases of schizophrenia and AIDS. *Social Sciences and Medicine, 67,* 398–408.

Yang, L. H., Kleinman, A., Link, B. G., Phelan, J. C., Lee, S., & Good, B. (2007). Culture and stigma: Adding moral experience to stigma theory. *Social Science and Medicine, 64,* 1524–1535.

Yang, L. H., Phelan, J. C., & Link, B. G. (2008). Stigma and beliefs of efficacy towards Chinese medicine and Western psychiatric treatment among Chinese-Americans. *Cultural Diversity and Ethnic Minority Psychology, 14,* 10–18.

Yang, L. H., & Wonpat-Borja, A. H. (2006). Psychopathology among Asian-Americans. In F.T. Leung, A. Inman, A. Ebreo, L. H. Yang, L. Kinoshita & M. Fu (Eds.), *Handbook of Asian-American psychology* (pp. 379–404). San Francisco: Sage.

Zhu, Y., Zhang, Li., Fan, J., & Han. S. (2007). Neural basis of cultural influence on self representation. *Neuroimage, 34,* 1310–1317.

Part V
Beyond Neural Correlates
Ecological Approaches to Psychiatry

15

Re-Socializing Psychiatry
Critical Neuroscience and the Limits of Reductionism

Laurence J. Kirmayer and Ian Gold

Contemporary neuroscience is advancing our understanding of the role of the brain in psychiatric disorders. These successes, allied with broader social forces, have allowed biological psychiatry to largely displace psychodynamic and social psychiatry, which emphasized the importance of meaning and experience in psychopathology. In contrast to these traditions, biological psychiatry tends to treat experience as an epiphenomenon of neural activity and the social world as an independent set of external stimuli or adaptive contexts. As a result, psychiatry reduces phenomenology to a list of symptoms and signs, and reduces the social world to a set of learned behaviors, attitudes, and social contingencies. In fact, the social world plays a fundamental role in human functioning and experience, with causal effects on mental health and illness. In this chapter we critically review the reductionist picture in contemporary psychiatry and provide illustrations of the importance of the social world in psychopathology from research in social neuroscience and psychiatric epidemiology.

In an editorial in *JAMA* (*Journal of the American Medical Association*) in 2005, Thomas Insel and Remi Quirion, the scientific directors of the US and Canadian national institutes that fund mental health research, argued that psychiatry is a discipline of "clinically applied neuroscience" (Insel & Quirion, 2005). Given their influential positions, this vision of psychiatry is important not only for the immediate future of funding psychiatric research, but for the direction of the whole field. The examples of neuroscience they described as providing a new foundation for psychiatry were drawn mainly from genetics and neuroimaging research. There is no question that these fields have made dramatic progress in recent decades. It is equally clear, however, that psychiatry as currently practiced includes a far more varied and complex array of human problems than can be neatly fitted into a biologically driven nosology, set of theoretical models and corresponding treatments. Twenty years ago, Leon Eisenberg warned of the stunting effects on psychiatry of ignoring either the brain

Critical Neuroscience: A Handbook of the Social and Cultural Contexts of Neuroscience, First Edition.
Edited by Suparna Choudhury and Jan Slaby.
© 2012 John Wiley & Sons, Ltd. Published 2016 by John Wiley & Sons, Ltd.

or the mind (Eisenberg, 1986). To this we must add the continuing tendency to downplay the social and cultural origins of disability and distress as well as resilience and healing. Defining psychiatry as applied neuroscience valorizes the brain but urges on us a discipline that is both mindless and uncultured. Critical neuroscience can work against this conceptual shrinkage to locate psychiatric research, theory, and practice in a wider social, cultural, and political world.

Critical neuroscience aims to trace the social origins and implications of claims like those of Drs Insel and Quirion. Behind their enthusiasm for neuroscience as a foundation for psychiatry is a reductionistic view of the origins and nature of human behavior and experience as rooted in neurobiology. This neuroreductionism seems attractive and even compelling for several reasons: (1) the technologies of neuroscience have made the activities of the brain visible in new and vivid ways; (2) in some instances, neuroscientific research has generated partial explanations for specific symptoms, diseases or disorders; (3) in the social sphere, neurobiological explanations for mental illness have been embraced by many because they shift causality away from human agency and so work to exculpate individuals and their families as the causes of their own suffering; (4) the biological turn has been heavily promoted with many inflated claims because this serves powerful interests in the pharmaceutical industry; and (5) more broadly, the emphasis on neurobiology diverts attention from social, structural, and economic factors that are politically contentious. Ultimately, neurobiological reductionism in psychiatry serves a larger ideology that locates human problems in our brains and bodies rather than in our histories and social predicaments.

In this chapter we want to challenge the logic of this neuroreductionist program, especially as it applies to psychiatry. Our position can be expressed simply: the social environment makes a difference to mental life and to mental illness. Therefore, a reductionist psychiatry which restricts itself to the processes inside the brain is doomed to be incomplete. We begin by surveying some types of reductionism and challenge its commitment to the idea that a single level of explanation of human behavior is possible. We then illustrate the importance of social processes in psychopathology through examples from social neuroscience and social psychiatric epidemiology. Finally, we consider why, despite the obvious importance of higher order cognitive and social processes in psychiatry, many continue to believe that the future of psychiatry rests with the discovery and clinical application of lower-level biological explanations.

Varieties of Reductionism

Reductionism has many forms or versions, encompassing methodological strategies, ontological claims, and epistemological commitments. Some forms of reductionism are useful while others may promote work that is profoundly misleading and potentially damaging to individuals, groups, and communities. Conflating the different forms of reductionism makes it hard to see the virtues and costs of each.

In the domain of psychiatry, there are at least three different versions of neuroreductionism to distinguish:

(1) *Methodological reductionism* assumes that it is a necessary and sufficient methodological strategy to break down complex systems and phenomena into simpler components or analogues to study. This includes focusing on animal models—even though these cannot address the more complex processes of narrative construction, reasoning, or imagination—and studying simple uni-directional or linear causal effects, even when it is clear that most biobehavioral systems involve circular feedback loops or mutual causality. Even when psychological processes are recognized as important, the assumption is that clinical science can advance by approaching such higher order phenomena (like pathological behavior and experience) in terms of lower-level (neurobiological) processes.

(2) *Ontological reductionism* claims that the higher order phenomena are constituted by the lower, that is, that there is no additional entity that is introduced to give rise to these higher order (mental) phenomena. Thus, mind is nothing other than the brain (or the brain at work) and we can, therefore, ultimately dispense with our folk language that treats the psychological (or social) domain as something distinct.

(3) *Epistemological reductionism* argues that there is no need for information about the higher order levels to explain human behavior and experience; everything that can be or needs to be known can be derived or deduced from our knowledge of lower order mechanisms. Hence, self-reports can ultimately be by-passed when we can measure what is going on inside the other person with a brain-imaging device like the philosopher's science-fictional "cerebroscope." In seeing that certain patterns of brain activation have occurred, we would have all the same information about the person we derive from statements like "I am in pain" or "I see red" or "The CIA has planted a bug in my brain."

Methodological reductionism has proved an enormously productive strategy for the advance of science—though it always risks losing sight of the crucial phenomena to be explained. In fact, successful reduction often depends on using the higher order phenomena to guide the search for lower level explanations and to recognize an adequate explanation when it has been found. The mathematical biologist Robert Rosen (1968) made this argument using the example of the relationship of statistical mechanics and thermodynamics. The kinetic molecular theory represents one of the best examples of a successful reduction; it shows how the macroscopic thermodynamic properties of a gas can be reduced to (explained by) the movements (dynamics) of the particles making up the gas. However, there are an immense number of ways to describe the ensemble of gas molecules and so, Rosen argued, the rules of statistical mechanics could only be discovered because an adequate description of the macroscopic properties existed against which to develop and test the lower level theory.[1] Therefore, even the most successful cases of reductionism in science argue against the adequacy of a program of research focused only on the simpler (lower-level) system as the sole methodological strategy.

---

[1] Indeed, the existence of molecules themselves was demonstrated through macroscopic phenomena like Brownian motion.

On the face of it, human beings are comprised of many systems at many different levels of organization: molecules, organelles, cells, tissues, organs, physiological systems, neural circuits and information processing systems, psychological faculties and functions (some of which have been called "modules"), memories, schemas and other knowledge structures, habits and dispositions to respond, patterns of interpersonal interaction, and so on. There may well be additional levels between these well-identified levels of structure. Clearly, there is no need to posit different types of substance to encompass these different levels of organization: they are all biological in the sense that there is an unbroken continuity of material constitution as one moves up and down the hierarchy. Ontological reductionism in the form of physicalism is widely accepted as part of the scientific worldview (though challenged in some traditions as still in conflict with religious values that insist on a fundamental dualism or supervenience of the spirit and the sacred in human existence). One version of this physicalism results in eliminative materialism: the idea that we can dispense with our notions of mind and experience and replace them with an empirically grounded vocabulary of neural or brain processes.

Nevertheless, explanations of human behavior employ multiple sets of conceptual models or descriptive languages that reflect different levels of organization: the social level of interpersonal relationships; the psychological level of cognitive schemas, motivations, and emotions; the neuropsychological notions of brain functions, regions, and circuits; the neurophysiological vocabulary of axons and synapses; the molecular language of neurotransmitters and receptors; and so on. Reductionism assumes that the higher levels in this list of descriptive languages either have no causal efficacy or else can be explained entirely in terms of the lower level descriptions. This means that we can dispense with the higher order language and replace it with a more fundamental conceptual vocabulary that will yield complete explanations.

The "decade of the brain" witnessed a thorough biologization of psychiatry, justified in part by this reductionist view. In psychiatric theory, reductionism amounts not only to a basic confidence in the adequacy of neurobiological mechanisms to explain psychopathology, but in a preference for lower level explanations. In this view, molecular biology represents the most basic descriptive and explanatory level of psychopathology. This reductionist view ignores the extent to which neurophysiology, psychology, interpersonal interaction, group and family process, and other social processes represent emergent levels of organization with their own structure and dynamics (Morowitz, 2002).[2] As such, these levels require their own languages of description and provide their own modes of explanation (Prosser, 1970). In a sense, they are all "biological" in so far as we are dealing with a single (material) world with many hierarchically structured levels of complexity. However, this is a systems biology that makes use of concepts and constructs from many disciplines to describe

---

[2] A thorough-going reductionism seems to require that one aim for reduction to the most fundamental of the sciences—physics. Even if one restricts oneself to those sciences that are most basic to the phenomena of interest—here, mental phenomena—then presumably one should aim for a reduction to molecular biology. But that seems absurd. A satisfactory theory of the mind given exclusively in terms of neurons (say) would surely count as a reductionist success. Whereas reductionism aspires to fundamental explanation, science aims at good explanations, at whatever level they can be found (see Fodor, 1997)

and explain basic processes. The systems involved are not only molecular or neurophysiological but also social and cultural.

Indeed, there is a substantial literature on neural networks that demonstrates how even simple systems can exhibit complex properties (Scott, 2002). However, the implications of this for psychiatry are not always drawn out. Instead, we follow a "neo-humoral" approach of treating disorders as the result of too much or too little of some neurotransmitter. Psychopathology, on this view, reflects a chemical imbalance. This is not only a way to simplify the complexities of neurochemistry for popular consumption—in a form that fits with prevalent metaphors of balance and harmony as intrinsic to well-being. It is used equally in clinical texts on psychopharmacology and in research on animal models of psychopathology. This model ignores the fact that neurotransmitters do not map in any simple way onto specific functions, behaviors, or disorders. Neurotransmitters are associated with pathways that perform different functions in different circuits and generally do not code for a specific type of information processing or adaptive system. As a result, a drug treatment that affects one type of neurotransmitter or receptor will have an enormous number of concurrent effects. However, the neo-humoral approach to partitioning psychiatric disorders into categories based on their putative association with disturbances of specific neurotransmitters fits with the technology of psychopharmaceuticals and so it serves powerful economic interests.

The architecture of current psychiatric diagnostic systems was underwritten by observations of the differential effects of certain classes of medication on psychiatric disorders (Healy, 2002; Wilson, 1993). In particular, the distinction between schizophrenia and bipolar disorder became very important when evidence accumulated that lithium had some specificity for bipolar disorder. Antipsychotic or neuroleptic medications, by contrast, were clearly effective at suppressing psychotic symptoms across a very wide range of different disorders. The simplification imposed by a psychiatric nosology organized according to drug classes works in part because manufacturers exaggerate the specificity of medications (most of which, in fact, work for a wide range of symptoms) and patients and clinicians are encouraged to focus on one salient therapeutic effect and ignore all of the other effects—or to view them as more or less troublesome "side-effects."

Accounts that try to explain behavior in terms of neurotransmitters often jump levels, leaving out the interaction of networks and circuits that traverse the brain—a highly differentiated "organ" with many anatomically distributed subsystems. This seems to represent a confusion between a reductionist viewpoint—which can and should make use of a wide range of biological data—and the (unargued for) view that there is a privileged biological level at which deep explanations of mental life are to be found. Similarly, attempts to correlate activity in specific regions of the brain with behaviors leave out the intervening processes of coordinating perception and activity over time. These leaps across levels sometimes work because some problems can be traced to a global problem at the level of neurotransmitters or other cellular or biochemical processes. However, the ultimate expression of most developmental problems depends on individuals' unique learning history (their character, personality, and idiosyncratic psychology) and the environmental contexts in which they live (their social world). These other levels can sometimes be minimized or ignored because: (1) there are developmental trajectories that are influenced by isolable changes in single genes or other crucial steps in epigenetic processes that persist over time and across diverse environments; (2) there are final common pathways or "attractor basins" such that the developmental history of the brain's networks do not matter much for the final forms of pathology; or (3) some of the "degrees of freedom" (types of behavior) associated with a new level of organization are held constant so that the dynamics of the system can be described in fewer dimensions or simpler terms (that is, the levels of a single neurotransmitter). This last simplification also can occur because we take our psychological constructions and social worlds as static and unchangeable.

On the other hand, it is easy to construct models of even a few interacting neurons (cell assemblies, circuits) that exhibit very complex behavior and a whole range of perturbations that could have various pathological effects. In particular, it is possible to construct a system in which the parameters associated with neurotransmission are all within "normal" limits at each location initially, but the effect of the overall pattern is to create instability or mutually amplifying interaction that is abnormal. The essential point is that systems have different dynamical properties than their components—and systems of systems have still other dynamics. As a result, each level can have its own pathological dynamics that arise from patterns of connection and coordination that are not reducible to the activity of single units—or even families of units grouped together on the basis of their use of a common neurotransmitter or other molecular characteristics. Reduction to a different level may fail to capture the patterns of interest. Systemic pathologies cannot always be reduced to problems with components of the system. The trouble may lie in the connectivity, circuitry, or activity of the system as a network—and, in the case of psychopathology, the relevant networks may include loops through the social environment of family, community, and society.

The picture given to us by biology then is of a hierarchy of systems with emergent levels of structure and dynamics at each level. Emergence, in the sense used here, refers to the appearance of new structures and dynamics in a system that were not present in the elements of the system (Bedau & Humphreys, 2008; Meehl & Sellars, 1996; Morowitz, 2002). The notion of emergence recognizes that systems have properties that are not present in their components. This is true in a trivial sense for most things: a house made of bricks gives shelter in ways that an individual brick, or even a heap of bricks, does not. But to count as an interesting case of emergence, the new level of systemic organization must have radically new properties that cannot be predicted from the properties of the components or from simpler systems. In fact, there are many examples of phenomena that occur only in the context of the larger system; and even when the rules of interaction of the components are known it may not be possible to predict the system's properties except through modeling or simulation of the system as a whole. Even when prediction is possible, it may not be the case that the more "fundamental" level of description is the most perspicuous; the emergent level may provide more illuminating explanations. Even if molecular genetics were reducible to fundamental physics, for example, it does not follow that physical genetics would be a better theory. The structure of the phenomena may be most clearly revealed at the molecular level.

Nonlinear dynamical systems display a wide range of emergent phenomena that are not obvious from the rules that govern the interaction of their components (Mainzer, 2004; Nicolis & Prigogine, 1989).[3] These dynamic properties require new languages of description. The processes of one level organize themselves to create the structures of the next higher level which then allow new processes to occur (Pattee, 1973). An example would be the assembly of the receptor proteins at a synapse that allow neural transmission. Looking from the top down, macro-structures like a synapse can be decomposed into molecular processes. But describing the synaptic arrangement of the molecular processes requires additional sorts of information, to characterize how multiple components are arranged in space and time in relation to each other to give rise to new processes. These arrangements then give rise to specific dynamics with new properties not present in (and, arguably, not even inherent in) the elements of the lower level. The properties inhere in the *arrangement* of the molecular components, not in the components alone. It is the arrangements, organization, and spatio-temporal pattern that supply the missing ingredients needed for processes to emerge and move from one level to the next. This arrangement may be spontaneously self-assembling (as we assume it was in the origins of life or in the developmental processes of embryogenesis) or receive top-down influences from previously constructed higher order structures of greater or lesser complexity (Kauffman, 1993). Even when it appears spontaneous or autonomous, such emergence always involves specific environmental circumstances—at least in terms of the energy supplied to an open system but often in terms of the ordering effects available from interactions with other external structures. Thus, the higher level of order or organization may not be exclusively constituted by or dependent on the lower level, local system but depends also on cooperative interactions with an emerging "macro" level or environmental context that surrounds the local system.

Against the assumptions of methodological and ontological reductionism that would direct scientific (and clinical) attention to the fundamental building blocks of nature as holding the ultimate causal efficacy and explanatory power, the hierarchical systems view of nature introduces orders of magnitude of complexity and requires that we consider a local system in its interactions with an environmental context that is partly shaped and constituted by the emerging system itself (Rosen, 1991). Studying these processes of *autopoeisis* requires specific methodologies that examine systemic properties that cannot be found in (or even predicted from) the isolated components (Kauffman, 1993; Maturana & Varela, 1980).

There is debate about the sense in which these emergent levels are really ontologically distinct. Certainly, they are all physically instantiated with the same raw materials that make up the rest of the world, but their new properties (complex behaviors,

---

[3] A linear system can be reduced to a weighted sum of its components, which leads to the notion of linear causality (Scott, 2002): if a certain cause C_1 leads to an effect, E_1 and another cause C_2 leads to effect E_2, then the co-occurrence of both causes will lead to a state that is a sum of the two independent effects, that is $C_1 + C_2 \rightarrow E_{1,2} = E_1 + E_2$. In contrast, nonlinear systems do not have such independent effects of causes on outcomes and hence we can speak of nonlinear causality in which $C_1 + C_2 \rightarrow E_{1,2} \neq E_1 + E_2$. The system is literally more (or other) than the sum of its parts. Much has been learned about a variety of nonlinear dynamical systems but many systems remain mathematically intractable and can only be studied through computer models or other analogues.

reproduction, self-repair, adaptation to new environments) seem substantially different in kind from those of simpler systems, in that they demand different theoretical formulations and may, in turn, be more or less informative about the mental phenomena of interest. It is more contentious whether this systems view demands a new epistemology of science (Maturana & Varela, 1987; Wolfram, 2002). However, the higher levels of organization of the nervous system do pose special problems for our notions of the nature of knowledge—of what can be known—and how we come to know it.

Ontologies of Mind

Social factors are implicated in the development of mental phenomena. Where does this leave the question of reduction? Methodological reductionism, as a set of prescriptions about how to do science, is largely a pragmatic question. How best to decompose a system for study is a question that continues to be addressed in the conduct of scientific research itself. No one doubts that along with taking a system apart, one must also be able to reconstruct its functioning within a successful theory. As a result, methodological reductionism can be taken as a family of techniques that are demonstrably effective for studying particular processes but that must be guided by theories of the higher order phenomena that the reductionist method aims to address.

Ontological reduction is, perhaps surprisingly, an area of continuing controversy. Leaving dualist options entirely to one side, the fact of the significant interactions between psychological processes and the environment raises the possibility that mental life requires more than the brain; it can include tools or aspects of the outside world crucial to mental life. There is a long tradition arguing that mental life extends into the environment through processes of embodiment and enactment. The anthropologist Gregory Bateson (1972) argued that tools were extensions of mind, which emerged from a social ecology. Maturana and Varela (1980) argued that cognitive processes could only be understood in terms of the organism's interactions with the environment. These theories do not claim that mind is a different substance than body, but that there are emergent processes that are new and different in substantial ways from the prior or lower level of organization from which they arise. In some sense, therefore, they represent new phenomena with distinct ontologies.

A controversial version of this "extended mind" hypothesis was articulated by Andy Clark and David Chalmers (1998; see also Clark, 2008) in a paper in which they offer the following simple thought experiment: seeing an advertisement for an exhibition at the Museum of Modern Art (MoMA) in New York City, Inga remembers that the MoMA is on 53rd Street and starts walking in that direction. Otto also sees the advertisement and decides to visit the exhibition. Unfortunately, Otto suffers from Alzheimer's disease and is losing his memory. In an effort to cope with the disability, he has begun to carry around a notebook in which he keeps various bits of information. Consulting his notebook, Otto finds that he has written in it the address of the MoMA. With the address now available, he too heads in the direction of the museum. Clark and Chalmers argue that there is no principled reason to think that Otto's notebook is not part of his mind, despite the fact that it is, of course, not part of his brain. Since it performs precisely the same function as Inga's memory, it is no more than prejudice in favor of the biological that leads us to exclude it from the domain of the mental.

It seems clear that the mind does not involve a new physical substance but there are nevertheless new sorts of processes, entities, and events that come into being as a result of social arrangements and interactions that may both augment and constrain brain activity (Hacking, 1999; Searle, 1995). Recognizing the importance of the social world could lead one to reject ontological reductionism even though one does not believe in non-physical entities like souls. The very words you are now reading emerged from a collaboration between two authors that has resulted—we would like to think, at any rate—in an intellectual product that is more than the sum of its parts. If one were inclined to see the mind as extended beyond the skin, then social interactions of an intellectual kind would regularly engender cognitive activity that would be *ontologically* different from the mental activity of a single person.

The social arrangements of interpersonal interaction can give rise to new sorts of cognitive and brain activity. Some of these interactions are governed by rules and institutions, others by the physical configuration of space and place. To the extent that we accept that the social world creates new sorts of things with their own structures and processes, we can speak of new ontologies, with a social and cultural history and a contemporary politics (Hacking, 2002).

Whether or not we grant the social world a distinct ontological status, it clearly can be decisive for individuals' health and illness (Wilkinson & Pickett, 2009). Psychiatric theory and practice therefore must include knowledge of social context. The crucial question then is what a social view of mental illness does to epistemological reduction. Commitment to this form of reduction is the most theoretically and practically significant because it is here that the question of the right approach to a science of mental illness must be decided. Leaving aside the question of the extended mind, there is broad agreement that the mind is ontologically nothing over and above the interactions of brain, body, and environment. But that fact does not constrain what theories of the mind or mental illness will turn out to be correct, any more than the fact that the universe is made up entirely of quarks implies that every scientific theory must be a theory of quarks. Larger-scale phenomena have their own dynamics and hence require their own languages of description of macro-level processes. "The world," as Jerry Fodor (1997, p. 162) puts it, "runs in parallel, at many levels of description."

Even with respect to the brain itself, there is controversy over the levels of description needed. While there may be wide agreement among neuroscientists that the emergent levels of organization seen in the nervous system do not require a different physical ontology, it is less clear whether they require a different epistemology. Cognitive systems are intentional—they refer to events in the world and can only be rightly understood as parts of loops that involve perception and action in the world. This leads to an epistemological problem when efforts are made to understand the cognitive system by isolating it from the environment. This dilemma is still more contentious when one considers the phenomena of consciousness and self-awareness. Whether or not subjectivity requires different ontology (following Chalmers (1996), Jackson (1982), and Nagel (1974)), it certainly requires a different epistemology. Moreover, this epistemology must not only respect the privileged (though also biased and distorted) perspective of the subject and the role of their agency in constructing both their own experience and the larger social world, but also the emergence of many aspects of mind and self through that self-constituting interaction with the social world.

Subjectivity and the Social Construction of the Self

A special type of emergent phenomenon characterizes the human brain: that is the ability to construct representations or descriptions and to operate at this logical level. Abstractly, this is what makes the brain utterly different than the liver or the lungs. The brain does not secrete or exchange information with the world the way other organs operate: it manipulates patterns. This has been captured in the notion of the brain as simultaneously a dynamical system and a cognitive or linguistic system (Pattee, 1977). Another analogy that leads to a similar distinction was introduced by von Neumann in his comparison of the brain and the digital computer: both require hardware and software to process information (von Neumann, 1958). In principle, these are distinct and dissociable.[4] For digital computers, the hardware may vary in speed and other characteristics but as long as it can carry out a basic set of computations it suffices to run any conceivable program. In reality, of course, knowing the characteristics and constraints of the hardware allows programmers to devise more efficient programs that run especially well on specific hardware. In the case of the brain, the software is instantiated as changes in the hardware; that is, the abstract manipulation of symbols and its physical realization in terms of neural networks are thoroughly intertwined. The structure of the brain exerts constraints on what is easy or difficult to compute—resulting not only in the limits of specific cognitive abilities but in the bounded nature of everyday rationality and our propensity for certain types of systematic biases, errors and akrasias.

The programs that are inscribed in the brain reflect our developmental histories and the demands of the contexts or environments in which we dwell. A unique set of these programs concern our abilities to monitor, represent, control, and reflect on our own behavior and experience. These control processes include efforts to match or reconcile our behavior to various standards we have, some of which are attached to a sense of our social personhood or to our subjective sense of selfhood. A lot goes on both in and around the construction and reconstruction of the sense of self as one or more images, plans, or narrative centers that include a sense of personal history (grounded in memory), agency, and subjectivity.

---

[4]   The links between hardware and software may include the ability of software to modify hardware—this lay behind von Neumann's notion of self-reproducing automata (von Neumann, 1951). Because any physical instantiation of a program requires energy to make order out of disorder in the course of its computations, running a program inevitably has physical effects on the substrate that conducts its computations. Thus, a program that runs in a rapid loop could overheat the processor and set the machine on fire or cause a meltdown. Even computers, therefore, have bidirectional causal pathways between hardware and software. Nevertheless, the functioning of the software (the linguistic level) can be described in hardware-free terms and has its own logic and "pathologies."

The sense of subjectivity and selfhood we experience from the inside interacts with a social construction of personhood seen from the outside. As persons then, we have emergent levels of organization of behavior associated with subjectivity and self-awareness and with our social roles and the corresponding responses of others. The self cannot be fully reduced to any lower level of structure or representation.[5] The self is not an arrangement of synapses and the cultural world is not an aggregate of individuals' cognitive or neural representations. The brain cannot stand in for the person and the person cannot stand in for society or culture. So, to achieve and maintain a person-centered viewpoint, we need to understand the ways in which people use and are used by their brain and their culture.

The recognition that, as subjective agents, we are not simply manifestations of brain activity but that we *use* our brains, reflects the supervenience of the self as an organizing system that can reflect on and work with the idiosyncrasies of the brain and the body it inhabits. Our brains are plastic and pluripotent and we can feed and nurture them or abuse them with chemical substances we ingest or experiences we seek out—indeed, we can choose to expose them to new environments where they are shaped, sculpted, and transformed (Malabou, 2008).

On the other hand, conscious self-direction is not the only determinant of behavior. Non-conscious cognitive processes and non-cognitive regulatory processes—like the activity of the cardiovascular system or the gut—constantly influence our behavior (and our experience of agency). Some of these non-conscious processes may organize behavior in a planful or purposive way. In a sense then, to the extent we identify the self with the conscious "I," we might think of the brain as using us for its own purposes, compelling us to do things we would rather not (Wilson, 2004). The awkward locution of "being used by one's brain" is not meant to misplace agency, but to counteract the tendency to exaggerate the autonomy and agency of the self that comes from a person-centered view of the world. It also opens the door to recognizing that our brains can betray us or can be hijacked by others—the domains of psychodynamic theory and the social psychology of persuasion, respectively, each with its own hermeneutics of suspicion.

Similarly, contemporary social sciences tend to exaggerate the agency of the individual against the constitutive and countervailing forces of the social world. Ascribing agency and purpose to society is not meant to personify impersonal networks (though for groups and communities this does make sense), but to acknowledge that we live

---

[5]   Though, to the extent that the self reflects distinctive patterns of responding to context, it may be partially inscribed in lower level dispositions to respond, that persist even when self-awareness is damaged or constricted. Consider, for example, the person with Alzheimer's who, while showing an alteration of personality or "loss of self" (Cohen & Eisdorfer, 2001), nevertheless, reveals flashes of their old self in certain turns of phrase, emotional responses, or other patterns of behavior. The self, like other complex representational processes, may be holographically distributed in the brain so that destruction of some areas does not simply eliminate its processes but degrades their specificity or detail; much as cutting up a hologram results not in a fragment of the original image but in a blurry version of the whole image. The notion of distributed networks in the brain has a long lineage that antecedes the invention of holography (Pribram, 1990). Of course, to the extent that the self resides in (or is sustained by) interactional processes, its preservation or loss—in Alzheimer's or other neurological disorders—depends on interpersonal processes (how others perceive and respond to the afflicted person) as well as on the neural machinery of memory and self-reflection (Herskovits, 1995).

in and among dense networks of interpersonal and institutional processes that shape our developmental trajectories. These processes are not expressions of a passive social matrix in which we can freely locate ourselves, but are themselves determined by political and economic interests. One way in which these larger political-economic formations influence us is by structuring the social worlds that afford us identities, power, and purpose. They underpin the collective notions of personhood that define our goals and aspirations. They influence the narratives that regulate our sense of autobiographical memory and identity and the forms of embodiment through which we acquire our sense of self. And, with technologies both old and new, they may reach past the self to directly manipulate the neural substrates that subserve the programs of the self.

In the face of this complex hierarchy of levels of organization and the emergence and supervenience of subjectivity and agency, the epistemology of biomedicine requires some rethinking. Biomedical practitioners generally assume that we can treat verbal reports as more or less accurate indices of bodily experience (Kirmayer, 2008). When a patient says "I am in pain," the assumption is that there is a specific physiological process (or one of a family of processes) going on in the body and the brain that yields a specific experience, which the person can then reliably report. Of course, patients may be "unreliable historians" and either exaggerate, minimize, or deny their experience. But this only reinforces the sense that the normal condition allows a direct link between bodily events, symptom experience, and clinical presentation. With such naive semiotics, biomedicine ignores the way in which experience is shaped by an array of psychophysiological and psychological processes that depend on past learning, cognitive schemas, memory, and attention. In addition to this cognitive and attentional mediation, both experience and its verbal report depend on context and may involve more or less conscious attempts at rhetorical self-fashioning and positioning. A symptom report, autobiographical story or response to a question, must then be understood not just in terms of the individual's history but also in terms of their relationship to the interlocutor, to unseen participants in their social world who wait beyond the doors of the consulting room, and indeed, in terms of the circulation of ideas and ways of construing oneself in local communities and global systems (Kirmayer, 2000).

The complexity, ambiguity, and indeterminacy of verbal reports is not simply a matter of "noise" in a communication channel confounding what would otherwise be a clear communication. There are aspects of experience that can only be known in and through language because they are made up of language in the first place and reside in cognitive structures and corresponding ways of thinking, or else are located in a conversation as a discursive formation or way of speaking. On this view, knowledge and experience are socially constituted and not reducible to an internal representation in the mind or brain of an individual (Bloor, 1983). Nor is it merely a matter of an "epidemiology of representations," each carried by an individual and distributed according to social position (Sperber, 1996). Rather, the discursive formations that constitute complex experiences of selfhood reside in culturally constituted forms of life.

This points to an important limitation of current work in social neuroscience which, despite its recognition of the importance of the social world in the evolution and development of the brain, tends to focus on lower-level biological phenomena (Insel & Fernald, 2004). For example, studies that show how important the neuro-hormones oxytocin and vasopressin are for our feelings of love and attachment, have important consequences, including alerting us to the possibility that psychiatric treatments like SSRIs might undermine romantic love and stable attachments in couples (Fisher, Aron, & Brown, 2006). But this model captures only a small part of the tapestry of thoughts and feelings, interactions, and interpersonal responses that go into the experience of different forms of love. Recognizing the power of a hormonal system may give us an understanding of some of our vulnerabilities and some leverage in responding to the human predicament—but it does not eliminate the choice of stance and strategy to pursue our lives. That requires a different level of analysis and a different language of description.

We can see this in studies on the psychobiological effects of an affectionate hug, in which holding another person close for a time stimulates the release of oxytocin, which in turn causes feelings of comfort, calm, trust, and, eventually, attachment to the other (Carter, 1998; Insel & Fernald, 2004). The more frequent the hugs, the greater the oxytocin release and the stronger the induced feelings of calm and trust, with health benefits in terms of reducing heart rate and blood pressure (Light, Grewen, & Amico, 2005). But the effects of oxytocin interact with contextual factors that shape the meaning of the embrace. Women in a warm, supportive relationship experience stronger oxytocin effects in response to physical contact with their partner (Grewen, Girdler, Amico, & Light, 2005). Of course, even before any contact, we have the opportunity to anticipate and interpret the meaning of an embrace, which may be desired or unwanted, socially appropriate, or transgressive. And during the embrace, thoughts and competing emotional responses can give the experience layers of reinforcing or contradictory meaning that may override any hardwired or previously learned propensity to respond.

On a larger temporal and social scale, love involves a refiguring of our personal identities, biographies, and life trajectories. We locate ourselves in relation to the loved one, and space itself is reoriented to define the familiar places of hearth and home and the unfamiliar spaces of the public realm, which are progressively more unfamiliar. So love involves cognitive maps as well, even if there are some contour lines drawn by gradients of comfort and response that are based on experiences linked to hormonal mechanisms of attachment. The affective systems revealed by social neuroscience interact with other biobehavioral systems, as well as cognitive and interpersonal processes to create a map of our local social worlds with hills and vales corresponding to places of safety and danger, comfort and distress. But this is only a sketch of a social world, with its own exigencies, that exceed in complexity any of our cognitive constructions. Love and marriage have their own interpersonal dynamics that are not reducible to psychological or biological processes (Gottman, Murray, Swanson, Tyson, & Swanson, 2002). In addition the local system of a marriage is embedded in larger social institutions that regulate its meaning and durability.

Social neuroscience certainly gives us insight into the dynamics of attachment in prairie voles and other animals and the same systems can be shown to be operating in humans. At the same time, it is unclear how far this takes us in an understanding of human love and attachment. As Insel and Fernald (2004) note, "Less clear is the relevance of these observations to the primate brain, where visual processing trumps vomeronasal signals and cortical networks may override the neuropeptide signals from the hypothalamus." (p. 715). It is not simply that visual processing or wider cortical networks have more influence in the primate brain but that, in humans, vision and cortical associations bring information about others in a complexly configured social world. In what sense, then can we view love as "an emergent property of the nervous system" (Porges, 1998)? The social meanings of love are only possible because of the autonomic and neurohormonal systems that enable certain types of strong emotional response, memory, and attachment to others. At the same time, the neural systems that contribute to feelings of comfort and attachment only become the processes we call "love" given the socially guided use of our cognitive capacities for desire, imaginative fantasy, and commitment (Griffiths & Scarantino, 2009; Gross, 2006; Reddy, 2001). Deprived of its biological substrate, love would be a weak or non-existent force in the world; deprived of its social history, embodiment, and enactment, it would be literally unimaginable.

Social Origins of Psychiatric Disorder

The failure of a reductive epistemology of the mind can best be seen when we reflect on the role of self and personhood in psychopathology. The social world allows us to recognize certain aspects of our self-fashioning and compels us to treat other aspects as natural or given. Cross-cultural comparison is important then not only to respect human diversity, but to look behind the curtain of our commonsense constructions of the person—which may not only serve vested interests but obscure the very processes that constitute mind itself. It is always easiest to see this by looking at other peoples' cultures. The field of cultural psychiatry uses such cross-cultural comparison to identify the role of social processes in the origins, course and outcomes of mental health and illness. One of the most striking recent findings in this area is evidence for social influences on the incidence of schizophrenia.

As some of the most severe forms of psychopathology, psychotic disorders tend to be viewed as the exemplars of biological psychiatric disorders. Indeed, after a period of interest in the importance of social factors in the causes, course, and outcome of schizophrenia in the 1950s and 1960s, there has been a decline of research on, and interest in, social factors in schizophrenia in North America (Jarvis, 2007). This de-emphasis of social determinants has gone hand-in-hand with a search for genetic causes—a goal which, to date, has proved elusive. At the same time, however, there is substantial evidence for profound social influences on the causes and course of schizophrenia.

Perhaps the most important source of relevant evidence for social effects on the etiology of psychosis comes from investigation of the effects of migration on the incidence of schizophrenia (Cantor-Graae, 2007; Coid et al., 2008). Over the last 30 years, a number of studies of African and Caribbean migrants to Britain have found higher rates of schizophrenia in these populations, ranging from rates that are twice to 14 times higher than the white population (Fearon & Morgan, 2006). A meta-analysis conducted by Cantor-Graae and Selten (2005; see also Bourque, van der Ven, & Malla, 2011) produced a mean weighted relative risk for developing schizophrenia of 2.7 (95% CI 2.3–3.2). Remarkably, the relative risk in the children of these immigrants (either born in the country of migration or brought up there from a young age) was higher still (mean weighted relative risk of 4.5, 95% CI 1.5–13.1). This demonstrates that the effect on mental health cannot be attributed exclusively to the stress of the process of migration itself. Whatever factors are operative seem to affect the second generation still more strongly.

While striking, there are many methodological challenges involved in conducting these studies, so the findings must be interpreted with caution (McKenzie, Fearon, & Hutchison, 2008). These studies do not usually distinguish between types of immigrant (for example, economic migrants versus refugees) whose psychological profile and reaction to the process of migration might be expected to vary considerably. Nor do these studies distinguish well among different ethnic groups. While most studies make use of first-admission or first-contact cases, it is known that members of different social groups typically come to the attention of the mental health system in different ways. It is thus not possible to be sure that the numbers of cases in different populations are being measured equally accurately. Moreover, if members of some groups are more likely to seek care than others, then the numbers of clinical cases may not be representative of the numbers in the general population. There are also concerns about comparing the incidence of schizophrenia in migrant groups with the incidence in the country of origin given that diagnostic methods are not uniform cross-culturally. Finally, there are questions about the accuracy of diagnoses across cultures and the possibility of ethno-racial bias in assessment.

Despite these difficulties, the size of the increase in the incidence of schizophrenia and the consistency of findings strongly suggests that the phenomenon is real and no mere artifact (McKenzie et al., 2008). In addition, the AESOP study carried out by Fearon and colleagues (2006) controlled for some of the relevant variables, and their findings confirmed those of the earlier studies. Incidence rate ratios (IRRs) were calculated for each ethnic group in comparison to the White population and were found to be very high for schizophrenia and manic psychosis in African-Caribbeans (9.1 and 8.0, respectively) as well as in Black Africans (5.8 and 6.2, respectively) in both men and women. Thus, whatever the stresses of migration, they act somewhat selectively, affecting some mental processes more than others, increasing vulnerability to—or undermining protective factors against—schizophrenia and mania in particular.

There is no consensus about what actually does the psychological damage either to immigrants themselves or to their children, but there is no evidence that the differential incidence of schizophrenia is genetic in nature; the incidence of schizophrenia in the countries from which most Caribbean migrants come is no higher than in the White population of the UK (Hickling & Rodgers-Johnson, 1995; Mahy, Mallett, Leff, & Bhugra, 1999). Whatever is increasing the vulnerability, or decreasing the efficacy of protective factors, seems to be social in nature. At the very least, genetic vulnerabilities are being manifest by changes in social conditions. Leading candidates include poverty and, more generally, socioeconomic disadvantage, racism, and living in an urban environment (McKenzie et al., 2008).

The effect of the urban environment has been studied in some detail and may constitute one of the strongest risk factors for the development of psychosis (Krabbendam & van Os, 2005). Studies over many decades have repeatedly shown that the rates of schizophrenia are influenced by exposure to urban environments and that there is also a dose effect: the larger the city, the higher the incidence of psychosis. Indeed there is evidence dating from the nineteenth century showing the same effect (Torrey, Bowler, & Clark, 1997). Furthermore, the effect is greater according to the number of years one spends in an urban region between birth and 20 years of age (Pedersen & Mortensen, 2001). In addition, the effect of urban life increases psychosis-like symptoms in non-clinical populations (van Os, Hanssen, Bijl, & Vollebergh, 2001). Most importantly, the urban effect seems to be specific to psychosis. Bipolar disorder, for example, is no more common in cities than in rural areas (Mortensen et al., 1999).

In order to assert the causal role of the urban environment, however, one must be able to exclude at least two alternative hypotheses: (1) that psychotic individuals, or those in the prodromal phase of psychosis, are more likely than non-psychotic or pro-dromal individuals to move to the city (the "social drift" hypothesis); and (2) that those who are mentally ill are less likely than those who are not psychotic to leave the city for more attractive (rural) communities—the "social residue hypothesis." Dauncey and colleagues (Dauncey, Giggs, Baker, & Harrison, 1993) investigated the place of residence of psychotic patients during the five-year period before admission and found no evidence for the social drift hypothesis. Mortensen and colleagues (Mortensen et al., 1999) argue that for this drift to have occurred in the previous generation would require an extremely high degree of movement from rural to urban areas.

A number of other potentially confounding factors have also been examined, including obstetric complications, adverse life events, and season of birth, and do not account for the effect of urban environment (Boydell & McKenzie, 2008). While socioeconomic disadvantage might be expected to account for at least some of the urbanicity effect, many of the relevant studies have been carried out in countries in which the standard of living is higher in urban than in rural regions. Drug use, in contrast, may constitute part of the explanation for the urbanicity effect in psychosis, though the effect remains even when adjusted for the use of cannabis.

It is worth noting that there seems to be a complex interaction between the effect of urban life and genetic predisposition to psychosis. Those with a genetic vulnerability to schizophrenia seem to be disproportionately affected by urban life, so that the urban environment constitutes a greater stress on vulnerable individuals than on those who are not (van Os, Pedersen, & Mortensen, 2004). A parallel synergy occurs at the social level. Van Os and colleagues (van Os, Driessen, Gunther, & Delespaul, 2000) found that people without partners are disproportionately at risk for psychosis if they are city-dwellers. We will return to this issue below.

Although it is at present unclear just what causes the urbanicity effect, it seems to be a function of human relations, an idea supported by the fact that within cities the effect is distributed differentially across neighborhoods (Kirkbride et al., 2006). The incidence of schizophrenia is higher in economically deprived areas with a high proportion of single-person households and high levels of population mobility (Boydell & McKenzie, 2008). This suggests that the effect is determined by the structure of particular communities and is thus fundamentally social. In the case of immigrants, there is evidence that the ethnic density of the neighborhood affects risk for psychosis (Veling et al., 2008). Those living in areas where there is a smaller proportion of their own ethnic group are at greater risk.

Taken together, these studies suggest that social factors are crucial determinants of the risk of schizophrenia. The nature of these factors and their differential distribution and impact on individuals from different backgrounds result from processes that can only be adequately described at the level of the social world, in terms of the impact of the histories of colonialism, migration, racism, and discrimination on social and economic inequalities. Although neuroscience can help us understand the proximate mediators of these social effects, it can never predict their spatial or geographic distribution and may misdirect attention away from crucial, modifiable social structural factors that demand remediation.

Socializing Biological Psychiatry

The evidence for social determinants of health—and of mental health in particular—is compelling. All of this might be granted, yet the biological psychiatrist could claim it lies outside the purview of psychiatry, which studies only the proximate neural mediation of the effects of the social environment. However, the whole thrust of our argument is that there should not be an either/or in considering brain–society interactions. Instead, psychiatry needs theories of social and cultural biology that recognize the fundamental role of social processes not only as determinants of health and illness but as the mediators and mechanisms of psychopathology as well as of healing and recovery.

We raised, in passing, the possibility that genetic factors could contribute to the increase in the incidence of schizophrenia seen in migrants, that arises as a result of uncovering of genetic vulnerability when protective factors—for example, the organization of family or social life in the home country—are no longer present in the destination country. We also noted the possibility of synergies between genetic susceptibility and the urban environment. This raises the possibility that social factors interact in some way with genetic mechanisms.

There are at least three ways in which this could be happening. The first is that genes could predispose to behavior in ways that feed back on mental life. Kendler and Prescott (2006, pp. 264–265) provide an apt, if hypothetical, example of the basic idea:

> A cancer geneticist has collected a sample of 400 patients with lung cancer and 400 control participants. She scans a chromosome looking for gene variants that differentiate the two groups and finds a gene that is much more common among the lung cancer patients. With great excitement, she writes up her results and submits them to a major scientific journal, claiming to have found a new oncogene (i.e., a gene that can cause cancer). However, unbeknownst to her, the gene has no effect on the risk for cancer at a physiological level. Instead, it exerts an indirect effect, through behavior, on the risk for chronic cigarette smoking. For example, genetically controlled variation in nicotine receptors, which stimulate the pleasure centers in the brain, might affect the chances that individuals will seek repeated exposure to carcinogenic compounds. Has this researcher really found a new oncogene? Yes and no. Traditional oncogenes act via inside-the-skin pathways (e.g., by influencing cell division), whereas this oncogene acts via an outside-the-skin pathway. This oncogene will have a few unusual properties not possessed by

traditional oncogenes. In a culture in which tobacco is not smoked, it will have no effect on cancer risk. Any social process that reduces the frequency of heavy tobacco smoking (such as reduced social acceptability or increased taxation) will reduce the impact of the oncogene on risk for lung cancer.

"Outside-the-skin" gene expression could of course also occur in psychiatric disorder. Consider another researcher who finds a gene that correlates with schizophrenia. She infers that the gene is likely to code for a protein that is implicated in dopamine function, which in turn is associated with the cardinal symptoms of schizophrenia. It turns out, however, that the gene is actually associated with temperament; people who have it tend to be unassertive and therefore are more likely to be bullied as children—and bullying may play a causal role in the later development of psychosis (Bebbington et al., 2004). Has this researcher found a gene for schizophrenia?[6] Not really. Like the putative oncogene, the effect of this gene has to be understood in the context of the environment in which it is expressed. The social environment may thus be part of a loop that affects mental life, and ignoring the potential role of the environment may lead to a misunderstanding of biological function of the gene.

A second possibility is that mentioned in relation to the effects of the urban environment on those disposed to schizophrenia. If a genetic disposition renders one individual more vulnerable to a social stressor than others, then this is evidence that there is a synergy between biological and social features that must be understood together. For example, individuals with a particular form of the serotonin transporter (5-HTT) gene are more susceptible to stress and, therefore, to depression and suicide than those without it (Caspi et al., 2003). This same sort of genetic polymorphism might confer adaptive advantages in other environmental and social contexts (Suomi, 2006).

A third way in which the social world may be interacting with our biology is via epigenetic processes—that is, processes in which the expression of genes, rather than the genes themselves, is altered. Research on epigenetics has begun to reveal how interactions of the genome with the environment over development lead to structural changes in the methylation patterns of DNA that regulate cellular function. These changes may be lasting so that experience remodels the functional genome. For example, there is compelling evidence in rodents and primates that early parenting experiences alter the regulation of stress response systems for the life of the organism via the hypothalamic-pituitary-adrenal stress response (Meaney, 2001; Meaney & Szyf, 2005; Zhang & Meaney, 2010). This process occurs in humans as well. In a recent paper, McGowan and colleagues (McGowan et al., 2009) reported a post-mortem study of hippocampal tissue that showed differences in glucocorticoid receptors' gene expression in suicide victims with a known history of abuse compared to suicide victims without such a history. Gene expression was reduced in individuals who suffered from abuse, but no difference was found between suicide victims without

---

[6] Kendler (2005) discusses the assumptions in the phrase "X is a gene for Y," pointing out that since psychiatric disorders have multiple causes and the causal pathway from any genetic variation to any specific type of behavioral disturbance is usually long, complicated, and context dependent, it will rarely if ever be appropriate to say that "X is a gene for psychiatric disorder Y."

a history of abuse and controls. It is reasonable to infer, therefore, that the changes in gene expression are correlated with the abuse itself and not with some aspect of the suicide behavior or its prodrome. This seems to be compelling evidence that the social world—in this case, home life—has a direct influence on gene expression and therefore, perhaps, on behavior in humans. This important finding shows that the nervous system is reshaped by experience not only at the synaptic level but in its underlying genetic regulation as well.

Recent work suggests that schizophrenia might be associated with specific epigenetic modulation of multiple systems (Mill et al., 2008). This points to a more refined way of thinking about the interactions between the brain and the social environment (Mill & Petronis, 2009; Petronis, 2004). The types of social adversity faced by immigrants, described above, may exert influences over the course of development through epigenetic processes that render individuals more vulnerable to schizophrenia. The epigenetic effect of social stressors will interact with ongoing social processes that constrain individuals' adaptation and expose them to prolonged and persistent stresses such as those associated with poverty, inequality, marginalization, and discrimination (Wilkinson & Pickett, 2009).

We thus need models and corresponding languages of description that allow us to recognize, study, and intervene in patterns and processes of adversity and resilience that are located outside the brain—even if, through learning and development, the social world comes to have shadows, refractions, or reflections in the functional genome and the circuitry of brain. The social world has its own organization—it is not comprised of isolated risk or protective factors but of coordinated systems with persistent effects over time that reflect dynamics that are irreducibly social.

Conclusion: Beyond Reductionism

We have tried to show that (1) as a methodological strategy, biological reductionism is useful but not sufficient to understand the origins of human behavior and experience in health and sickness; (2) as an ontological position, biological reductionism is undermined by the higher level of organization at which mental life must be understood, which includes interactions between the brain and the social world; and (3) partly in consequence of these first two conclusions, epistemological reductionism will never be adequate as a comprehensive understanding of human behavior and experience. In fact, promoting such reductionism in psychiatry does real violence to our conceptual models and the production of knowledge and, ultimately, to clinical practice that aims to be person-centered and integrative.

Given that the non-reductionist view we have described has a long lineage and is grounded in solid observation and argument about the nature of hierarchical systems—and more specifically about the nervous system—the persistent enthusiasm for reductionist epistemologies requires some explanation. This is a task for critical neuroscience. We think the answers for this bias will be found not only in the methodological advantages of reductionism for scientists seeking to design experiments, or their desire to argue for the utility of simple models to address important mental health problems. We believe that they will also be discovered in the ways in which biological explanations draw attention away from highly contested social and political issues—issues that would demand much political consensus and will to address—and focus instead on a level of explanation distant from everyday experience, that can be framed as a politically neutral arena for scientific explanation and technical mastery. This neutralization of the politically loaded issue of the social origins of mental health disparities goes hand-in-hand with the economic exploitation of biological theory by pharmaceutical companies.

When Insel and Quirion express the view that psychiatry is "clinically applied neuroscience," they are expressing a form of epistemological reductionism—a form of reductionism according to which mental illness will ultimately be understood and treated by a successful theory of the brain. If, however, as we have argued, one cannot understand mental illness without reference to social causes of mental illness, then no theory that is exclusively about the brain can be complete. At best, a neuroscientific theory can articulate the end result of the complex interactions of the organism with its environment. Even if it turns out that a disorder of dopamine, for example, is a necessary and sufficient condition for the symptoms of schizophrenia, it would be a profound error to ignore the social world that contributes to the causes, course and outcome of that disorder as scientifically insignificant. A successful theory of the brain will undoubtedly explain a great deal about mental life and mental illness, but on its own it will provide no more than a keyhole view of the mind. It seems likely, therefore, that unless economic forces conspire to shrink it to a narrow technical domain in the future psychiatry will become not just behavioral neurology or applied neuroscience but also clinically applied social science.

References

Bateson, G. (1972). *Steps to an ecology of mind*. New York: Ballantine Books.

Bebbington, P. E., Bhugra, D., Brugha, T., Singleton, N., Farrell, M., Jenkins, R., Lewis, G., & Meltzer, H. (2004). Psychosis, victimization and childhood disadvantage: Evidence from the second British National Survey of psychiatric morbidity. *British Journal of Psychiatry, 185*, 220–226.

Bedau, M. A., & Humphreys, P. (Eds.). (2008). *Emergence: Contemporary readings in philosophy and science*. Cambridge, MA: MIT Press.

Bloor, D. (1983). *Wittgenstein: A social theory of knowledge*. London: Macmillan.

Bourque, F., E. van der Ven., & Malla, A. (2011). A meta-analysis of the risk for psychotic disorders among first- and second-generation immigrants. *Psychological Medicine, 41*, 897–910.

Boydell, J., & Mckenzie, K. (2008). Society, place and space. In Morgan, C., Mckenzie, K., & Fearon, P. (Eds.), *Society and psychosis*. New York: Cambridge University Press.

Cantor-Graae, E. (2007). The contribution of social factors to the development of schizophrenia: A review of recent findings. *Canadian Journal of Psychiatry, 52*, 277–286.

Cantor-Graae, E., & Selten, J.-P. (2005). Schizophrenia and migration: A meta-analysis and review. *American Journal of Psychiatry, 162*, 12–24.

Carter, C. S. 1998. Neuroendocrine perspectives on social attachment and love. *Psychoneuroendocrinology, 23*, 779–818.

Caspi, A., Sugden, K., Moffitt, T. E., Taylor, A., Craig, I. W., Harrington, H., Mcclay, J., Mill, J., Martin, J., Braithwaite, A., & Poulton, R. (2003). Influence of life stress on depression: moderation by a polymorphism in the 5-HTT gene. *Science, 301*, 386–389.

Chalmers, D. J. (1996). *The conscious mind*. New York: Oxford University Press.

Clark, A. (2008). *Supersizing the mind: Embodiment, action, and cognitive extension.* New York: Oxford University Press.

Clark, A., & Chalmers, D. (1998). The extended mind. *Analysis, 58,* 7–19.

Cohen, D., & Eisdorfer, C. (2001). *The loss of self: A family resource for the care of Alzheimer's disease and related disorders.* New York: Norton.

Coid, J. W., Kirkbride, J. B., Barker, D., Cowden, F., Stamps, R., Yang, M., & Jones, P. B. (2008). Raised incidence rates of all psychoses among migrant groups: Findings from the East London first episode psychosis study. *Archives of General Psychiatry, 65,* 1250–1258.

Dauncey, K., Giggs, J., Baker, K., & Harrison, G. (1993). Schizophrenia in Nottingham: Lifelong residential mobility of a cohort. *British Journal of Psychiatry, 163,* 613–619.

Eisenberg, L. (1986). Mindlessness and brainlessness in psychiatry. *British Journal of Psychiatry, 148,* 497–508.

Fearon, P., Kirkbride, J. B., Morgan, C., Dazzan, P., Morgan, K., Lloyd, T., & Murray, R. M. (2006). Incidence of schizophrenia and other psychoses in ethnic minority groups: Results from the MRC AESOP Study. *Psychological Medicine, 36,* 1541–1550.

Fearon, P., & Morgan, C. (2006). Environmental factors in schizophrenia: The role of migrant studies. *Schizophrenia Bulletin, 32,* 405–408.

Fisher, H. E., Aron, A., & Brown, L. L. (2006). Romantic love: A mammalian brain system for mate choice. *Philosophical Transactions of the Royal Society of London B Biological Sciences, 361,* 2173–2186.

Fodor, J. (1997). Special sciences: Still autonomous after all these years. *Noûs, 31,* (Suppl 11), 149–163.

Gottman, J. M., Murray, J. D., Swanson, C. C., Tyson, R., & Swanson, K. R. (2002). *The mathematics of marriage: Dynamic nonlinear models.* Cambridge, MA: MIT Press.

Grewen, K. M., Girdler, S. S., Amico, J., & Light, K. C. (2005). Effects of partner support on resting oxytocin, cortisol, norepinephrine, and blood pressure before and after warm partner contact. *Psychosomatic Medicine, 67,* 531–538.

Griffiths, P. E., & Scarantino, A. (2009). Emotions in the wild: The situated perspective on emotion. In P. Robbins & M. Aydede (Eds.), *Cambridge handbook of situated cognition.* Cambridge: Cambridge University Press.

Gross, D. M. (2006). *The secret history of emotion: From Aristotle's rhetoric to modern brain science.* Chicago: University of Chicago Press.

Hacking, I. (2002). *Historical ontology.* Cambridge, MA: Harvard University Press.

Hacking, I. (1999). *The social construction of what?* Cambridge, MA: Harvard University Press.

Healy, D. (2002). *The creation of psychopharmacology.* Cambridge, MA: Harvard University Press.

Herskovits, E. (1995). Struggling over subjectivity: Debates about the "self" and Alzheimer's disease. *Medical Anthropology Quarterly, 9,* 146–164.

Hickling, F. W., & Rodgers-Johnson, P. (1995). The incidence of first contact schizophrenia in Jamaica. *British Journal of Psychiatry, 167,* 193–196.

Insel, T. R., & Fernald, R. D. (2004). How the brain processes social information: Searching for the social brain. *Annual Review of Neuroscience, 27,* 697–722.

Insel, T. R., & Quirion, R. (2005). Psychiatry as a clinical neuroscience discipline. *Journal of the American Medical Association, 294,* 2221–2224.

Jackson, F. (1982). Epiphenomenal qualia. *Philosophical Quarterly, 32,* 127–136.

Jarvis, G. E. (2007). The social causes of psychosis in North American psychiatry: A review of a disappearing literature. *Canadian Journal of Psychiatry, 52,* 287–294.

Kauffman, S. A. (1993). *The origins of order: Self-organization and selection in evolution.* New York: Oxford University Press.

Kendler, K. S. (2005). "A gene for...": The nature of gene action in psychiatric disorders. *American Journal of Psychiatry, 162,* 1243–1252.

Kendler, K. S., & Prescott, C. A. (2006). *Genes, environment, and psychopathology: Understanding the causes of psychiatric and substance use disorders.* New York: Guilford.

Kirkbride, J. B., Fearon, P., Morgan, C., Dazzan, P., Morgan, K., Tarrant, J., & Jones, P. B. (2006). Heterogeneity in incidence rates of schizophrenia and other psychotic syndromes: findings from the 3-center AeSOP study. *Archives of General Psychiatry, 63,* 250–258.

Kirmayer, L. J. (2008). Culture and the metaphoric mediation of pain. *Transcultural Psychiatry, 45,* 318–338.

Kirmayer, L. J. (2000). Broken narratives: Clinical encounters and the poetics of illness experience. In C. Mattingly & L. Garro (Eds.), *Narrative and the cultural construction of illness and healing.* Berkeley, CA: University of California Press.

Krabbendam, L., & van Os, J. (2005). Schizophrenia and urbanicity: A major environmental influence conditional on genetic risk. *Schizophrenia Bulletin, 31,* 795–799.

Light, K. C., Grewen, K. M., & Amico, J. A. (2005). More frequent partner hugs and higher oxytocin levels are linked to lower blood pressure and heart rate in premenopausal women. *Biological Psychology, 69,* 5–21.

Mahy, G. E., Mallett, R., Leff, J., & Bhugra, D. (1999). First-contact incidence rate of schizophrenia on Barbados. *British Journal of Psychiatry, 175,* 28–33.

Mainzer, K. (2004). *Thinking in complexity: The computational dynamics of matter, mind, and mankind.* New York: Springer.

Malabou, C. (2008). *What should we do with our brain?* New York: Fordham University Press.

Maturana, H. R., & Varela, F. J. (1987). *The tree of knowledge: The biological roots of human understanding.* Boston: Random House.

Maturana, H. R., & Varela, F. J. (1980). *Autopoiesis and cognition: The realization of the living.* Dordrecht, Holland: Reidel Publishing Company.

McGowan, P. O., Sasaski, A., d'Alessio, A. C., Dymov, S., Labonte, B., Szyf, M., & Meaney, M. J. (2009). Epigenetic regulation of the glucocorticoid receptor in human brain associates with childhood abuse. *Nature Neuroscience, 12,* 342–348.

McKenzie, K., Fearon, P., & Hutchison, G. (2008). Migration, ethnicity and psychosis. In C. Morgan, K. McKenzie & P. Fearon (Eds.), *Society and psychosis.* New York: Cambridge University Press.

Meaney, M. J. (2001). Maternal care, gene expression, and the transmission of individual differences in stress reactivity across generations. *Annual Review of Neurosciences, 24,* 1161–1192.

Meaney, M. J., & Szyf, M. (2005). Maternal care as a model for experience-dependent chromatin plasticity? *Trends in Neuroscience, 28,* 456–463.

Meehl, P. E., & Sellars, W. (1966). The concept of emergence. In H. Feigl & M. Scriven (Eds.), *Minnesota studies in the philosophy of science.* Minneapolis, MN: University of Minnesota Press.

Mill, J., & Petronis, A. (2009). The relevance of epigenetics to major psychosis. In A. C. Ferguson-Smith, J. M. Greally & R. A. Martienssen (Eds.), *Epigenomics.* New York: Springer SBM.

Mill, J., Tang, T., Kaminsky, Z., Khare, T., Yazdanpanah, S., Bouchard, L., & Petronis, A. (2008). Epigenomic profiling reveals DNA-methylation changes associated with major psychosis. *American Journal of Human Genetics, 82,* 696–711.

Morowitz, H. J. (2002). *The emergence of everything: How the world became complex.* New York: Oxford University Press.

Mortensen, P. B., Pedersen, C. B., Westergaard, T., Wohlfahrt, J., Ewald, H., Mors, O., & Melbye, M. (1999). Effects of family history and place and season of birth on the risk of schizophrenia. *New England Journal of Medicine, 340,* 603–608.

Nagel, T. (1974). What is it like to be a bat? *Philosophical Review, 83,* 435–450.

Nicolis, G., & Prigogine, I. (1989). *Exploring complexity: An introduction.* New York: W. H. Freeman.

Pattee, H. H. (1977). Dynamic and linguistic modes of complex systems. *International Journal of General Systems, 3,* 259–266.

Pattee, H. H. (1973). *Hierarchy theory: The challenge of complex systems.* New York: George Braziller.

Pedersen, C. B., & Mortensen, P. B. (2001). Evidence of a dose-response relationship between urbanicity during upbringing and schizophrenia risk. *Archives of General Psychiatry, 58,* 1039–1046.

Petronis, A. (2004). The origin of schizophrenia: Genetic thesis, epigenetic antithesis, and resolving synthesis. *Biological Psychiatry, 55*(10), 965–970.

Porges, S. W. (1998). Love: An emergent property of the mammalian autonomic nervous system. *Psychoneuroendocrinology, 23,* 837–861.

Pribram, K. H. (1990). From metaphors to models: The use of analogies in neuropsychology. In D.E. Leary (Ed.), *Metaphors in the history of psychology.* New York: Cambridge University Press.

Prosser, C. L. (1970). Levels of biological organization and their physiological significance. In J. A. Moore (Ed.), *Ideas in evolution and behavior.* New York: Natural History Press.

Reddy, W. M. (2001). *The navigation of feeling: A framework for the history of emotions.* Cambridge: Cambridge University Press.

Rosen, R. (1991). *Life itself: A comprehensive inquiry into the nature, origin, and fabrication of life.* New York: Columbia University Press.

Rosen, R. (1968). Hierarchical organization in automata theoretic models of the central nervous system. In K. N. Leibovic (Ed.), *Information processing in the nervous system* (pp. 21–35). New York: Springer.

Scott, A. (2002). *Neuroscience: A mathematical primer.* New York: Springer.

Searle, J. R. (1995). *The construction of social reality.* New York: Free Press.

Sperber, D. (1996). *Explaining culture: A naturalistic approach.* Oxford: Blackwell Publishers.

Suomi, S. J. (2006). Risk, resilience, and gene x environment interactions in rhesus monkeys. *Annals of the New York Academy of Science, 1094,* 52–62.

Torrey, E. F., Bowler, A. E., & Clark, K. (1997). Urban birth and residence as risk factors for psychoses: An analysis of 1880 data. *Schizophrenia Research, 25,* 169–176.

van Os, J., Driessen, G., Gunther, N., & Delespaul, P. (2000). Neighbourhood variation in incidence of schizophrenia. Evidence for person-environment interaction. *The British Journal of Psychiatry, 176,* 243–248.

van Os, J., Hanssen, M., Bijl, R. V., & Vollebergh, W. (2001). Prevalence of psychotic disorder and community level of psychotic symptoms: An urban-rural comparison. *Archives of General Psychiatry, 58,* 663–668.

van Os, J., Pedersen, C. B., & Mortensen, P. B. (2004). Confirmation of synergy between urbanicity and familial liability in the causation of psychosis. *American Journal of Psychiatry, 161,* 2312–2314.

Veling, W., Susser, E., van Os, J., Mackenbach, J. P., Selten, J. P., & Hoek, H. W. (2008). Ethnic density of neighborhoods and incidence of psychotic disorders among immigrants. *American Journal of Psychiatry, 165,* 66–73.

von Neumann, J. (1958). *The computer and the brain.* New Haven: Yale University Press.

von Neumann, J. (1951). The general and logical theory of automata. In L. Jeffress (Ed.), *Cerebral mechanisms in behavior.* New York: John Wiley & Sons, Inc.

Wilkinson, R. G., & Pickett, R. (2009). *The spirit level: Why more equal societies almost always do better.* London: Penguin.

Wilson, E. A. (2004). *Psychosomatic: Feminism and the neurological body.* Durham: Duke University Press.

Wilson, M. (1993). DSM-III and the transformation of American psychiatry: A history. *American Journal of Psychiatry, 150,* 399–410.

Wolfram, S. (2002). *A new kind of science.* Champaign, IL: Wolfram Media.

Zhang, T. Y., & Meaney, M. J. (2010). Epigenetics and the environmental regulation of the genome and its function. *Annual Review of Psychology, 61,* 439–466, C1–3.

16

Are Mental Illnesses Diseases of the Brain?

Thomas Fuchs

This chapter offers a systemic and ecological account as an opposing view to the naturalist idea that mental illnesses can be reduced to dysfunctions of the brain. Mental illness is regarded, on the one hand, as inseparable from the living organism and on the other, as inseparable from the patient's lifeworld or social environment. In order to grasp mental disorders in their context, the notion of monolinear causation has to be replaced by the notion of *circular causality*. In this view mental illnesses are marked by a disruption of *vertical circular causality*; that is, the interplay between lower-level processes and higher faculties of the organism. This primarily affects a mentally ill person's relation to themself which continually co-determines the course of the illness. On the other hand, mental illnesses are characterized by a disruption of *horizontal circular causality*; in other words of social relationships and the ability to respond adequately to the demands and expectations of others. This leads to negative feedback loops in socio-functional cycles that influence the course of the illness from the very beginning. Both kinds of circular causal processes are tied to mediation by the brain, but cannot exclusively be located within it. For this reason reduction of mental illnesses to diseases of the brain is in principle not possible.

The basic research program of the neurosciences consists in naturalizing consciousness, subjectivity, and also intersubjectivity—in other words explaining them in neurobiological terms. Even though this program is far from being realized, the impression is created that subjective experience can be imaged in the brain and in this way, as it were, materialized. This has far reaching effects on our image of the human being in general. The use of "brain language" is increasingly permeating our self-conception. In the wake of a popularized neurobiology, we are beginning to regard ourselves not as persons having wishes, motives, or reasons, but as agents of our genes, hormones, and neurones. Consequently, our problems and sufferings are often no longer considered existential tasks that we must face, but results of malfunctioning neuronal circuits and hormonal metabolism.

Critical Neuroscience: A Handbook of the Social and Cultural Contexts of Neuroscience, First Edition. Edited by Suparna Choudhury and Jan Slaby.

Biological psychiatry for its part aims to find the cause of mental illness in deviant functioning of the brain, according to the dictum commonly ascribed to Griesinger: "Mental illnesses are diseases of the brain."[1] The—as yet—poor attempts towards the end of the ninetweenth century by Theodor Meynert (1884), for instance, to subsume mental illnesses under the "diseases of the forebrain" were derided by Jaspers (1913/1973, p. 16) at the time as "brain mythologies." Today, however, it seems only a matter of time until specific genetic and neurophysiological correlates of all mental illnesses are found and allow us to causally trace them back to neuronal substrates. If anxiety disorders, depression, and schizophrenia are actually brain disorders, psychiatry finally becomes a branch of neurology and the psychiatrist a brain specialist.[2] Against such a background, there is a risk that therapeutic interventions in psychiatric practice will increasingly be oriented towards brain-centered procedures—pharmacological or directly stimulating modes of influencing brain functions—at the loss of psychotherapeutic or systemic approaches that consider the patients in their biographical and environmental context.

In what follows, I want to provide an opposing systemic-ecological view of mental illnesses. It is based on the assumption that, from birth on, the brain is embedded in interrelations between the person and the environment and is best seen as an organ of mediation and transformation for biological, mental, and social processes that are bound up in circular interplay. In this interplay, subjectivity—a person's experience and their relation to themselves—plays a central role, no less than the person's social interactions with others. For this reason, I claim that mental illnesses are not just brain diseases in the sense in which, for instance, we can trace back an angina pectoris to a coronary heart disease. The patient's altered subjective experience and disturbed relation to others are not mere epiphenomena of an effective organic process; much rather, they are essential elements of the illness itself. However, in order to grasp mental disorders in their subjective and intersubjective context, we first need to consider the notion of *causality* in living systems. Only by challenging the one-way causation that leads from the brain to the mind will we advance an ecological view of mental disorders and, through this, a person-oriented psychiatry.

Circular Causality of Living Systems

In order to embed the brain in the relations of organism and environment, I want to introduce, in what follows, the notion of *circular causality* as a property of living systems (Fuchs, 2009; Haken, 1993). It characterizes the systemic processes of interplay and feedback that were also foundational for Jakob von Uexküll's model of the *functional cycle* (1920/1973) and Viktor von Weizsäcker's theory of the *Gestalt cycle* (1940/1986; see also Fuchs, 2008, p. 121 et seq.). Both concepts refer to the

---

[1] Note that Griesinger himself in no way held a purely biological view. He was concerned with opposing a contemporary view according to which mental illnesses could not only be located in the brain, but in the entire body (see Schott and Tölle, 2006).

[2] See Insel & Quirion (2005): "The recognition that mental disorders are brain disorders suggests that psychiatrists of the future will need to be educated as brain scientists."

inseparable interconnection of perception and movement: what an organism senses is a function of how it moves, and how it moves is a function of what it senses. Thus, the touching hand anticipates and selects what it feels by its movements, whereas the shape of the object reciprocally guides the hand's touch. Through this, organism and environment co-constitute each other. Similar concepts have been developed more recently in enactivist theories of perception and cognition, as put forward by Varela, Thompson, and Rosch (1991), O'Regan and Noë (2001), Thompson (2007), and others. The feedback cycles between an organism and its biological as well as social environment may be termed *horizontal circular causality*. Examples are the aforementioned cycles of perception and action, but also the interactive processes in social systems, as they are, for instance, analyzed in family systems therapy (see also Fuchs & De Jaegher, 2009).

But there are also circular relations *within* the organism, namely between the whole and its parts, or between lower and higher systemic levels. I characterize them as *vertical circular causality*. Thus, a living being may be regarded as a system that continuously reproduces the components of which it consists (organs, cells), while these components reciprocally sustain and regenerate the system as a whole.[3] The whole is the condition of its parts, but is in turn realized by them. Such a structure, for instance, characterizes the relations between genes and the organism: the genetic structure of an individual cell nucleus controls the necessary production of specialized cellular organs and functions ("upward" causality). Conversely, the configurations and functions of the entire organism determine which genes are even given relevance for the development and regulation of a certain individual cell ("downward" causality).

This type of causality is often regarded as problematic or obscure, for two main reasons. First, since the whole consists of the parts itself, cause and effect cannot be assigned here to separate agents acting externally on each other. Second, the causal effect of higher systemic levels seems to presuppose unknown physical forces, thus either contradicting the laws of physics or falling prey to Occam's razor (see Craver & Bechtel, 2007 for a criticism). However, there is no need to restrict the notion of causality to *efficient* causality, according to the paradigm of billiard balls acting on each other. Macro-structures can well have *formative* causal influences on the micro-elements by which they are structurally realized. This formative causality does not imply the emergence of novel natural forces that are at odds with the laws of physics. Rather, macro-structures, by their particular form and configuration, are able to "select" certain properties of their components, and "block" others (Campbell, 1974; Moreno & Umerez, 2000). Moreover, the components may also acquire new, emergent properties. For example, iron molecules integrated into haemoglobin become able to reversibly bind oxygen, which is an extremely improbable state in anorganic nature. No physical "miracle" is required to accomplish this, but only a higher order structure (in this case haemoglobin) which "enslaves" its own constitutive elements (Haken, 1993) and involves them in specific patterns of behavior.

---

[3] Accordingly, Varela has defined an autopoietic system, or the minimal living organization, as "one that continuously produces the components that specify it, while at the same time realising it (the system) as a concrete unity in space and time, which makes the network of production of components possible" (Varela, 1997, p. 75).

Similarly, mental processes may have a formative impact on the physical behavior of living beings without being reducible to the physical events by which they are realized. When I am speaking, the muscles of my tongue and larynx show organized movement patterns. Their immediate efficient cause is the neuronally triggered release of acetylcholine at the motor endplates causing the muscle fibres to contract. However, it is equally adequate to say that my tongue and larynx move the way they do "because I am speaking these sentences." The over-arching, formative or organizing cause of the muscle actions is my speaking ("downward"), which in turn is realized by a series of combined physiological mechanisms ("upward"). The cause of my speaking, however, is neither my tongue nor my brain (though both are necessary to realize it)—it is me. Thus, in any conscious performance (speaking, writing, running, or thinking, for example), the living being itself acts as a downward formative cause, or in other words, the achievement in question is realized by vertical causality.

Accordingly, vertical causality also characterizes the functions of the brain. Pain stimuli from the periphery, for instance, through central processing in the brain, lead to pain experience ("upward"); conversely only the overall situation of attention and affectivity determines whether an impulse is "admitted" as a painful experience or whether it is suppressed by descending, inhibitory tracts ("downward"), as may be the case in a state of intense affective excitement. To give another example, an emotional state can, on the one hand be treated pharmacologically, by influencing the transmitter metabolism in the brain (upward). On the other hand, this can also be achieved psychotherapeutically, by changing the subjective perception of one's personal situation (downward). In this sense, anxiety can be influenced by sedatives as well as by a calming talk. As such, subjectivity represents a high or integral systemic level of the organism that feeds back into lower-level physiological processes. The brain functions as a *transformer* for this vertical circular causality, by converting higher- and lower-level influences on the organism and "translating" them into the other levels in the hierarchy (see Fuchs, 2009, 2008, p. 158 et seq.).

Mental Illness as Circular Process

Having introduced these terminological clarifications, let us now try to characterize mental illnesses as circular processes. I begin with vertical circular causality.

Vertical circular causality

Other than in the case of somatic conditions, in mental illnesses the patient does not succeed in attributing the condition to their body; for the condition primarily affects their experience of themself. Put differently, the subjective side of the illness does not consist merely in a secondary reaction to physiological dysfunctions. To a certain extent, it always involves a self-alienation or a "splitting" of the self. Something "in me" confronts me, defies my control, or dominates me while I desperately try to take control again, be it a panic attack, a depressive mood, a compulsion, or audible thoughts. Impulses or functions that have so far been integrated, take on a life of their own or become particularized and defy control. Mental illness hereby affects the person centrally, namely in their experience of themself and in their autonomy. What is more, the altered experience of and relation to themself, as such, is an effective factor in the further course of the illness. It follows that, independent of its origin, *vertical circular causality* always plays a decisive role in the illness.

Take the example of a depressive disorder. In whatever way different causal conditions—be they genetic, neurobiological, biographical, or interpersonal—interact in the particular case, as soon as the depression becomes manifest, it is per se an *illness of the person*. The disorder is accompanied by a fundamental change in bodily experience (inhibition, restriction, anxiety, heaviness, and loss of motivation); hardly any other illness has a comparable effect on a person's bodily subjectivity. But it also gives rise to negative perceptions and evaluations of oneself (self-reproach, feelings of guilt) and typical, depressive patterns of thought—negative assessments of their situation by the patients. These negative self-assessments, as self-fulfilling prophecies, increase the likelihood of further failures and contribute to the depression. Similar vicious circles are well known in anxiety disorders. They have the following pattern: the occurrence of physiological features of stress (activation of the sympathetic nervous system, increased pulse rate, and so forth) leads in turn to the perception of the physiological symptoms as "threatening," catastrophic cognitions and evaluations, increased physiological stress and so on. The subjectivity of experience, as a relation to oneself, thus becomes an important component affecting the course of the illness.

Every psychopathological experience is characterized by a personal meaning that the patient attributes to it, and a certain stance that they take towards it—suffering passively, giving in, acting out, fighting against it, or detaching oneself from it. This position taking is a relevant clinical feature in itself. Of course, these subjective modes of experience and behavior are enabled by neuronal processes, otherwise they could not be effective within the organism. The brain here functions as a transformational organ that converts peripheral and central, lower- and higher-level components of the previously mentioned "vicious circles" into one another. However, the phenomena of subjectively ascribing meaning, assessing a situation, and relating to oneself cannot be equated with processes in the neuronal substrate, as these lack acts of meaning making or intentionality. That all thought is realized in neuronal activity does not make it the case that it is identical with brain processes. Intentional content and directedness is inseparable from a subject's relation to the world. If neuronal processes function as "carriers" of intentional acts, they can do so only as part of an over-arching life process that includes the organism as a whole and its environment. In this way, mental processes are enabled or realized by neuronal processes, but are not localizable in the brain.

In a similar vein, it is not possible to reduce mental illness to circumscribed neurobiological dysfunctions—no matter how reliably correlated dysfunctions of the substrate can be identified. For, on the one hand, the subjective experience of the illness in its specific quality—its "what-it-is-like-ness" and its intentional contents—is not reducible to physiological descriptions. No imaging of brain activities can provide a psychiatrist with an understanding of what it is like to be depressive, to experience a panic attack or to hear voices. In fact, imaging methods themselves do not even provide criteria for what counts either as a pathological or as an ordinary physiological process—this can only be known from clinical practice, that is, from the patient's experience and behavior. Moreover, no description of the biological markers of anxiety or depression, however detailed, will tell us whether the patient in question is worried about a failure in the past, a threatening loss of his or her job, a public speech the patient has to give, or a current illness of his or her child. Obviously the biological data will be of very limited value as long as it remains isolated from its experiential context.

On the other hand, an even more crucial reason for this irreducibility is given by the patient's relation to themselves, which is continually involved in the illness process, influences it positively or negatively, and, as such, bars us from seeing mental illness as purely biological. The perception and assessment of one's own condition are genuinely personal phenomena that also limit the transferability of animal models to particular components of the illness. They give rise to a unique, specifically human kind of vertical circular causality, namely the feedback from subjective perceptions and evaluations into more fundamental processes of the illness. Not least, the possibility of suicide—which only humans have—bears witness to the fact that the relation to oneself can significantly influence the course of the illness, though, in this case, fatally.[4]

Horizontal circular causality

Just as mental illnesses cannot be detached from the person and be ascribed exclusively to the neuronal substrate, it is also not possible to see them as purely individual dysfunctions; in other words as detached from their *interpersonal aspects*. Irrespective of their causes, mental illnesses are always disturbances of the patient's interactions and relationships. They are accompanied by various impairments of the freedom to flexibly and autonomously respond to situations, offers, and demands of the social environment. As such, one can characterize them as impairments of a person's *responsivity* (Fuchs, 2007): certain abilities of the patient to shape social relationships according to their needs are either inhibited due to the illness or have not been developed in the first place. Thus, a significant part of psychopathology cannot be assessed in isolated patients, let alone their brains, but only as interactional dysfunctions.

As soon as social responsivity is impaired, feedback effects necessarily occur in the socio-functional cycles and, from the very beginning, influence or even determine the progression of the illness. The gestalt cycle of social perception and action is impaired or interrupted; the patients lose the usual resonance of their environment. Therefore, one can also characterize mental illnesses as communicative dysfunctions in the broadest sense. Symptoms of the illness evoke these dysfunctions, but they, in turn, are sustained, promoted, or even generated by the communicative impairments.

In the case of depression, for instance, a loss of emotional resonance occurs; that is a severe dysfunction of the responsivity to and exchange with the environment (Fuchs, 2000, 2001). This dysfunction in turn intensifies the patient's depressive self-perception. However, it also has an effect on their social system. Family and friends

---

[4]   This is not to portray suicide as a freely chosen action, for it is almost always based on a severe cognitive and emotional narrowing of situative perception. Nonetheless, it presupposes an assessment of the situation by the patient and cannot just be seen as a manifestation of a neurobiochemical dysfunction.

usually react at first by giving support, but over time with an increasing sense of helplessness and feeling of guilt as well as with latent or open annoyance. Their mostly inconsistent behavior and the patient's depression amplify one another in a vicious circle (Ruf, 2005, p. 178). The crucial influence of partnership interaction on depression has repeatedly been confirmed (Backenstrass et al., 2007; Barbato & D'Avanzo, 2006; Mundt, Kronmüller, Backenstrass, Reck, & Fiedler, 1998). Further factors aggravating the illness are negative consequences in the workplace, the feared or actual stigmatization of the patient, but also a possible secondary gain. All of these influences are certainly not generated by the brain, but are continuously processed and transformed into altered dispositions of experience and behavior.

Circular Causality in Pathogenesis

Let us take a look at the aetiology of mental illnesses, once again, in the case of depression. Here, too, we find the above-mentioned dysfunctions in vertical and horizontal functional cycles. A look at the epidemiology of the condition is sufficient to show the inadequacy of purely biological explanations, for, in recent years, a significant increase in depressive disorders can be observed in highly industrialized societies, which most certainly cannot be traced back to genetic or neurobiological causes, but is rather due to social and cultural causes.[5] The fact that the brain functions as the biological "final common path" for the various influences does not make the resulting illnesses brain diseases.

Nevertheless, epidemiological observations aside, we can also easily clarify the role of subjective and intersubjective processes in the aetiology of the illness. The manifestation of a depression is usually preceded by a personal situation that is perceived as a severe loss or threat by the person under the assumption that they do not have the resources for coping with it ("learned helplessness;" Seligman, 1975). Subjective perception and evaluation is, therefore, the decisive triggering factor. At the neuronal level, mediated by linking of prefrontal and limbic centers, and with significant involvement of the amygdala, this is accompanied by physiological stress which consequently leads to massive dysfunctions of the organismic functional cycles. This primarily affects the CRH-ACTH-cortisol, respectively the sympathetic nervous system as well as the serotonin-transmitter regulation in the limbic system. The self-perception of this altered organismic state, as a negative feedback loop, intensifies the physiological symptoms of stress. As a result, the organismic reaction becomes detached from its integration in superordinate feedback cycles and eludes

---

[5] According to studies in the US, the frequency of depressive disorders increased tenfold between 1945 and 1990; certainly, altered diagnostic habits had a considerable share in this (Cross-National Collaborative Group 1992; Weissman & Klerman, 1978). Data from other countries, however, points in a similar direction. Even the incidence of schizophrenia is not independent of cultural influences, as has often been assumed. Studies in several European countries, among these the United Kingdom and the Netherlands, have shown that the frequency of schizophrenia among immigrants in the new environment is 4–10 times higher than in the native population as well as in the original population in the country of origin (Cantor-Graae, 2007; Fearon et al., 2006).

the person's control. Negative horizontal feedback loops connecting to the social environment then influence the further course as described above.

It thus emerges that, for the development of depressive disorders, subjective experience in no way merely plays an epiphenomenal role. Rather, the illness originates in a specific perception of the situation, in an "individual act of meaning-ascription" that is not, as an intentional relation to the environment, reducible to neuronal processes. Depression results from a perceived loss of meaning and social resonance, not from a lack of serotonin. Moreover, it is not the objective features of the situation, but their subjective evaluation as insurmountable, which is decisive for the depressive reaction. Consequently, biographically acquired dispositions such as lack of self-worth or self-efficacy become highly influential factors in pathogenesis. Only secondarily do the physiological reactions take on a life of their own as a sustained regulatory dysfunction affecting the entire organism. Granted, in later stages depressive episodes may result from minor events or even from somatic triggers. But even then the organismic dysfunction always remains circularly connected to the patient's subjective perceptions as well as to their illness-related behavior in interpersonal relations.

Though the weighting of the factors involved differs in other disorders, we can generalize the paradigm of depression insofar as we always find in mental illnesses a *complex interplay of circular processes* both at the vertical, organismic level and at the horizontal, interpersonal level. In each of these internal and external circularities, the brain functions as an organ of transformation or mediation—as the carrier of the biological component of pathogenesis. Its structure is, however, continually shaped and modified in turn by psychosocial interactions. In this way subjective experience, as a significant component of the interaction of environment and organism, exerts a structuring influence on the neuronal substrate—an insight that is of no little relevance for psychotherapeutic practice.

Asserting this general basic structure does not imply that all mental illnesses need to be considered in the same way. It is by all means necessary to distinguish whether an illness is to be traced back to a comprehensible relation between a personal learning history and experience of the environment (as in the case of anxiety disorders), to a neurosystemic dysfunction affecting the constitution of the self (as in the case of schizophrenia), or to a macroscopically identifiable lesion of the brain (as in the case of an apoplectic stroke). Depending on the illness, psychosocial and biological aspects have to be weighted differently. Intentional and psychosocial explanations remain indispensable for neurotic disorders that are derived from dysfunctional patterns of perception, behavior, and relationships (Henningsen & Kirmayer, 2000). Even if dysfunctions of neuronal systems are involved here as well, these are usually epiphenomena that necessitate pharmacological treatment only in the event that they become independent and chronic. Neurophysiological approaches are generally more relevant for those disorders that can be seen as defects in ordinary functioning. But even psychiatric or neurological defects are always connected to adaptive coping processes that are accessible to intentional modes of understanding and treatment—and this even applies to the formation of delusions (Kern, Glynn, Horan, & Marder, 2009; Solms, 2004).

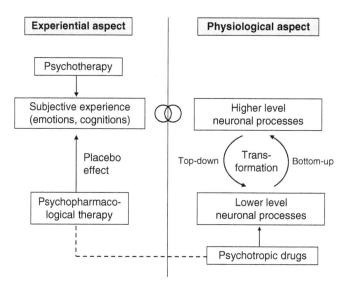

Figure 16.1 Effects of psychotherapy and drug therapy as seen from an experiential aspect (left) and from a physiological aspect (right). The two circles in the middle (∞) signify concomitant or concordant changes within both aspects; there is no "efficient causality" between them. Thus, the effect of psychotherapy on the brain is mediated by concomitant higher level neuronal processes being transformed into changes on lower levels. Conversely, the physiological effect of psychotropic drugs is transformed into higher level changes that realize altered subjective experiences (for example, decreased anxiety). However, drugs appear on the left side as well, because they are also efficient within the dimension of experience and meaning, this being known as placebo effect.

Circular Causality in Therapy

Finally, an ecological conception of mental illness also suggests a pluralistic understanding of treatment. The dualistic distinction between somatic therapies acting on the brain and psychological therapies having elusive, purely subjective effects is no longer tenable. The circular interactions of self, body, brain, and environment may be approached at various levels or turning points, since any mode of treatment will be transformed by the brain and hereby contribute to a holistic effect. Psychosocial influences on the level of meaning and intentionality are transformed into altered patterns of neuronal activity on lower levels, and vice versa: pharmacological effects are transformed into changes of brain activity at higher levels, resulting in altered affective or cognitive experience (see Figure 16.1). This means that any therapeutic intervention is of a physiological as well as of a psychological nature.

Psychotherapy addresses the patient as an experiencing, self-conscious, and self-relating subject. Yet its long-term impact is mediated by its effect on brain functions, as has been shown in a number of neuroimaging studies. Psychotherapy produced lasting effects mainly on prefrontal and frontal brain metabolism (for an overview, see Beauregard, 2007; Fuchs, 2004). Thus, in Positrone Emission Tomography (PET) studies of depressive patients, Brody and colleagues (2001) and Martin and colleagues (2001) found significant decreases in prefrontal lobe activity following treatment with Interpersonal Psychotherapy (IPT). These and other comparable studies strongly support the view that the subjective nature and the intentional content of mental processes (thoughts, feelings, beliefs, expectations, and volitions) significantly influence the various levels of brain functioning (molecular, cellular, and neural circuits) as well as brain plasticity. The transformation runs "top-down" that is, it starts from subjective experience that is realized by (though not localized in) higher level neuronal processes (mainly in cortical networks), and results, on the lower level, in altered synaptic transmission, altered gene expression, and rewiring of neuronal networks.

On the other hand, effects of psychotropic drugs start from influencing the transmitter metabolism at lower levels, mainly in subcortical regions, and are transformed "bottom-up" into higher level processes, resulting in a modification of subjective experience. In a particularly interesting PET study of depressive patients, Goldapple and colleagues (2004) found differential target areas of successful Cognitive Behavioral Therapy (CBT) versus pharmacotherapy: CBT primarily produced changes in the medial frontal and cingulate cortex, whereas drug treatment changed metabolism in limbic-subcortical regions (brainstem, insula, subgenual cingulate). This fits the idea of CBT interventions focussing mainly on modifying dysfunctional cognitions, and leading to an alleviation of vegetative symptoms and inhibition, while pharmacotherapy rather takes the opposite course.

However, direct subjective effects of pharmacological treatment must not be overlooked: each drug administration also operates on the intersubjective level of shared meaning and emotional relationship between doctor and patient, commonly known as placebo effect (see left side of Figure 16.1). The resulting changes in brain metabolism have also been demonstrated by neuroimaging: in an fMRI study on major depression, Mayberg and colleagues (2002) again found mainly cortical effects of placebo treatment, as against more subcortical-limbic and brainstem effects of antidepressant drugs.

This underlines that there is no separation, but rather a circular interaction of psychological and biological processes, and accordingly, no "merely biological" or "merely psychological" treatment. This interaction, however, cannot be expressed in terms like "the mind acting on the body" or "the brain producing the mind." Instead, the brain acts as a mediator and transformer which may be addressed through input on different hierarchical levels and which converts it in both directions: neurobiochemical changes become mood changes on the subjective level, but subjectivity in turn influences the plasticity, structuring, and functioning of the brain. Vertical circular causality allows for both approaches equally.

This illustrates that both ways of treatment may also interact synergically. On the one hand, beyond a certain point, the neurobiological and endocrine dysfunctions involved in depression, for example, may be too advanced to be accessible to interventions on the psychological level. Pharmacological ("bottom-up") treatment may then enable the patient to re-engage in his relationships and, therefore, will indirectly further his or her social well-being. On the other hand, psychotherapy can help the individual to reframe their beliefs, for example, so that they align with the actual nature of events to which they are directed (Glannon, 2008). This can alter the patient's misperception of events or social situations as well as his or her corresponding behavior in a beneficial way. Moreover, as we have seen, psychotherapy not only changes the patients' implicit relational patterns, attitudes, and behavior, but also the functions and structures of their brains. Mental states are not epiphenomenal to brain states but can have a causal influence on them. In view of the limited effectiveness of medication, especially in chronic illness, it would be wrong to neglect these "top-down" options of treatment.

Drug therapies targeting neuronal pathways and transmitter systems treat only one dimension of mental disorders (Glannon, 2008). Moreover, a mere biological view still tends to isolate the individual patient and to make his illness seem separated from its interconnections with his environment. However, the intentional and qualitative aspects of beliefs and emotions cannot be explained in terms of physical processes in the brain. Nor can we forego (inter)subjective experience if we want to change the patient's maladaptive cognitive, emotional, and behavioral dispositions that have led to his or her illness and may lead to a relapse in the future. Only conscious, embodied experience is able to correct the corresponding dysfunctional patterns of neural activity. And only repeated interactions with the environment—in other words, processes of interpersonal learning—can stabilize new attractors of perception and behavior in the brain. Since the neural structures that underly our personal dispositions are shaped by embodied experience, there will probably never be a way to create new views of the self and the world by brain manipulation directly. Any psychotherapeutic and social approach to psychiatry is thus based on a holistic, ecological view of life.

Conclusion

The brain is not the sole producer of the mind but a relational organ that mediates the interaction between the organism and its complementary environment (Fuchs, 2008). Our mental states are the emergent products of circular causation consisting of neuro-physiological, environmental, and social influences continuously interacting with each other in a series of feed-forward and feedback loops (Fuchs, 2004, 2009; Glannon, 2008). Disordered states of mind result when these circular processes are disturbed in some way. I have distinguished two dimensions which characterize this disturbance:

- On the one hand, mental illnesses can be characterized as *dysfunctions in vertical feedback cycles*. The central integration of partial functions or impulses fails; the latter take on a life of their own and elude the person's control, for instance, in the form of neurotic symptoms, compulsions, panic attacks, disorders of impulse-control, self-disorders, hallucinations, and so forth. These particularized processes, in turn, affect the person's relation to themself. They lead to various attempts at coping and reintegration, but also to secondary reactions and symptoms ("fear of fear", self-reproaches, for example) that make the illness worse; in other words, they are a significant component in its progression.
- On the other hand, mental illnesses can also be seen as *dysfunctions in horizontal feedback circuits*, for they are connected to more or less severe impairments of

responsivity and interactions with the social environment. In relationships to significant others, negative feedback loops and vicious circles occur that sustain or further intensify the symptoms. These feedback loops are tied to (inter)subjective perception and evaluation, to the patient's and their relatives' experience and behavior. Though they certainly influence brain functioning, they may not be described on the brain level alone.

Having mentioned both aspects, it follows that a reductionist description and explanation of mental illness based on neurophysiological facts alone does not do justice to its actual complexity. No mental illness can be diagnosed, described, or explained without taking account of the patient's subjectivity and their interpersonal relationships. Mental illnesses are always illnesses of the person and their relationships to other persons. The brain, with its functions, is centrally implicated in them, but a narrowly neurobiological perspective is never sufficient to describe and explain all facets of the illness. The final disorder is the product of a cascade of subjective, neuronal, social, and environmental influences continuously interacting with each other. Within these circular interactions the brain acts as a mediating, transforming, and also amplifying organ, but not as "the monolinear cause."

While advances in neurobiology have contributed to overcoming dualistic models of mental illness, one would be throwing out the baby with the bathwater if one wanted to trace all forms of mental illness back to brain processes in an undifferentiated manner. Neurophysiologically (by means of imaging technologies) determinable anomalies in themselves are not more than correlative in character. No such findings could be identified as pathological at all without being related to subjective suffering and intersubjective disturbances. They only become aetiologically relevant if they are embedded in the overarching circular processes that include the organism–environment system as well as the patient's interpersonal relationships.

In the case of obsessive-compulsive disorder (OCD), for instance, hyperactivity of the caudate nucleus provides no indication as to the cause of the disorder. Local activations of the brain's metabolism only correlatively reflect the function that is being activated; they are only a partial component of the illness. Depressive and anxiety disorders are not solely caused by the amygdala, just as OCD is not solely caused by the caudate nucleus, even if these brain regions are implicated in the illnesses. To the extent that neurophysiological changes are to be found, these are correlates, adaptive processes, or biological scars that have emerged in the context of repeated perceptions of situations as dangerous or threatening. Even if neurosystemic developmental impairments in schizophrenia or amygdaloid hyperactivity in posttraumatic stress disorder clearly act as restricting factors, such dysfunctions never become monolinear causes.[6]

Integrally viewing mental illnesses as relational dysfunctions, however, is also a precondition for treating them adequately. The complexity of the circular processes is not best captured either by an opposition between or a mere summation of various

---

[6] Monolinear causation may only be attributed to brain lesions (for example, apoplexy, brain tumor) that result in a failure of functions. However, even failures of this nature are followed by manifold processes of adaptation and coping that imply circular interactions of person, brain, and environment.

therapeutic approaches. What is called for is rather a bi- or polyperspectival approach. Here various, especially somatic and psychotherapeutical approaches, can be combined to influence circular causalities. However, psychosocial descriptions and interventions will remain indispensable, for a purely neurobiological explanation or treatment of mental illnesses is not in principle possible. What psychiatry needs is a systemic or ecological view of the brain in order to better understand the interplay of biological, psychological, and socio-cultural processes and to do justice to the complexity of its subject matter. This is not the brain in isolation, but the embodied human being living in relationships.

References

Backenstrass, M., Fiedler, P., Kronmüller, K. T., Reck, C., Hahlweg, K., & Mundt, C. (2007). Marital interaction in depression: A comparison of structural analysis of social behavior and the Kategoriensystem für Partnerschaftliche Interaktion. *Psychopathology, 40*, 303–11.

Barbato, A., & D'Avanzo, B. (2006). Marital therapy for depression. *Cochrane Database Syst Rev. 2006 April 19*(2), CD004188. Review.

Beauregard, M. (2007). Mind does really matter: Evidence from neuroimaging studies of emotional self-regulation, psychotherapy, and placebo effect. *Progress in Neurobiology, 81*, 218–236.

Brody, A. L., Saxena, S., Stoessel, P., Gillies, L. A., Fairbanks, L. A., Alborzian, S., Phelps, M. E., Huang, S. C., Wu, H. M., Ho, M. L., Ho, M. K., Au, S. C., Maidment, K., & Baxter, Jr., L. R. (2001). Regional brain metabolic changes in patients with major depression treated with either paroxetine or interpersonal therapy. *Archives of General Psychiatry, 58*(7), 631–640.

Campbell, D. (1974). "Downward causation" in hierarchically organized biological systems. In F. J. Ayala & T. Dobzhansky (Eds.), *Studies in the philosophy of biology* (pp. 179–186). Berkeley: University of California Press.

Cantor-Graae, E. (2007). The contribution of social factors to the development of schizophrenia: A review of recent findings. *Canadian Journal of Psychiatry, 52*, 277–278.

Craver, C. F., & Bechtel, W. (2007). Top-down causation without top-down causes. *Biology and Philosophy, 22*, 547–563.

Cross-National Collaborative Group (1992). The changing rate of major depression. Cross-national comparisons. *Journal of the American Medical Association, 268*, 3098–3105.

Fearon, P., Kirkbride, J., Morgan, C., Dazzan, P., Morgan, K., Lloyd, T., Hutchinson, G., Tarrant, J., Fung, W., Holloway, J., Mallett, R., Harrison, G., Leff, J., Jones, P., & Murray, R. (2006). Incidence of schizophrenia and other psychoses in ethnic minority groups: Results from the MRC AESOP Study. *Psychological Medicine, 36*, 1541–1550.

Fuchs, T. (2009). Embodied cognitive neuroscience and its consequences for psychiatry. *Poiesis and Praxis, 6*, 219–233.

Fuchs, T. (2008). *Das Gehirn – ein Beziehungsorgan. Eine phänomenologisch-ökologische Konzeption.* Stuttgart: Kohlhammer.

Fuchs, T. (2007). Psychotherapy of the lived space. A phenomenological and ecological concept. *American Journal of Psychotherapy, 61*, 432–439.

Fuchs, T. (2004). Neurobiology and psychotherapy: An emerging dialogue. *Current Opinions in Psychiatry, 17*, 479–485.

Fuchs, T. (2001). Melancholia as a desynchronization. Towards a psychopathology of interpersonal time. *Psychopathology, 34*, 179–186.

Fuchs, T. (2000). *Psychopathologie von Leib und Raum*. Darmstadt: Steinkopff.

Fuchs, T., & De Jaegher, H. (2009). Enactive intersubjectivity: Participatory sense-making and mutual incorporation. *Phenomenology and the Cognitive Sciences, 8*, 465–486.

Glannon, W. (2008). The blessing and burden of biological psychiatry. *Journal of Ethics and Mental Health, 3*, 1–4.

Goldapple, K., Segal, Z., Garson, C., Lau, M., Bieling, P., Kennedy, S., & Mayberg, H. Modulation of cortical-limbic pathways in major depression. Treatment-specific effects of cognitive behavior therapy. *Archives of General Psychiatry, 61*(1), 34–41.

Haken, H. (1993). *Advanced synergetics* (3rd ed.). Berlin: Springer.

Henningsen, P., & Kirmayer, L. J. (2000). Mind beyond the net: Implications of cognitive neuroscience for cultural psychiatry. *Transcultural Psychiatry, 37*, 467–494.

Insel, T. R., & Quirion, R. (2005). Psychiatry as a clinical neuroscience discipline. *Journal of the American Medical Association, 294*, 2221–2224.

Jaspers, K. (1973). *Allgemeine Psychopathologie* (9th ed.). Berlin, Heidelberg, New York: Springer (originally published 1913).

Kern, R. S., Glynn, S. M., Horan, W. P., & Marder, S. R. (2009). Psychosocial treatments to promote functional recovery in schizophrenia. *Schizophrenia Bulletin, 35*, 347–361.

Martin, S. D., Martin, E., Rai, S. S., Richardson, M.A., & Royall, R. (2001). Brain blood flow changes in depressed patients treated with interpersonal psychotherapy or venlafaxine hydrochloride. *Archives of General Psychiatry, 58*(7), 641–648.

Mayberg, H. S., Silva, J. A., Brannan, S. K., Tekell, J. L., Mahurin, R. K., McGinnis, S., & Jerabek, P. A. (2002). The functional neuroanatomy of the placebo effect. *The American Journal of Psychiatry, 159*, 728–737.

Meynert, T. (1884). *Psychiatrie. Klinik der Erkrankungen des Vorderhirns begründet auf dessen Bau, Leistungen und Ernährung*. Wien: Braumüller.

Moreno, A., & Umerez, J. (2000). Downward causation at the core of living organisation. In P. B. Andersen, C. Emmeche, N. O. Finnemann & P. V. Christiansen (Eds.). *Downward causation: Minds, bodies and matter* (pp. 99–117). Aarhus: Aarhus University Press.

Mundt, C., Kronmüller, K. T., Backenstrass, M., Reck, C., & Fiedler, P. (1998). The influence of psychopathology, personality, and marital interaction on the short-term course of major depression. *Psychopathology, 31*, 29–36.

O'Regan, J. K., & Noë, A. (2001). A sensorimotor account of vision and visual consciousness. *Behavioral and Brain Sciences, 24*, 939–1011.

Ruf, G. D. (2005). *Systemische Psychiatrie*. Stuttgart: Klett-Cotta.

Schott, H., & Tölle, R. (2006). Magna Charta der Psychiatrie: Leben und Werk von Wilhelm Griesinger. *Sozialpsychiatrische Informationen, 4*, 2–9.

Seligman, M. E. P. (1975). *Helplessness: On depression, development and death*. San Francisco: Freeman and Comp.

Solms, M. (2004). Is the brain more real than the mind? In A. Casement (Ed.) *Who owns psychoanalysis?*, (pp. 323–342). London: Karnak.

Thompson, E. (2007). *Mind in life: Biology, phenomenology, and the sciences of the mind*. Cambridge, MA: Harvard University Press.

Uexküll, J. V. (1973). *Theoretische Biologie*. Frankfurt: Suhrkamp (originally published 1920).

Varela, F., Thompson, E., & Rosch, E. (1991). *The embodied mind: Cognitive science and human experience*. Cambridge, MA: MIT Press.

Varela, F. (1997). Patterns of life: Intertwining identity and cognition. *Brain and Cognition, 34*, 72–87.

Weissman, M. M., & Klerman, G.L. (1978). Epidemiology of mental disorders. Emerging trends in the United States. *Archives of General Psychiatry, 35*, 705.

Weizsäcker, V. V. (1986). *Der Gestaltkreis. Theorie der Einheit von Wahrnehmen und Bewegen*. (5th ed.). Stuttgart: Thieme (originally published 1940).

17

Are there Neural Correlates of Depression?

Fernando Vidal and Francisco Ortega

Since the decade of the brain, a number of projects with names such as neurotheology, neuroaesthetics, neuropsychoanalysis, neuroeducation, neuroeconomics, or social neuroscience have rapidly developed with the goal of bringing knowledge about the brain to bear on questions hitherto dealt with by the human and social sciences. For some, their ultimate goal is comprehensively to reform the human sciences on a neuroscientific basis. Thus, Semir Zeki (2002, p. 54), professor of neuroaesthetics at University College London, declared, "My approach is dictated by a truth that I believe to be axiomatic—that all human activity is dictated by the organization and laws of the brain; that, therefore, there can be no real theory of art and aesthetics unless neurobiologically based." Insofar as the usual subjects of the human and social sciences are intimately connected with culture, we shall refer to the *neuro* disciplines that have developed since the 1990s as being "neurocultural."[1]

Driven by the availability of brain imaging technologies, these fields share a quest for "neural correlates" of behaviors and mental processes. The recent debate, prompted by the uncovering of *voodoo correlations* in social-neuroscientific research (Abbot, 2009; Margulies, Chapter 13, this volume; Vul, Harris, Winkielman, & Pashler, 2009), has not dampened the enthusiasm for the imaging-correlational approach. Since the early 1990s brain imaging studies have increasingly dealt with topics of potential ethical, legal, social, and policy implication, such as attitudes, cooperation and competition, violence, or religious experience. Commercial enterprises like neuromarketing have developed concomitantly, and the media, both popular and specialized, has given much room to these new fields. This chapter will

---

[1] This is not the place to discuss the notoriously elastic concept of culture. In the light of statements such as Zeki's and of the more systematic treatment offered by a Spanish neuroscientist in *Neuroculture. A culture based on the brain* (Mora, 2007), it makes sense, for the purposes of this chapter, to call fields such as neuroaesthetics or neurotheology, their vision, approach, claims, findings, publications, institutions, and so forth, "neurocultural." See Ortega and Vidal, 2011.

Critical Neuroscience: A Handbook of the Social and Cultural Contexts of Neuroscience, First Edition.
Edited by Suparna Choudhury and Jan Slaby.
© 2012 John Wiley & Sons, Ltd. Published 2016 by John Wiley & Sons, Ltd.

examine the notion of "neural correlate," and discuss its application in the field of depression research, with a focus on imaging as a method, and on neural correlates as an investigative and interpretive tool.

The vast majority of studies in the new neurocultural fields use functional magnetic resonance (fMRI). It has been shown that from a handful of papers in 1991 and an initial focus on sensory and motor tasks, the number of publications exploring the use of fMRI in the human sciences increased to 865 in 2001, with an average increase of 61 % per year during that period (Illes, Kirschen, & Gabrieli, 2003). The database was updated at the end of 2004 and showed an astounding increase: in the three years since the original study, an additional 3,824 papers had been published (Racine, Bar-Ilan, & Illes, 2005), and the same trend was found in the coverage of those studies in the print media (Illes, Racine, & Kirschen, 2006). The same period saw a proliferation of studies with potential ethical, legal, social, and policy implications, covering several broad thematic areas: altruism, empathy, decision making; cooperation and competition; judging faces and races; lying and deception; meditating and religious experience (Illes et al., 2006).

Neurotheology wishes to establish the neurological bases of spiritual and mystical experience. Similarly, neuroaesthetics, neuropsychoanalysis, neuroeducation, neuro-economics, or social neuroscience present themselves as searching for the neuro-biological underpinnings of processes and phenomena studied and described by aesthetics, psychoanalysis, education, economics, or social psychology. Although these neurocultural fields are dominated by the neurocorrelational approach, they do not elaborate the very notion of *neural correlate*, and give no consideration to the contested nature of the concept. In contrast, philosophers of mind and scientists researching the neural correlates of consciousness (NCC) have devoted considerable energy to debating the notion. It may well be that, since neural correlates seem legitimate for studying consciousness, they have been seen as a fortiori appropriate for investigating comparatively simpler processes: hence, the need to begin here with NCCs.[2]

The Neural Correlates of Consciousness

"Most neuroscientists believe that experience happens in the brain," write philosophers Alva Nöe and Evan Thompson, and they add, "most scientists assume that if we could understand what is going on at the neural substrate of an experience, then we would understand how the brain's action produces states of consciousness" (Noë & Thompson, 2004a, p. 87). Their observation captures the main tendency of consciousness research since the 1990s. As Sabine Maasen shows in a quantitative survey of journal articles from 1974 to 2000, the academic concern with consciousness has become increasingly cognitive, neurological, and pharmacological throughout the years (Maasen, 2003, p. 127). The growing interest in consciousness research within

---

[2] We are aware that we should tie our discussion to locality and contexts—two elements we do not examine. The fact, however, is that, whatever their home country and institution, and regardless of local differences, scientists doing the kind of research reviewed here share (at least in the restricted realm of their profession) basic aims, values, beliefs, criteria, interests, and investigative tools, topics, and strategies.

neuroscience contrasts with the decreasing philosophical and sociological elaboration of the topic. Maasen also notes that despite neuroscientists' interest in conscious experiences, they refrain from using "consciousness," as if they wished to avoid any association with philosophical connotations of the idea. On the other hand, NCCs have become the object of a large research program. To what extent, however, do they concern consciousness?

For Christof Koch, a major figure in the field, the "C word" evokes such powerful aversive reactions among scientists that "you're better off with some other word in grant applications and journal submissions. 'Awareness' usually slips under the radar" (Koch, 2004, p. 320). When Koch writes "awareness," he claims "consciousness;" and yet, the referent remains awareness, and not, as he believes, consciousness. As Maasen notes, neuroscientific efforts tell us something about awareness, "yet nothing about consciousness in its experiential aspect" (2003, p. 148). This is evident from the two main objects of the neuroscientific study of consciousness: wakefulness and visual awareness (Maasen, 2003, 2007; Zeman, 2001).

Research on consciousness-as-wakefulness aims at revealing interconnected neural, psychological, and behavioral functions and their control systems. It consists largely of the study of the visual system in animals and humans and investigates fine-grained correlations between cerebral activity and conscious experience in the fields of visual perception, distinctions between declarative and procedural memory, and action, through the study of changes in cerebral activity divorced from conscious control (Zeman, 2001). Research on visual perception in this field concerns mainly visual experiences without changes in the external stimuli, as in hallucinations or ambiguous figures. It also deals with the distinction between "explicit" neural processes that give rise to the conscious awareness of seeing, and "implicit" neural processes responsible for visuomotor performance without awareness (for example, blindsight). The assumption is that the distinction will help delineate the key neural substrates of awareness. A large part of the neurobiological theories of consciousness assume that structures in the upper brainstem core play a critical role in arousal, and that thalamic and cortical activity supplies much of conscious content (Zeman, 2001).

Neuroscientific research into awareness and consciousness largely draws on and overlaps with the search for the neural correlates of consciousness, which "is arguably the cornerstone of the recent resurgence of the science of consciousness" (Chalmers, 2000, p. 17). Although the notion of NCC has been around since the 1980s, and the first printed use of the term is usually credited to Francis Crick and Christof Koch (1990), the expression "neurological correlates of conscious mental activity" appeared in 1961, in an article on disorders of attention and perception in early schizophrenia: "Although it seems unlikely that we can ever define consciousness in purely physiological terms, work in recent years has certainly provided us with evidence for some neurological correlates of conscious mental activity" (McGhie & Chapman, 1961, p. 111). The authors were psychoanalytically-oriented, and wished to develop a theory of schizophrenia based on clinical observations and interviews with patients; they were not concerned primarily with the theory of consciousness.[3]

---

[3] We would like to thank Nicolas Langlitz for this reference.

NCC is usually defined as "the minimal set of neuronal events and mechanisms jointly sufficient for a specific conscious percept" (Koch, 2004, p. 16; see also Crick & Koch, 1995, 1998, 2003; Metzinger, 2000; Tononi & Koch, 2008). Such a definition embodies the "minimal substrate thesis" (Noë & Thompson, 2004b). In a frequently quoted article, philosopher David Chalmers undertook a sophisticated analysis of various definitions, from the most general to the most elaborate. At one end, an NCC is a "specific system in the brain whose activity correlates directly with states of conscious experience;" at the other end, it "is a minimal neural system N such that there is a mapping from states of N to states of consciousness, where a given state of N is sufficient, under conditions C, for the corresponding state of consciousness" (Chalmers, 2000, p. 18, 31).

As for the nature and location of NCC, numerous candidates have been proposed: 40-hertz oscillations, intralaminar nuclei in the thalamus, reentrant loops in thalamo-cortical systems, neural assemblies bound with N-methyl-D-aspartate (NMDA) receptors, certain neurons in the inferior temporal cortex, or processing within the ventral stream involved in object recognition and form representation (Chalmers, 2000; Tononi & Koch, 2008). Even if everyone agreed that NCCs are "neural assemblies" responsible for the genesis of consciousness, there would be room for disagreement about such specifics as the regions involved (Zeman, 2001). Hence the focus on empirical hypotheses about where to look and which technology to use (for example, fMRI versus EEG), or whether to focus on neural assemblies or on re-entrant processing (Hohwy, 2007).

Neural correlates of conscious vision have become a major research target. The main reasons are that the anatomy and function of the primate visual system are particularly well understood, and that visual experience is amenable to experimental manipulation. The quest for the neural correlates of vision is the field where NCC research has claimed its most significant results. Visual consciousness is here synonymous with "seeing," characterized as the perceptual, phenomenological, or qualia descriptions of visual experiences, devoid of linguistic, intentional, or self-referential implications (ffytche, 2000, p. 221). Research in this field has involved mapping the area responsible for color perception in the macaque's brain, describing the neurons whose activity is correlated with what is "seen" (rather than with what is presented) in experiments of binocular rivalry, stimulating visual neurons specialized in the detection of motion in the macaque, and studying lesions in brain areas associated with deficits in color and movement perception in humans (Ffytche, 2000). A number of techniques, such as masking, binocular rivalry, continuous flash suppression, and various forms of induced blindness are used to study how the seemingly simple and unambiguous relationship between a physical stimulus in the external world, the resulting neural activity, and its associated percept is disrupted (Tononi & Koch, 2008). One significant characteristic of this research is that, as Tononi and Koch (2008) recognize, it basically equates consciousness with awareness.

Work on the neural basis of binocular rivalry has prompted speculations on the existence of a content NCC for visual consciousness (Logothetis, 1999; Logothetis & Schall, 1989). The interest here is not limited to the neural states that produce visual consciousness, but extends to those that determine its specific contents. The ideal aim is to identify a neural system from whose activity one could establish the precise contents of a visual experience, or at least its content with regard to some dimensions, such as shape or color.

Binocular rivalry takes place when two visual patterns are presented simultaneously to each eye. After an initial period of frequent exposure to superimposition, one comes rapidly to experience the two patterns as if they were alternate, in a sort of perceptual dominance. This phenomenon provides a means of dissociating stimulus-driven neural activity from the neural activity that corresponds to the subjective visual experience. It is assumed that the neural activity most closely associated with the percept is likely to represent what the subject is actually seeing; it is therefore considered the neural correlate of the perceptual experience.

Noë and Thompson (2004b) argue that studies on the neural basis of binocular rivalry do not support the notion of a "content match" between neural activity and perceptual experience, but only provide evidence for some sort of "content agreement." A particular neuron may fire when a vertical line appears in its receptive field (RF). However, the perceptual experience of a vertical line does not concern the vertical line alone, but the line against a background and occupying a certain relation to the embodied perceiver (Noë & Thompson, 2004b, p. 12). Moreover, neuronal RF content could not be established without taking into account the sensorimotor context of the individual as a whole. Since RF contents lack features—such as being active and attentional—that are the hallmark of perceptual experience, there can be no content NCC for visual perception (p. 14). In short, perceptual experience is incommensurable with neural activity.

While perceptual content is intrinsically experiential and is, therefore, necessarily experienced from a first person perspective, neural systems as such do not experience anything and have no point of view. Furthermore, the active and attentional character of first-person experience calls into question the existence of a "minimal substrate"—a minimal set of neurobiological properties that would suffice to activate a certain conscious content. In the NCC perspective, however, perceptual experiences take place independently of other neural processes, as well as of the rest of the body and even of the external world. In other words, the "entire brain is sufficient for consciousness" (Koch, 2004, p. 87). That is why cognitive neuroscientist Antti Revonsuo assumes that "subjective phenomenal consciousness is a real, natural, biological phenomenon that *literally resides* in the brain" (Revonsuo, 2000, p. 59, our emphasis). Since, for Revonsuo, the brain and the phenomenal level " 'are deep inside the skull', never actually in direct contact with external objects," consciousness, "within the brain" (p. 65, p. 62). Revonsuo criticizes the philosophers who doubt that "a fully functional brain is sufficient for a fully realized subjective phenomenology," and accuses them of endorsing an "antibiological metaphysics of consciousness" (p. 61).

Revonsuo's internalist and "embrained" conception of consciousness and experience looks like yet another version of "the myth of interiority;" in other words, the attempt to "solve the psychophysical problem by substituting the ethereal and elusive soul of the philosopher by the material and tangible soul of the savage, namely the brain" (Bouveresse, 1976, p. 677). His position explains the choice of the "dreaming brain" and "virtual reality" as metaphors for consciousness and subjective experience. Since the dreaming brain, he writes, "creates the phenomenal level in an isolated form," it provides insights into the processes that are "*sufficient* for producing the phenomenal level." Similarly, virtual reality is, for him, a useful metaphor for consciousness, insofar as experienced reality or the sense of our presence in the world is merely an "out-of-the-brain-experience," a "*telepresence* for the brain," an illusion that there is an extra-cerebral real world (Revonsuo, 2000, p. 65).

Dreaming for Revonsuo can be a metaphor for conscious experience because both in dreaming and in consciously experiencing we are under the illusion that perceptual events "do not take place inside the brain" but in an "externalized perceptual world." Of course, it is one thing to say that in dreaming we are, as in virtual reality, disconnected from the world, another to claim that such an experience can be induced by the corresponding NCC, and yet another to argue that our experience of reality and of dreaming are brain-generated illusions. Even in its less solipsistic versions, NCC research tends to illustrate the "Foundation Argument" (Noë, 2009), according to which the fact that we dream and that some events can be produced in consciousness by directly stimulating the brain proves that the brain suffices for consciousness.

Even in dreams, however, we are not totally disconnected from the world. The scope of dreams is limited by the dreamer's past experiences. Moreover, while dreaming we are not fully decoupled from the world; we breathe, move, snore, make noises, even have orgasms. Sometimes dreams correspond to internal states or external stimuli; brain, body, and world are present when we dream. The same applies to vision. NCC research into visual experience excludes a sensorimotor context involving eye, head, and body movement. As Noë points out, seeing is not something that happens "in us" or "to us in our brains," but something we do, "a kind of skillful activity" (Noë, 2009, p. 60; see also Gallagher & Zahavi, 2008, pp. 99–100). To conceive of consciousness as a biological phenomenon does not imply that it is located in the brain in the form of NCCs. The brain is always in an organism and an organism is always involved in self-regulating relationships with the environment; consciousness is therefore coupled with the world through sensorimotor and intersubjective interactions (Thompson & Varela, 2001). Even a brain in a vat would need something like a living body to sustain its metabolic activity and flush away waste products, and consciousness is a live experience that spans the nervous system, the body, and the environment (Cosmelli & Thompson, 2008; Noë, 2009; Vaas, 1999).

The purported isolation of consciousness in the head goes hand in hand with the assumed isolation of the experimental subject in the lab; both consciousness and the subject in which it is studied are totally decontextualized. And yet, as Simon Cohn (2008a, p. 153) has shown in an ethnographic study of the neuroscientific mapping of pleasure, the social dimensions that researchers insist on excluding from their experiments and theories "are not only instrumental in the experiments, but also remain embedded in the final conceptualizations." The artificiality both of the situation and of the scientists' way of seeing it is epitomized in the assumption that subjects actually respond to the stimuli as they should for the purposes of the experiment. In fact, as a neuroscientist put it to Cohn (2004, p. 64), "you just can never be sure that they're not thinking about the weekend, or something."

There is a tension between the procedures for insuring that the imaging experiment is carried out in appropriately neutral conditions, and the persistence of subjectivity in the lab at several levels. A subjective alliance between the scientists and their volunteer subjects is required for establishing the "script" that makes the experiment work and allows for a co-production of scientific facts that can be considered objective (Marcel, 2003; Roepstorff & Frith, 2004). The rhetoric of direct mind reading through brain scans contrasts not only with the complexity of the image-production process, but also with the dependence of imaging studies on a "relational necessity," on "lines of intimacy" between subjects and experimenters that are crucial for experimental success (Cohn, 2008b, p. 100).

In short, for NCC critics, neuroscience should give up the framework of psychophysical correlation it embodies, and consider that neurobiological processes enable, but do not constitute, mental life (Noë & Thompson, 2004b, pp. 18–19). Nevertheless, it is not only such critics as Alva Noë (2009) who note that NCC research does not illuminate the mechanisms of qualia and the subjective quality of experience. Revonsuo also recognizes that "none of the current methods in cognitive neuroscience can be expected to reveal the phenomenal level in the brain" (2000, p. 72), and philosopher Valerie Hardcastle (2000, p. 264) remarks that NCC is a Hard Problem (in other words, a problem that persists after the relevant functional or causal mechanisms have been established) "with no solution in sight."

Such warnings have not had the sobering effect that might have been expected. Metzinger (2000, p. 7), for example, affirms that our "awareness of mortality will be greatly enhanced as we—especially people outside the academic world and in nondeveloped countries—learn more and more about the neural correlates of consciousness." The neuroscientifically-updated image of the human would put who choose to "live their lives outside the scientific image of the world" under intense social and emotional pressure. Thus will the unfortunate millions who do not inhabit a German university perhaps incorporate into their beliefs elements of a new "folk neurology" (Vrecko, 2006) or even of a "reductionistic neuroanthropology" (Metzinger, 2000, p. 6). But this would not result directly from scientific evidence. A brain-based image of human action can displace other current notions only when the environment in which we live and in which science is produced shifts so that such a picture makes cultural sense (Martin, 2000, p. 575).

Neuroimaging-based neural correlations have become a major instrument for the development of a brain-centered picture of the person. This applies not only to the field of conscious experience. As defined above, a neural correlate of consciousness is the set of neuronal events that suffice for a specific conscious percept, or the minimal neural system sufficient for the corresponding state of consciousness. In emotion, cognition, or mental pathology, a neural correlate consists of the neuronal events that are enough for a specific phenomenon in the relevant area. The correlates' cerebral location and psychological referents are different, but the concept and the method are the same in NCC research as in other neurocorrelational investigations. Thus, any neurocultural field may illustrate the promises and shortcomings of the neurocorrelational method. Nonetheless, not every psychiatric area offers an equally good instance for this purpose. Probably more than any other psychiatric condition among those being investigated by means of neuroimaging, depression is torn between biomedical and psychological accounts, between a search for neurological (including chemical and anatomical) causes, and the quest for contextualized explanations. Depression research thus constitutes a particularly valuable field for inquiring into the rationale and significance of the neurocorrelational approach.

Depression: Neuroimages and Neurocorrelates

Even a superficial look at the successive editions of the *Handbook of Depression* (Beckham & Leber, 1985, 1995) demonstrates that, like most other psychological and psychiatric entities, depression is not a single entity, and that, as for most other topics in psychology and psychiatry, there is no single approach that may be deemed simultaneously necessary and sufficient for understanding and treating the condition. Accordingly, in the *Handbook*'s most recent edition (Gotlib & Hammen, 2008), 62 authors of 29 chapters address four main areas. Part 1 deals with "descriptive aspects" such as the epidemiology, course, outcome, and assessment of depression, as well as issues in methodology, classification, and diagnosis (for example, the relations between personality and mood disorders, or the comparison of unipolar and bipolar depression). Part 2 moves from the genetics of major depression to the interpersonal and social environment of the condition, dealing along the way with the contributions of neurobiology and affective neuroscience, depression and early adverse experience, children of depressed parents, and the cognitive aspects of depression. Part 3 examines depression in specific populations (with a chapter on understanding the condition across cultures), and Part 4 considers prevention and treatment—not only pharmacological, but also cognitive, behavioral, and psychosocial.

The neurocorrelational and neuroimaging approach is obviously used in only a few of these areas. Nevertheless, as we shall see, neurocorrelationists implicitly claim for their method a foundational status, with neurocorrelational imaging research conducting with live humans the ethically impossible tasks of experimental anatomo-pathology, which should in the long run reveal the most essential causal mechanisms of mental disorder.

Given the vastness of the field of depression, the condition's high degree of comorbidity with other psychiatric conditions, the heterogeneity of the nosological category and the methods and interpretations at work, neuroimaging approaches to depression should be studied in connection with the wider economy of investigative, therapeutic, and economic practices and interests. We shall here proceed more narrowly, by focusing on some major reviews that throw light on the neurocorrelational mindset and expectations.

Our choice to explore depression was largely dictated by the contrast between the hopes stated by neurocorrelationists and the nature of the overall discourses about the condition. In the case of schizophrenia, for example, social and experiential indicators (social adversity, stressful life events, childhood abuse, or trauma) have been correlated with chances of developing the disorder; conversely, psychological and social interventions play a role in its management. Nevertheless, more than biopsychosocial models, which emphasize factor interdependence, it is the diathesis-stress model—according to which a stressor may trigger an initial illness episode in persons with a genetic predisposition ("diathesis")—which seems to have become the predominant framework for thinking about schizophrenia (see Jones & Fernyhough, 2007, for a discussion of the neural version of this model). Although there is considerable research into possible stressors, thinking in terms of vulnerability implies a focus on genetic and neurobiological factors. Moreover, treatments are predominantly pharmacological, with psychotherapy used as an adjunct. In any case, psychodynamic and behavioral explanations, such as the "schizophrenogenic mother" theory proposed in the late 1940s, are no longer considered valid.

The diathesis-stress model is also central in depression research. The etiology of depression, both unipolar ("major" depression) and bipolar (the former "manic" depression), is generally thought to include a significant genetic component in the determination of risk; moreover, the condition is correlated with changes in neuro-transmitter systems (whose exact role is not clear) involving serotonin, norepinephrine, and dopamine. Thus, like other mental illnesses, depression has become an object of biological psychiatry and neuroscientific inquiry. Nevertheless, while giving considerable weight to biological and vulnerability factors, depression studies tend to underline the interdependence of a multiplicity of risk and etiological mechanisms, from the genetic to the cultural. At the same time, they generally ignore such works as David Healy's *Let Them Eat Prozac* (2004) and his *Mania: A short history of bipolar disorder* (2008), Edward Shorter's *Before Prozac* (2008), and Emily Martin's *Bipolar expeditions* (2007), a study of mania and depression in American culture. Neither do they seem to acknowledge Allan Horwitz and Jerome Wakefield's *The Loss of Sadness* (2007), a critique of how "normal sorrow" is pathologized, which, like Martin's book, illuminates the contexts in which the diagnosis of depression arises, is applied, and is experienced (see also Bentall, 2009 and Kirsch, 2009, for recent critiques of pharmacological treatment).

The scientific literature explored here nonetheless suggests how much more difficult it is to detach depression from a complex of biopsychosocial factors than it is to isolate the purported biological causes or neural correlates of autism or schizophrenia. Indicative of such a phenomenon, and perhaps inherently connected to the experience of depression, is the fact that the condition is often accompanied by an exceptionally penetrating reflexivity; probably no other form of mental suffering has generated so many autobiographical accounts. In diverse, often contradictory ways, narratives by hitherto anonymous patients, as well as by movie stars, famous writers, diagnosed mental health professionals, and academics such as the author of *Bipolar Expeditions*, have contributed to the modern persona of the depressive and the public image of the condition. These personal narratives neither counterbalance nor contradict neuro-biological explanations (Dumit, 2003). Nevertheless, the evocation of contexts, moments, relationships, and inner life gives depression a meaning that constitutes a kind of causal interpretation. For their authors (admittedly a minority of the depressive population) such elucidations, which explore reasons rather than causes, make more existential sense than the demonstrations of biological psychiatry. Self-reflexive depressed persons may be fascinated by brain scans but, as autobiographical writings show, they wish primarily to understand contextual and relational factors that images and correlations have no chance of revealing or illuminating. While organic explanations of autism or schizophrenia may satisfy the persons concerned, they seem intrinsically insufficient to those directly or indirectly touched by depression. Given the limitations of a purely biomedical approach, what specifically would the brain imaging pursuit of depression's neural correlates contribute?

In 2005, an article in *The New York Times* noted that brain scans, celebrated as "snapshots of the living human brain," had been long expected to help cut through the mystery of mental illness, but that the promise had not been fulfilled (Carey, 2005). The neuroscientists' response, expressed in that article by Steven Hyman, a Harvard professor of neurobiology and former director of the National Institute of Mental Health (NIMH), is that those who oversold the technology forgot that "the brain is the most complex object in the history of human inquiry." He implied that the solution does not include a change in perspective but consists in further pursuing the same line of research.

In other respects, however, the *New York Times* article suggested that neuroimaging had not lived up to its promise because it is intrinsically inadequate to clarify some of the issues it is expected to address. Variability among brains is such that there might be no way of using morphological and functional findings to diagnose or classify; as far as etiology is concerned, the technology provides no means for answering the underlying question it has itself raised: "which comes first, the disease or the apparent difference in brain structure or function that is being observed?" Contrary to what is suggested by the existence, in the United States, of a prosperous neuroimaging market, especially for ADHD (attention deficit hyperactivity disorder), there is no evidence that brain scans add significantly to standard individual psychiatric examinations. Nonetheless some specialists are sure that the volumetric—and perhaps also functional information—provided by scans, will someday acquire diagnostic value.

Before showing that the situation has not fundamentally changed since the *New York Times* article appeared in October 2005, let us take a look at its scientific background. In 1998, Wayne C. Drevets, who has since become Senior Investigator at the Neuroimaging Section of the NIMH Mood and Anxiety Disorders Program in Washington, DC, published a review of functional neuroimaging studies aimed at elucidating the pathophysiology of major depression, and thereby establishing its "anatomical correlates" (Drevets, 1998, p. 341). He hoped that such neurocorrelational studies "may ultimately localize specific brain regions for histopathological assessment, elucidate anti-depressant treatment mechanisms, and guide pathophysiology-based classification of depression" (p. 342).

In 1998, as Drevets noted, the capabilities of neuroimaging to determine diagnosis or guide treatment had not yet been established. Functional imaging seemed nonetheless a promising research approach to depression because some depressive symptoms can be experimentally induced in non-depressed subjects. This opened the way for depressed-control comparisons of the changes in cerebral blood oxygenation and glucose metabolism "associated with" depression. But the exact nature of the association is nebulous. For example, functional brain imaging measures can be affected by non-depressive conditions sometimes present in depressed patients; regional blood oxygenation or metabolic differences between depressives and control subjects "may thus reflect either the physiological correlates" of depression "or pathophysiological changes that predispose subjects to or result from affective disease" (Drevets, 1998, p. 342). In short, as put in a 2008 review of biological vulnerability factors in early-onset depression, the quest for the "neurobiological roots" of the condition is obscured by the fact that, when assessing differences in brain function or activity between patients and controls, "it is unclear whether we are measuring causal factors making an etiological contribution to the illness, or, conversely, consequences or associated factors of the illness" (Nantel-Vivier & Pihl, 2008, p. 105).

The dominant language remains epistemically ambiguous (is "may" freely conjectural or more or less rigorously hypothetical?), reveals undecidability ("either-or") and avoids causal connectives ("predispose" and "result" are dissolved into a purely speculative remark), while trying to escape its own neurocorrelational framework through the use of metaphors ("reflect"). On its first page, the review we just quoted explains that the "[p]utative biological, psychological, and environmental etiological mechanisms" of pediatric depression are "intrinsically linked, interactive, and complementary;" starting with the second page, however, it becomes clear that the reviewed research concerns "biological correlates" which should lead to a better understanding of "etiological roots" (Nantel-Vivier & Pihl, 2008, pp. 103–104). The authors claim that, by studying pediatric populations, they "significantly decrease the likelihood of the occurrence of confounding factors and can therefore more clearly investigate causative neurobiological forces by getting closer to their etiological roots" (p. 105). One of the main goals of "[d]isentangling the neurobiological factors" is to develop a "biological etiology," and, on that basis, a taxonomy of illness that will yield "more homogenous diagnostic categories" (p. 106). But if some factors are "confounding" and others can be "disentangled" in the way proposed, then they are not "intrinsically" linked. As far as we can tell, such ambiguities are commonplace in neuroimaging depression research, and characterize the field of psychiatric neuroimaging as a whole (Boyce, 2009).

The same can be noticed about the prevalent attitude towards the variability of research results. The clinical heterogeneity of depression and the anatomical differences across individuals are major sources of a variability that, as Drevets explained, also implies that "diverse signs and symptoms may exhibit distinct neurophysiological correlates" (Drevets, 1998, p. 343). "Localization," he wrote, "is now limited as much by the anatomical variability across individuals as by the spatial resolution of imaging technologies" (p. 345). A related source of confusion comes from the fact that imaging results do not differ significantly between subjects with primary depressive syndromes, and those whose similar syndromes arise secondary to neurological conditions such as Parkinson's or Huntington's diseases (p. 353).

The two chief explanations for the inconsistency of the data (there are others, mainly methodological) are placed on the same level. Yet, while imaging resolution can improve, as it indeed has, variations in anatomy and brain circuitry are not limitations to be overcome. It is nevertheless hoped that they will cease being an obstacle when the nosography that still frames neuroimaging studies is replaced by a "pathophysiology-based classification." The stated hope is to refine "our understanding of the anatomical correlates" of depression (p. 358); the ultimate goal, however, is to integrate imaging, neurochemical, and anatomical data so as to move from physiological correlates to anatomo-pathological localizations. At the same time, the data Drevets reported seemed to support a "circuitry model in which mood disorders are associated with dysfunctional interactions between multiple structures, rather than increased or decreased activity within a single structure" (p. 355). The coexistence of a vocabulary of localization with an emphasis on brain circuitry characterizes neuroimaging and depression literature.

In 2002, a shorter overview of the same area noted the lack of a "general theory" capable of integrating the findings about functional abnormalities in the amygdala and hippocampus, and reached conclusions of confounding generality: since the medial prefrontal cortex is connected to areas where neuroimaging identifies structural and functional abnormalities in depression, "dysfunction in this region may be fundamental to depression ... These results thus support a neural model of depression in which dysfunction in regions that modulate emotional behavior may result in the emotional, motivational, cognitive and behavioral manifestations of depressive disorders" (Erk, Walter, & Spitzer, 2002, p. 67). The ambiguous, evocatively rather than assertively causal language is the same as in Drevets, but adds an element of self-evidence, since dysfunction in regions that modulate emotion necessarily affect emotion. Insofar as the nosography of depression includes emotional signs, depression necessarily involves brain areas implicated in emotion.

That same year, an extensive review was co-authored by Richard J. Davidson, the high profile director of the Laboratory for Affective Neuroscience at the University of Wisconsin-Madison. As a scientist with extensive media coverage and a well-publicized connection to the Dalai Lama; as "a veritable rock star in the world of neuroscience" (Smith, 2009) and one of the world's 100 most influential people according to *Time*'s 2006 ranking, Davidson is not your average laboratory scientist. But in other respects he is typical of the neurocultural universe.

One of Davidson's best-known messages is that meditation alters the brain. This is obvious, since any human activity whatsoever involves and affects the brain. It is of course interesting to know what exactly appears to be altered (for example, increases in left-sided anterior activation, a pattern associated with positive affect, as well as increases in antibody titers following influenza vaccination in meditators compared with a non-meditators control group; Davidson et al., 2003). Davidson, however, ultimately means to demonstrate that meditation can be put to useful social and psychological uses, such as reducing stress in all of us or making life easier in maximum-security prisons. This may be a valuable insight—but one whose demonstration does not require spending hundreds of thousands of dollars in brain scans (Davidson et al., 2003, used EEG, electroencephalography, but many other studies, and in growing numbers, use neuroimaging; see Davidson & Lutz, 2008, for a recent short discussion).

Davidson declares that the best way to study the mind is to study the brain (Redwood, 2007); and yet, neither the neurosciences in general, nor neuroimaging in particular, can tell us anything about the psychological or social effects of meditation. Thus, when asked about "the link between compassion for others and a sense of personal happiness," Davidson relied on psychological, not neuroscientific, data and cited a well-known experiment "in which participants were given $50 to spend. Half were instructed to spend it on themselves, half to spend it on others. Those who bought gifts for others reported feeling happier after the exercise" (Smith, 2009). Illustrating claims for neuroscience by discussing psychological rather than neuroscientific results is a widely shared strategy among neurocultural actors. Chris Frith's enthusiastic and graceful *Making up the Mind: How the Brain Creates our Mental World* (2007) provides an extensive example. My brain, Frith writes, "can act perfectly without me," it "builds models" and "discovers" things by itself. This, however, the author knows mostly from the psychological experiments that constitute the bulk of his book, not from the relatively small amount of neuroscientific research he also cites (see Tallis, 2007, for an insightful critique of the book).

Davidson's review of affective neuroscience perspectives on depression focuses on research about the representation and regulation of emotion (Davidson, Pizzagalli, Nitschke, & Putnam, 2002; almost identical to Davidson, Pizzagalli, & Nitschke, 2002). First, it corroborates the emphasis on "brain circuitry" (in this case, "underlying" mood, emotion, and affective disorders) that has become increasingly popular in fields sometimes depicted as turning the neuroimaging approach to mind into a new phrenology (Uttal, 2001, 2008). Second, the primacy of circuitry and connectivity coexists with a focus on brain structures (prefrontal cortex, anterior cingulate cortex, hippocampus, and amygdala), which are treated separately in different sections of the review; such structure is typical of the publications under consideration, and reflects the organization of research. Third, Davidson both presupposes and intends to show that the study of brain circuitry will open the way for parsing the "heterogeneity" of affective disorders, and generate a new approach to subtyping "that does not rely on the descriptive nosography of psychiatric diagnosis but rather is based on a more objective characterization of the specific affective deficits in patients with mood disorders" (Davidson et al., 2002, p. 546).

Like cognitive neuroscience, affective neuroscience aims at decomposing complex processes into "elementary constituents that can be studied in neural terms" and "examined with objective laboratory measures" instead of self-reports (p. 546). One of the "crucial issues" which are thought to plague the field and which affective neuroscience wishes to resolve by neurologizing clinical concepts, is the heterogeneity of mood disorders. Symptoms are broadly similar, but the proximal causes can be extremely varied, and even "the underlying mechanisms may differ." (p. 546) Indeed, symptoms come in clusters whose specific features "are likely mediated by different neural circuits despite the fact that they culminate in a set of symptoms that are partially shared" (p. 547). Since descriptive phenomenology does not yield a "clean separation of underlying neural circuitry," the goal of affective neuroscience is to move beyond it, "toward a more objective, laboratory-based parsing of affective processing abnormalities" (p. 547).

The claim to "objectivity," here identified with what happens in a laboratory, bolsters the ultimate goal of affective neuroscience: to reevaluate the relationships between etiology and nosography (of depression in the present case) by defining symptom clusters "that may arise as a consequence of dysfunctions in specific regions," and thus "to offer suggestions for different ways of parsing the heterogeneity of depression in ways that more directly honor the circuitry of emotion and emotion regulation in the brain" (p. 547). In short, the refashioning of nosography on neuroscientific-causal foundations and through the identification of biomarkers is a major ambition of the field. "The specific [depression] subtype, symptom profile, and affective abnormalities should vary systematically with the location and nature of the abnormality" (p. 565). The "delineation of brain-based illness models ... is seen as a promising strategy for redefining our depression nosology" (Mayberg, 2007, p. 729), and "[n]eural markers of at-risk individuals may prove to be more sensitive predictors of subsequent depression and sensitivity to treatment than the clinical predictors we have at present" (Keedwell, 2009, p. 97).

From a developmental viewpoint, "[i]dentifying depression subtypes based on age of onset and neurobiological characteristics may provide us with more etiologically consistent and uniform diagnostic categories" (Nantel-Vivier & Pihl, 2008, p. 111). The language, floating between the normative and the expectant ("should"), the permissible and the hoped-for ("may"), contrasts with the methodological and empirical technicalities of the research, implicitly favors biological causality over integrative models, and expresses the limits of what is—and probably can—be known.

The empirical evidence demonstrates that the coveted "clean separation" of brain circuits is either far from being achieved (and, therefore, that more research and increasingly powerful technologies are needed) or constitutes a misguided goal (in which case the entire project ought to be reconsidered). Davidson and his colleagues' expression, "may arise as a consequence of" is as far as they advance toward understanding the causal mechanisms of depression. In connection with the prefrontal cortex, for instance, they observe that some types of depression "may be caused" by certain abnormalities in the circuitry which implement positive affect-guided anticipation; similarly, anatomical differences in the brain of patients with mood disorders "might account" for some of the detected functional differences (Davidson et al., 2002, p. 548, 550). The existence of hippocampal-dependent Pavlovian conditioning (in the form of an association between fear responses and places) "has important implications for our understanding of the abnormalities that may arise as a consequence of hippocampal dysfunction" (p. 556). Indeed, patients with mood and anxiety disorders are known to display "normative affective responses" in inappropriate contexts; these patients "may be characterized by hippocampal dysfunction," as suggested by MRI (magnetic resonance imaging) studies showing smaller hippocampal volumes in patients with major depression, bipolar disorders, PTSD (post-traumatic stress disorder) and borderline personality disorder.

Davidson, however, notes that "[w]hether hippocampal dysfunction precedes or follows onset of depressive symptomatology is still unknown" (p. 557). "We do not know," he adds, if any of the discussed functional and structural abnormalities "precede the onset of the disorder, co-occur with the onset of the disorder, or follow the expression of the disorder" (p. 565). Such remarks are characteristic of neurocorrelational research, which is by definition unable to fulfill its own goal of differentiating between causes and consequences. Neither the updated version of the same review (Davidson, Pizzagalli, & Nitschke, 2009) nor any of the brain-related articles in the new *International Encyclopedia of Depression* (Ingram, 2009) offer a different view or evidence of progress toward a knowledge of causes and causal mechanisms, even in a probabilistic framework.

While the scientific literature invariably underlines the advancements made in knowledge of the brain structures said to "subserve" or be involved in depression, it also acknowledges the lack of advancement towards causality and localization. In 2008, for example, an article in *Current Directions in Psychological Science* reviewed the status and unresolved issues in neuroimaging and depression. It summarized neurocorrelational research, assessing the role of several brain structures in major depression and concluded that heightened activity in the limbic structures engaged in emotional experience and expression dampens activation in the dorsal cortical structures involved in affect regulation. Following the usual form, the article devoted different sections to distinct structures or systems (the amygdala, the subgenual anterior cingulate cortex, and the dorsolateral prefrontal cortex), and pointed out that identifying "the patterns of functional connectivity that characterize the depressive neural network" is still a challenge for future work (Gotlib & Hamilton, 2008, p. 161).

As the authors made clear, the fact that "neural abnormalities" play a role in depression was known before the advent of neuroimaging. But they also recognized that determining the timing of those abnormalities, as can be done by means of activation patterns, (for instance, greater-than-normal amygdala reactivity to affective stimuli during a depressive episode) has so far not illuminated their role in the disorder. The results concerning the temporal relation between neural activation and depression, as well as the etiological role of neural dysfunction "are complex and do not cohere to tell as clear a story as we would like" (p. 162). Indeed, anomalies can be present in a diagnosed person's brain or precede the onset of the disease "without being involved in its development" (p. 162).

Moreover, as we saw with Davidson, the findings discussed in *Current Directions in Psychological Science* "underscore the fact that 'depression' refers to a heterogeneous group of disorders that are not carved at their neurobiological joints in DSM-IV;" hence the desire to define depression subtypes and symptom profiles "that are related systematically to neural functional and structural abnormalities" (p. 162). In other words, one should go beyond correlations, establish causal connections, and amend the nosology of depression on the basis of what appear to be the disorder's neural substrates. The goal of deconstructing present diagnostic entities in that way is widely shared among researchers in psychiatric neuroimaging (Abou-Saleh, 2006; Hyman, 2007). It is in this respect revealing that the metaphor of "parsing" is applied to the "heterogeneity" of depression. It implies that depression should not be heterogeneous—or not in the present manner—but, rather, that it should be reconceptualized so as to facilitate its breakdown into brain-based nosographic types and components (for example, patterns of brain activation that correspond to individual differences in severity, accompanying symptoms, or treatment response).

The main conceptual action always consists in *correlating* (though the results in this respect are not consistent), but the ultimate aim is to *relate causally*. Hence the problem of what to do with the observation (one among hundreds of similar ones) that positive correlations between increased functional connectivity in the amygdala network and Geriatic Depression Scale scores in elderly patients with amnestic mild cognitive impairment "suggest" that connectivity in those areas "is related to the degree of depression." It seems impossible to go beyond hazy general conclusions—in this case, that there is an "interactive neural mechanism" between the dysfunction of emotional processing (supported by the amygdala) and cognitive and memory functions (Xie et al., 2008). Although the predominant "functional connectivity" strategy aims at extracting patterns of covariance, it is assumed that the "activity changes in different locations influence one another" (Mayberg, 2007, p. 729).

The same language characterizes the "dernier cri" in psychiatric imaging research, namely diffusion tensor imaging (DTI) studies of white matter hyperintensities. White matter hyperintensities appear on magnetic resonance images as ultra-white patches that indicate injury to axons. DTI is a magnetic resonance imaging method that produces neural tract images on the basis of the diffusion of water in tissue (such as the axons in white matter). The variation of diffusion along different spatial directions provides information about diffusion anisotropy (the direction preference of the diffusion process); the results are couched in terms of "fractional anisotropy" (FA), that is to say in degrees of anisotropy (from 0 for isotropic, that is, homogenous in all directions, to 1 for fully anisotropic). The technique is used to investigate tissue structure and connectivity between regions or points in the brain. While DTI is different from fMRI and other imaging technologies, its basic goal—to correlate pathologies with cerebral locations and circuits—illustrates just as well the assumptions, promises, and limitations of the neurocorrelational/neuroimaging logic.

In the field of depression, white matter hyperintensities have been found consistently in elderly unipolar patients. A recent DTI study found that, in comparison with controls, patients with major depressive disorder (MDD) tend to show lower FA values in the left sagittal stratum; the implied structural changes "may contribute" to the previously detected dysfunction in the limbic-cortical network in depressive patients (Kieseppä et al., 2009, p. 5). An equally recent meta-analysis of MRI studies of brain volume in MDD observes that some of the areas "involved in" emotion regulation and stress responsiveness exhibit volume reduction. The authors conclude that the integration of MRI and DTI measurements "may improve our understanding of the neural circuitry involved in MDD," and that their own meta-analytic results "strongly suggest that studying brain structure in MDD will contribute to understanding the pathogenesis of this disease" (Koolschijn, van Haren, Lensvert-Mulders, Hulshoff Pol, & Kahn, 2009, p. 11, 13). They do not explain, however, how pathogenesis can be inferred or demonstrated without some sense of causality or at least temporal direction.

A recent meta-analysis of structural imaging studies remarks that after 25 years of scanning bipolar disorder patients, and generating over 7,000 MRIs, brain regions "affected in" the disorder remain ill defined. Given the number of studies considered, significant findings are surprisingly few—in fact three—and all "regionally nonspecific." First, bipolar disorder is "associated" with lateral ventricle enlargement and (second) with increased deep white matter hyperintensities; third, lithium use is "associated" with increased total gray matter volume. "There may be genuinely limited structural change in bipolar disorder, or between-study heterogeneity may have obscured other differences" (Kempton, Geddes, Ettinger, Williams, & Grasby, 2008, p. 1026). The high inter- and intra-study heterogeneity, and the fact that individual investigations prove to be chronically underpowered are crucial for understanding such limited results. Nevertheless, however much they may result from deficient sampling, type I and type II errors, insufficient control of intervening variables (medication for example), or discrepant nosologies, it is likely that the results also express a variability that is a characteristic feature of the objects and phenomena studied, rather than a methodological artifact.

Understandably, authors shy away from speaking in terms of causes. Abnormalities "play a role," are "involved in," or "may contribute" to mental disorder; functional and anatomical differences, or the activation of brain structures do not signal the cause of depressive symptoms, but only have "temporal relations" with their expression, or are "significantly positively associated" with them. There is much to commend in a cautious attitude towards causal connections. Yet the intentionally imprecise language not only reveals ambivalence regarding causality, but also betrays a historical situation. Although the existence of a link between neurotransmitters and mood disorders has been known since the 1950s, when drugs that alter neurotransmitters (specifically chlorpromazine) were found to relieve those disorders, it is still unknown if changes in neurotransmitter levels cause depression or the other way around; the same is acknowledged in connection with volumetric, anatomical, and neuroimaging data. Given how little our authors say, and seem to know, about the meaning of the correlations they establish, as well as their recurrent use of an ambiguous "may" to evoke causality without asserting it, their caution sustains the neurocorrelational enterprise, while at the same time it inhibits reflection on the lack of expected progress after half a century of research.

Concluding Remarks

We began by discussing neurocultural disciplines that have been developing since the 1990s and brought to light several common features. From neurotheology to neuromarketing, they deal with topics defined by the human and social sciences, rely on neuroimaging as a research tool, look for the neural correlates of the phenomena they study—and do not elaborate the notion of neural correlate. The last point is particularly significant, since, starting in the 1980s, NCC research gave rise to considerable debate, from the technical and methodological to the epistemological and metaphysical. On the one hand, professional philosophers and a few philosophically-driven neuroscientists engage in highly technical discussions of the notion, both theoretical and empirical. On the other hand, the majority of researchers in the new neuro fields look for neural correlates in apparent isolation from the debates that take place in consciousness studies. We have attempted here to bring these two lines of work closer together, by outlining first the NCC debates and then by illustrating the neurocorrelational turn of mind in depression research. While psychiatry offers many possibilities, we chose a condition that has been the subject of extensive neuro-correlational investigation, yet involves an existential dimension hardly amenable to neurocorrelational analysis.

The case of depression and its nosography highlights the hopes conveyed by the use of neural correlates: to discipline multifaceted phenomenological realities that are difficult to sort out and classify, as well as to create disciplines constitutively defined by the use of brain imaging techniques and a neurocorrelational framework. The NCC discussion, as conducted by both critics and practitioners, demonstrates that there is no empirical support for the metaphysical assumption that conscious experience is exclusively neural. Consciousness, like all the other phenomena studied by the neuro-cultural disciplines, requires a functioning brain and involves its anatomy, physiology, and chemistry. Like most of those other phenomena, however, it does not happen exclusively inside the head, but entails two-way connections between neuronal and extra-neuronal systems (including the body as well as socio-cultural and historical environments) and emerges from their interaction.

Depression research is emblematic of the limitations of the neurocorrelational perspective. Neuroscientists themselves describe these limitations in detail, but that does not make them (at least not in print) doubt their approach. Quite the contrary, they insist on it, making the most of the slippery nature of neurocorrelational discourses, of the conceptual and rhetorical porosity between correlations and causes. Neuroscientists know that correlations are no more than correlations. For many of them, however, correlations prove the sufficiency (rather than merely the necessity) of the neural correlate in the production of the phenomena under study, and thus open the way for the establishment of material causes localized inside the skull. Yet the neurocorrelationists' mantra that their research demonstrates that human activity changes the brain is trivial. For example, it has been known for centuries, if not millennia, that regulating emotions has positive cognitive and social effects. In various forms that knowledge is, at least in the Western world, one of the foundations of moral and educational systems; the tests to which it has been subjected are pragmatic, and the criteria whereby it has been judged depend on historically contingent values. Thus, contrary to what neuroscientists may suggest, showing that voluntary emotional regulation modulates magnetic resonance signal change in the amygdala, or that there are connections between this structure and the prefrontal cortex, or that "social-emotional learning changes the brain," does not make ethical or pedagogical principles more or less effective, real, or legitimate.

The goal of education never has been and hopefully never will be to "change the brain by training the mind," but to educate a person.[4] The point is obviously not to deny that the brain is involved in all human behavior, but to suggest that neurocorrelational research, as it is being marketed by its producers (with ample pharmaindustry support in the case of mental disorder) and bought by funders, policy makers, and large swathes of public opinion, is diverting enormous financial and human resources from action and thought in social and psychological arenas where they would have a greater chance of making a difference.

References

Abbott, A. (2009). Brain imaging under fire. *Nature, 457,* 245.

Abou-Saleh, M. (2006). Neuroimaging in psychiatry: An update. *Journal of Psychosomatic Research, 61,* 289–293.

Beckham, E., & Leber, W. (Eds.). (1995). *Handbook of depression: Treatment, assessment, and research* (2nd ed.). New York: Guilford Press.

Beckham, E., & Leber, W. (Eds.). (1985). *Handbook of depression: Treatment, assessment, and research*. Homewood, Ill.: Dorsey Press.

Bentall, R. (2009). *Doctoring the mind: Why psychiatric treatments fail.* New York: New York University Press.

Bouveresse, J. (1976). *Le Mythe de l'intériorité. Expérience, signification et langage privé chez Wittgenstein.* Paris: Minuit.

---

[4] Quotations in this and the previous sentence come from a caricatural instance of neurocorrelational claims in education, influential neuroscientist Richard Davidson's lecture "The Heart-Brain Connection: The Neuroscience of Social, Emotional, and Academic Learning" (New York, December 10, 2007), http://www.edutopia.org/richard-davidson-sel-brain-video.

Boyce, A. (2009). Neuroimaging in psychiatry: Evaluating the ethical consequences for patient care. *Bioethics, 23,* 349–359.

Carey, B. (2005, October 18). Can brain scans see depression? *The New York Times.* Retrieved from http://www.nytimes.com.

Chalmers, D. (2000). What is a neural correlate of consciousness? In Metzinger, T. (Ed.), *Neural correlates of consciousness: Empirical and conceptual issues* (pp. 17–39). Cambridge, MA: MIT Press.

Cohn, S. (2008a). Petty cash and the neuroscientific mapping of pleasure. *BioSocieties, 3,* 151–163.

Cohn, S. (2008b). Making objective facts from intimate relations: The case of neuroscience and its entanglements with volunteers. *History of the Human Sciences, 21,* 86–103.

Cohn, S. (2004). Increasing resolution, intensifying ambiguity: An ethnographic account of seeing life in brain scans. *Economy and Society, 33,* 52–76.

Cosmelli, D., & Thompson, E. (2008). Embodiment or Envatment? Reflections on the Bodily Basis of Consciousness. Retrieved from http://www.individual.utoronto.ca/evant/EnactionChapter.pdf.

Crick, F., & Koch, C. (2003). A framework for consciousness. *Nature Neurosciences, 6,* 119–126.

Crick, F., & Koch, C. (1998). Consciousness and neuroscience. *Cerebral Cortex, 8,* 97–107.

Crick, F., & Koch, C. (1995). Are we aware of neural activity in primary visual cortex? *Nature, 375,* 121–123.

Crick, F., & Koch, C. (1990). Towards a neurobiological theory of consciousness. *Seminars in the Neurosciences, 2,* 263–275.

Davidson, R., Kabat-Zinn, J., Schumacher, J., Rosenkranz, M., Muller, D., Santorelli, S., Urbanowski, F., Harrington, A., Bonus, K., & Sheridan, J. (2003). Alterations in brain and immune function produced by mindfulness meditation. *Psychosomatic Medicine, 65,* 564–570.

Davidson, R., & Lutz, A. (2008). Buddha's brain: Neuroplasticity and meditation. *IEEE Signal Processing Magazine, 25*(1), 171–174.

Davidson, R., Pizzagalli, D., & Nitschke, J. (2002). The representation and regulation of emotion in depression: Perspectives from affective neuroscience. In I. Gotlib & C. Hammen (Eds.), *Handbook of depression* (pp. 219–244). New York: Guilford Press.

Davidson, R., Pizzagalli, D., & Nitschke, J. (2009). Representation and regulation of emotion in depression: Perspectives from affective neuroscience. In I.Gotlib & C. Hammen (Eds.), *Handbook of depression* (pp. 218–248). New York: Guilford Press.

Davidson, R., Pizzagalli, D., Nitschke, J., & Putnam, K. (2002). Depression: Perspectives from affective neuroscience. *Annual Review of Psychology, 53,* 545–574.

Drevets, W. (1998). Functional neuroimaging studies of depression: The anatomy of melancholia. *Annual Review of Medicine, 49,* 341–361.

Dumit, J. (2003). Is it me or my brain? Depression and neuroscientific facts. *Journal of Medical Humanities, 24,* 35–47.

Erk, S., Walter, H., & Spitzer, M. (2002). Functional neuroimaging of depression. *Advances in Biological Psychiatry, 21,* 63–69.

ffytche, D. (2000). Imaging conscious vision. In Metzinger, T. (Ed.). *Neural correlates of consciousness: Empirical and conceptual issues* (pp. 221–230). Cambridge, MA: MIT Press.

Frith, C. (2007). *Making up the mind: How the brain creates our mental world.* Hoboken, NJ: John Wiley & Sons.

Gallagher, S., & Zahavi, D. (2008). *The phenomenological mind: An introduction to philosophy of mind and cognitive science.* London: Routledge.

Gotlib, I., & Hamilton, J. (2008). Neuroimaging and depression: Current status and unresolved issues. *Current Directions in Psychological Science, 17*, 159–163.

Gotlib, I., & Hammen, C. (Eds.). (2008). *Handbook of depression* (2nd ed.). New York: Guilford Press.

Hardcastle, V. (2000). How to understand the N in NCC. In Metzinger, T. (Ed.), *Neural correlates of consciousness: Empirical and conceptual issues* (pp. 259–264). Cambridge, MA: MIT Press.

Healy, D. (2008). *Mania: A short history of bipolar disorder*. Baltimore: Johns Hopkins University Press.

Healy, D. (2004). *Let them eat Prozac: The unhealthy relationship between the pharmaceutical industry and depression*. New York: New York University Press.

Hohwy, J. (2007). The search for neural correlates of consciousness. *Philosophy Compass, 2/3*, 461–474.

Horwitz, A., & Wakefield, J. (2007). *The loss of sadness:. How psychiatry transformed normal sorrow into depressive disorder*. New York: Oxford University Press.

Hyman, S. E. (2007). Can neuroscience be integrated into the DSM-V? *Nature Reviews Neuroscience, 8*, 725–732.

Illes J. (Ed.). (2006). *Neuroethics: Defining the issues in theory, practice and policy*. New York: Oxford University Press.

Illes, J., Racine, E., & Kirschen, M. (2006). A picture is worth a thousand words, but which one thousand? In J. Illes (Ed.), *Neuroethics: Defining the issues in theory, practice and policy* (pp. 49–168). New York: Oxford University Press.

Illes, J., Kirschen, M., & Gabrieli, J. (2003). From neuroimaging to neuroethics. *Nature Neuroscience, 6*(3), 205.

Ingram, R. (Ed.). (2009). *International encyclopedia of depression*. New York, Springer.

Jones, S., & Fernyhough, C. (2007). A new look at the neural diathesis-stress model of schizophrenia: The primacy of social-evaluative and uncontrollable situations. *Schizophrenia Bulletin, 33*(5), 1171–1177.

Keedwell, P. (2009). Brain circuitry. In R.Ingram (Ed.), *International encyclopedia of depression*. New York, Springer.

Kempton, M., Geddes, J., Ettinger, U., Williams, C., & Grasby, P. (2008). Meta-analysis, database, and meta-regression of 98 structural imaging studies in bipolar disorder. *Archives of General Psychiatry, 65*(9), 1017–1032.

Kieseppä, T., Eerola, M., Mäntylä, R., Neuvonen, T., Poutanen, K., Luoma, K., Tuulio-Henriksson, A., Jylhä, P., Mantere, O., Melartin, T., Rytsälä, H., Vuorilehto, M., & Isometsä, E. (2009). Major depressive disorder and white matter abnormalities: A diffusion tensor imaging study with tract-based spatial statistics. *Journal of Affective Disorders, 120*(1), 240–244. doi:10.1016/j.jad.2009.04.023.

Kirsch, I. (2009). *The emperor's new drugs: Exploding the antidepressant myth*. London: Bodley Head.

Koch, C. (2004). *The quest for consciousness: A neurobiological approach*. Englewood, CO: Roberts and Company.

Koolschijn, C., van Haren, N., Lensvert-Mulders, G., Hulshoff Pol, H., & Kahn, R. (2009). Brain volume abnormalities in major depressive disorder: A meta-analysis of magnetic resonance imaging studies. *Human Brain Mapping, 30*(11), 3719–3735. DOI: 10.1002/hbm.20801.

Logothetis, N. (1999). Vision: A window on consciousness. *Scientific American, 281*, 68–75.

Logothetis, N., & Schall, J. (1989). Neuronal correlates of subjective visual perception. *Science, 245*, 761–763.

Maasen, S. (2007). Selves in turmoil: Neurocognitive and societal challenges of the self. *Journal of Consciousness Studies, 14*(1–2), 252–270.

Maasen, S. (2003). The conundrum of consciousness: Changing landscapes of knowledge at the turn of the millennium. In B.Joerges & H.Nowotny (Eds.), *Social studies of science and technology looking back, ahead* (pp. 117–143). Dordrecht: Kluwer.

Marcel, A. (2003). Introspective report: Trust, self-knowledged and science. In A. Jack & A Roepstorff (Eds.), *Trusting the subject? The use of introspective evidence in cognitive science*, Vol. 1. Exeter, Devon: Imprint Academic.

Martin, E. (2007). *Bipolar expeditions: Mania and depression in American culture.* Princeton: Princeton University Press.

Martin, E. (2000). Mind-body problems. *American Ethnologist, 27*, 569–590.

Mayberg, H. (2007). Defining the neural circuitry of depression: Towards a new nosology with therapeutic implications. *Biological Psychiatry, 61*, 729–730.

McGhie, A., & Chapman, J. (1961). Disorders of attention and perception in early schizophrenia. *British Journal of Medical Psychology, 34*, 103–16.

Metzinger, T. (Ed.). (2000). *Neural correlates of consciousness: Empirical and conceptual issues.* Cambridge, MA: MIT Press.

Metzinger, T. (2000). Introduction: Consciousness research at the end of the twentieth century. In Metzinger, T. (Ed.), *Neural correlates of consciousness: Empirical and conceptual issues* (pp. 1–12). Cambridge, MA: MIT Press.

Mora, F. (2007). *Neurocultura. Una cultura basada en el cerebro.* Madrid: Alianza.

Nantel-Vivier, A., & Pihl, R. (2008). Biological vulnerability of depression. In J. Abela & B. Hankin (Eds.), *Handbook of depression in children and adolescents* (pp. 103–123). New York: The Guilford Press.

Noë, A. (2009). *Out of our heads: Why you are not your brain, and other lessons from the biology of consciousness.* New York: Hill and Wang.

Noë, A., & Thompson, E. (2004a). Sorting out the neural basis of consciousness. Authors' reply to commentators. *Journal of Consciousness Studies, 11*(1), 87–98.

Noë, A., & Thompson, E. (2004b). Are there neural correlates of consciousness? *Journal of Consciousness Studies, 11*(1), 3–28.

Ortega, F., & Vidal, F. (Eds.). (2011). *Neurocultures. Glimpses into an expanding universe.* New York: Peter Lang.

Racine, E., Bar-Ilan, O., & Illes, J. (2005). fMRI in the public eye. *Nature Reviews Neuroscience, 6*(2), 159–164.

Redwood, D. (2007). Meditation, positive emotions and brain science: Interview with Richard Davidson Ph.D. Retrieved from http://www.healthy.net/scr/Interview.asp?Id=306.

Revonsuo, A. (2000). Prospects for a scientific research program on consciousness. In Metzinger, T. (Ed.), *Neural correlates of consciousness: Empirical and conceptual issues* (pp. 57–75). Cambridge, MA: MIT Press.

Roepstorff, A., & Frith, C. (2004). What's at the top in the top-down control of action? Script-sharing and "top-top" control of action in cognitive experiments. *Psychological Research-Psychologische Forschung, 68*(2–3), 189–198.

Shorter, E. (2008). *Before Prozac: The troubled history of mood disorders in psychiatry.* New York: Oxford University Press.

Smith, J. (2009, July 27). Building a better brain. *Isthmus.* Retrieved from http://www.isthmus.com/isthmus/article.php?article=25405.

Tallis, R. (2007). Not all in the brain. *Brain, 130*(11), 3050–3054.

Thompson, E., & Varela, F. (2001). Radical embodiment: Neural dynamics and consciousness. *Trends in Cognitive Sciences, 5*, 418–425.

Tononi, G,. & Koch, C. (2008). The neural correlates of consciousness: An update. *Annals of the New York Academy of Sciences, 1124,* 239–261.

Uttal, W. (2008). *Distributed neural systems: Beyond the new phrenology.* Cornwall-on-Hudson, NY: Sloan Educational Publishing.

Uttal, W. (2001). *The new phrenology: The limits of localizing cognitive processes in the brain.* Cambridge, MA: MIT Press.

Vaas, R. (1999). Why neural correlates of consciousness are fine, but not enough. *Anthropology and Philosophy, 3*(2), 121–141.

Vrecko, S. (2006). Folk neurology and the remaking of identity. *Molecular Interventions, 6,* 300–303.

Vul, E., Harris, C., Winkielman, P., & Pashler, H. (2009). Puzzlingly high correlations in fMRI studies of emotion, personality, and social cognition. *Perspectives on Psychological Science, 4,* 274–290.

Xie, C., Wu, Z., Li, W., Jones, J., Antuono, P., & Li, S et al. (2008). Neural correlates of depression in subjects with amnestic mild cognitive impairment. *Alzheimer's and Dementia, 4*(4), Supplement 1, T259–T260.

Zeki, S. (2002). Neural concept formation & art: Dante, Michelangelo, Wagner. *Journal of Consciousness Studies, 9*(3), 53–76.

Zeman, A. (2001). Consciousness. *Brain, 124,* 1263–1289.

18

The Future of Critical Neuroscience

Laurence J. Kirmayer

Progress in neuroscience is providing us with new technologies and novel metaphors for mind. With this language and technology come new ways of being and construing our selves. Contemplating images of our brains evokes both a sense of awe and a measure of self-estrangement because we believe it is the brain that enables and somehow contains our individuality, subjectivity, and agency—the essence of who we are as persons. The prospect of radical transformations of the brain through pharmaceutical, surgical, or other forms of intervention raises anxieties about potential ruptures in the continuity of self or violations of what it is to be a human being. By changing the substrate of the most basic processes of thinking, feeling, and identity, we may change the very basis on which we make choices, exercise will, and judge the outcomes of our actions. Neurotechnology may transform us in ways that undermine the ability to remember what things were like before we changed our minds—so that we lose the platform for any awareness and analysis of the meaning of these changes. Moreover, the sense of exigency surrounding these neurotechnologies can overshadow their actual capacities to bring about change so that we lose the ability to make realistic judgments of just what is (or isn't) at stake.

Critical perspectives on neuroscience are needed to explore the social, moral, and political implications of these new technologies for imaging and manipulating the brain but equally to examine the ways in which the promise and practice of neuroscience is changing our everyday sense of self and personhood well in advance of any actual applications of technology. A wide array of possible transformations of the self have been explored over the last 50 years of speculative fiction—from portraits of "wire-heads" plugged into an electric wall socket to relentlessly stimulate pleasure centers of the brain until they die from lack of food or water through self-neglect, to the many macabre versions of disembodied brains floating in vats dreaming of omnipotence, to cyborg brains placed in mechanical bodies, to brains augmented by computer chips that link the person to an invisible electronic web, or minds decanted from their brains

Critical Neuroscience: A Handbook of the Social and Cultural Contexts of Neuroscience, First Edition.
Edited by Suparna Choudhury and Jan Slaby.
© 2012 John Wiley & Sons, Ltd. Published 2016 by John Wiley & Sons, Ltd.

to dwell entirely in cyberspace.[1] These thought experiments have mapped some possible "posthuman" futures and in so doing invite us to imaginatively inhabit new worlds with radically different versions of personhood. In some ways, though, reality has caught up with speculation—for example, with widespread use of psychopharmaceuticals not only for treating mental disorders but for the enhancement of performance and well-being (Elliott & Chambers, 2004; Gold & Olin, 2009; Kirmayer & Raikhel, 2009; Rose, 2005; Stein, 2008), or through the pervasive but still largely uncharted effects of electronic media and networking on neurodevelopment and sociality (Anderson, 2007). Visions of the future are invoked to support the momentum of neuroscience with claims about its potentially transformative effects on our lives.

Critical neuroscience differs from the imaginative excursions of speculative fiction by insisting on systematic observation and analysis of the development, cultural meanings, and social ecology of the objects of neuroscience. Philosophically, this requires unpacking the theoretical and methodological assumptions of neuroscience research, particularly its representations of brain, mind, and person. Ethnographically, this involves close examination of the ways that neuroscientific knowledge and technologies are used in specific social and cultural contexts. Sociologically, this involves tracing the networks, interests, power relations, agents, and actors that present and promote these new forms of knowledge and configurations of personhood— which are always selected from a range of options—in ways that may ignore or conceal their wider consequences. So, true to the legacy of critical theory, we can engage in a critique of neurotalk as part of popular culture as well as neuroscientific and technical rationality and their economic and political motivations (Honneth, 2009).

Critical neuroscience, though, is not bound by the methods or agenda of critical theory. It borrows from the range of philosophical and social science traditions to interrogate the practice of neuroscience—not to distinguish "good" neuroscience from "problematic" neuroscience, but rather to understand how (from laboratory practices through to the circulation and application of knowledge) "facts" about the brain have come to be so salient in clinical, educational, and commercial settings and in the popular imagination. Crucially, it aims to go beyond critique to contribute directly to the scientific enterprise itself. Increasingly, there is convergence in the interests and activities of philosophers and cognitive neuroscientists, with philosophers developing theoretical models and suggesting experiments and neuroscientists examining the broader implications of their work. Social science perspectives offer added value to philosophical and psychological analyses, by grounding critique in the close observation and analysis of social interactions and institutions, and, at the same time, offering rich descriptive data about the social phenomena neuroscience is now setting out to study. Thus, not only can anthropology, history of science, and sociol-

---

[1] "Wire-heads" is from Niven (1969); the brain in the vat is a staple of horror movies and its isolation and vulnerability is usually compensated by a wild megalomania; augmenting the brain by computer chips and dwelling half-in cyberspace are the central devices of Gibson (1984) who spawned the genre of cyberpunk; accounts of entirely replacing embodied existence with virtual identities are found in countless recent novels (e.g. Wright, 2003), which almost always find ways to re-install the person in a feeling or sensing body, at least for a time, perhaps because this is essential to insure the reader's continued empathy with the characters. Each of these fictions has a corresponding philosophical literature exploring in more systematic ways the implications of these thought experiments for mind and identity.

ogy contribute to modes of critique; they can also contribute methodological tools and conceptual clarity towards multidimensional models of behavioral phenomena. The emergence of social and cultural neuroscience as active fields of inquiry makes this critical analysis and interdisciplinary engagement especially important (Choudhury & Kirmayer, 2009; Seligman & Brown, 2010). Though the aspiration to contribute directly to scientific work might lead to a loss of critical distance, it also serves to ground critical neuroscience in the dilemmas of research and clinical practice and move it toward a creative engagement that can contribute to solutions of the problems—both scientific and social—it uncovers.

Varieties of Critical Neuroscience

Although critical theory has a specific historical provenance and aim, emphasizing the political economic origins of our social arrangements, the use of the term by Choudhury and her colleagues (Choudhury, Nagel, & Slaby, 2009) has a much wider ambit. Their notion of critical neuroscience involves not only analysis of scientific practice but also systematic questioning of taken-for-granted, tacit, implicit, routine, conventional, background knowledge. This social background may be hard to see and to question precisely because of its everydayness and the ways in which it provides the commonsense or authoritative frameworks against which we recognize anything that is notable or out of the ordinary. Difficulty in identifying this tacit framework, whether in science or everyday discourse, may also occur because there are agents and interests that actively work to distract, deny or suppress efforts to become conscious of the background because to do so might lead to challenges to current social, economic, or political arrangements. A critical perspective then, aims not only to analyze and lay bare the undergirding of our conceptual frameworks but may also have a subversive or liberatory function, working against various forces and vested interests that would urge us to leave the status quo unquestioned.

Critical neuroscience depends on establishing a position outside the activity of neuroscience itself, from which one can question and critique the models, assumptions, and accounts of biology and experience that constitute neuroscience and its public image. This is not "a view from nowhere" but necessarily a view through lenses supplied by another discipline that offers a cultural or historical perspective, or simply a different metaphoric framework with which to interpret scientific activity. Depending on the critical goal and the frameworks mobilized, this step outside neuroscience may distance us from standard methods, categories, epistemologies, and tacit values not only of neuroscience research but of larger social ideologies and practices as well.

Among the many current forms of critical analysis applied to neuroscience, well-illustrated by the diverse contributions to this volume, are studies that seek to: (1) locate neuroscience in its historical, social, and cultural contexts; (2) re-situate the theories, style of reasoning, and taken-for-granted knowledge of neuroscience in the activities of researchers, clinicians, and others who generate and apply such knowledge; (3) understand the evolution of the field as influenced by social, economic, political, and other interests; (4) explore the origins, implicit meanings, and limitations of the dominant (or root) metaphors that guide research, theory, and practice; (5) examine the ontological and epistemological assumptions of neuroscience research and its clinical and social applications; (6) trace the journey of brain facts from the laboratory into the clinic, courtroom, and other social arenas (and, sometimes, back to the laboratory in a loop that has no fixed beginning or end); (7) examine the uses of neuroscience as a rhetoric of justification and legitimation in diverse domains of social life; and (8) identify what gets left out of neuroscientific explanations, particularly the texture of lived experience and the social and cultural origins, meanings, and consequences of brain processes, including the ways we understand and use (or are used by) our brains.

Although it differs in important ways, critical neuroscience shares some common concerns with neuroethics, a field that has recently received much institutional support, in recognition of the importance of emerging neurotechnologies (Marcus, 2002). As part of the broader field of bioethics, neuroethics aims to provide ethical guidelines for research and clinical care and explore the wider moral implications of neurotechnologies for society (Giordano & Gordijn, 2010; Glannon, 2007; Illes, 2006). Neuroethics offers ways to think about the applications of new technologies of the brain in medical, forensic, commercial, and political contexts. Critical neuroscience adds to this enterprise by widening attention to include the process of generating scientific knowledge and by deploying empirical observations, analytical tools, and interpretive frameworks drawn from the range of social sciences and neuroscience itself. Critical neuroscience is not explicitly a normative practice although it inevitably turns up issues on which we all must take a stand.

Rather than teasing out moral conundrums or generating ethical guidelines for practice, critical neuroscience aims to clarify the social meanings and consequences of neuroscientific "facts," claims, and practices. The cogency of the critique follows not from the consistency of a moral argument but from systematic exploration of the consequences of neuroscience knowledge and practice for modes of reasoning or forms of life. The shift from a focus on conceptual analysis to ethnographic observations of actual practice is central to the social sciences.

Critical neuroscience then works in two main directions: downward in a conceptual analysis of the underpinnings of our models and metaphors of the brain, and upward toward a political economic analysis of the uses of neuroscientific knowledge. What ties these two interpretive movements together is attention to the middle realm of interactions in networks or assemblages that include personal relationships, work groups, communities, or local worlds. So, critical neuroscience needs to bring together the analysis of discursive systems with the microsocial analysis of interpersonal interaction and institutional practices (Hacking, 2004).

Locating Nervous Systems

The nervous systems that are the focus of critical neuroscience are located in many places: in the imagination of individuals, in laboratories designing and conducting experiments, scientific literature, mass media, and other vehicles of popular culture, medical clinics, courts, and other arenas of everyday life and even in the private soliloquies of individuals struggling to make sense of their own identities or afflictions.

Each of these sites exists in a complex web of relations with the others. Tracing the circuits through which knowledge and practices related to neuroscience circulate is one crucial task for the sociological arm of critical neuroscience. What passes along these circuits may be metaphors or modes of interpreting or attributing experience, recipes for generating or applying knowledge, or other social practices involving money, power, and authority.

The language we use to describe the nervous system has changed substantially over time. New technologies have introduced new metaphors for the brain as telegraph, telephone switchboard, cybernetic control mechanism, digital computer, or chaotic dynamical system (Daugman, 2001). Cybernetic and computer metaphors present images of the brain as a hierarchically organized control system (Arbib, 1972; Miller, Galanter, & Pribram, 1960; Wiener, 1961), but neural network theory raises the alternative possibility of distributed processing with no central node or master controller in the brain except, perhaps, for a limited set of "executive" functions related to higher order plans and conscious awareness. Malabou (2008) has argued that the choice of metaphors for the brain does not simply reflect available technological models but mirrors the dominant social and political ideologies. She suggests that the notion of the brain as a distributed system with a high degree of plasticity fits with neoliberal ideas about the effective organization of capital in terms of flexible, decentralized networks.

Each of these metaphors has surplus meaning, specific connotations that were not intended by those who created or chose the metaphor to capture some intuitions about the nature of brain functioning and perhaps to build a model on that analogy.[2] This surplus meaning may do a lot of work for those who are captivated by—or who seize the opportunities presented by—any specific metaphoric image of the brain. For this reason, as Malabou insists, "any vision of the brain is necessarily political. It is not the identity of cerebral organization and socioeconomic organization that poses a problem, but rather the unconsciousness of this identity" (Malabou, 2008, p. 52).

To understand how neural models and metaphors reshape our experience of self we must understand: (1) how the ways of construing, inhabiting, and enacting the self

---

[2] Some of the methodological problems raised by tracing the metaphoric connotations of keywords are also revealed by one of Malabou's interpretations of brain plasticity. Beyond the connotations of modifiability, flexibility, and impermanence, Malabou suggests that "plastic" is also related to "plastique," that is, a type of explosive, and hence it points to the potential for catastrophic transformations of the brain:

> But it must be remarked that plasticity is also the capacity to annihilate the very form it is able to receive or create. We should not forget that *plastique*, from which we get the words *plastiquage* and *plastiquer*, is an explosive substance made of nitroglycerine and nitrocellulose capable of causing violent explosions. We thus note that plasticity is situated between two extremes: on the one side the sensible image of taking form (sculpture or plastic objects), and on the other side that of the annihilation of all form (explosion) (Malabou, 2008, p. 5).

This is a good example of argument by idiosyncratic or extraneous connotation. The word "plastique" was recruited to name the explosive because of its malleable (plastic) qualities, not its capacity to explode. While plasticity means that one form can give way to (be effaced by, replaced by, impressed by) another—the notion that this occurs through violent explosion is made only through the (recent) French etymological chain—a connotation not present in English (where the shift from plastic to *plastique* is also an audible change in language).

associated with an emphasis on certain metaphors constrain our lives and experience from the inside (as part of a reconfiguration of the self or other modes of experience, action, and control); (2) how the practices associated with an emphasis on certain metaphors or models fit with larger social practices and institutions that constrain our lives and experience from the outside.

Uncovering Looping Effects

The models and metaphors of the brain along with the technologies and practices of neuroscience are part of larger cultural systems. Laboratory life, clinical settings, and social institutions work together to generate, deploy, ratify, disseminate, and stabilize particular concepts of mind and brain. Actors at the levels of individuals, social groups, or communities, professions, institutions, and global systems may embrace or challenge these ways of understanding and intervening. Critical neuroscience can give us the tools to discern the influence of culture at each of these levels both in our theories and in the causal pathways by which social and cultural context influence neurobiology. We need critique precisely because each of these levels involves circular processes of self-construction, rationalization, and self-confirmation. These cycles give the emergent realities a sense of inevitability and obscure their historical contingency.

Hacking (1995a, 1995b, 1998, 1999) has approached this aspect of human experience through the notion of looping. Briefly, looping effects occur because human actions and institutions transform conceptual categories (which may initially be transitory, tentative, improvised, experimental, and purely rhetorical) into social realities. Hence, individual and collective cultural construals create new social entities. This occurs through social dynamics in which specific "vectors" or influences give the impetus to deploy a concept and to organize a social niche (including populations, practices, and institutions) that stabilizes the new construct.

These social constructs have material force on every aspect of our biology and experience because they configure our physical environments and largely determine our access to basic resources as well as our social position. In the case of neuroscience, loops may involve predominately social labeling and cognitive interpretation or may include physiological effects (what Hacking calls "bio-looping") in which the effects of a cognitive construal or social practice change bodily processes in a way that sustains the social practice. For example, if individuals who take SSRI (selective serotonin reuptake inhibitor) antidepressants feel better—perhaps, because the medication makes them feel more assertive and self-confident or less concerned about others— they may label their pre-existing condition "depression" and persist in the practice of self-medication whether or not their problems really fit the original template of depression as defined by psychiatry. In so doing, they contribute to creating and stabilizing a broader definition of "depressive disorder."

Every category of human experience—whether we trace its origins to a physical fact in the world that would exist independently of human cognition and perception (a natural kind) or to a socially constructed way of life (a human kind)—emerges from the interaction of culture and biology. This general analysis of the social construction of experience is crucial for understanding our everyday concepts of the person or folk psychology as well as the nature of psychiatric theory and practice. Psychiatric diagnoses are a blend of "natural kinds" (categories based on the biophysical characteristics of the world), human kinds (categories constructed by human cognition and social convention), and interactive kinds (categories emerging out of the interaction of biology and culture). In fact, because it is impossible to give a coherent account of natural kinds without invoking some aspects of human cognition, Hacking (2002, 2007) has abandoned this distinction between human kinds and interactive kinds. Looping effects are ubiquitous and inescapable. Indeed, given the likelihood of bio-looping over the time course of individual development, the distinction between human kinds and interactive kinds is hard to make in any given instance of human experience, including psychiatric disorders. All psychiatric disorders emerge not just from the human brain but from the biosocial loops in which we are embedded and which include our culturally mediated self-construals.

In some respects, Hacking's work re-states earlier insights of sociological labeling theory, but following Foucault, he emphasizes the implications of social categories for epistemology and "historical ontology." This gives his argument great power and generality. The limitation is that it gestures toward a social ecological view of knowledge and practice without fully engaging it empirically at the level of the social interactions of agents and institutions through which it is embodied and enacted. Ultimately, both Foucault and Hacking are less concerned with individual experience or social process than with salient examples that reveal potential ruptures, turning points, or transformations in our understanding of self and person. Hacking has acknowledged the importance of a complementary microsocial analysis of the ways in which interactions constitute local forms of life as exemplified by the work of the sociologist Erving Goffman (Hacking, 2004).

Critical neuroscience can go beyond the methodological limitations of philosophical or historical accounts by carefully tracing the circulation of knowledge and practice from their sites of production through to their deployment—from laboratory science, clinical encounters, public discourse, and other social settings and institutions, back to our self-understanding and internal psychological or microsocial regulation of behavior and experience. These circuits can be traced for most neurological and psychiatric disorders, even the least contested and most putatively universal. For example, the recent history of the diagnosis of depression is a tale of pharmaceutical companies' aggressive marketing, selective presentation and concealment of negative research findings, and huge monetary payoffs to leading figures in psychiatry (Angell, 2004; Healy, 2004). This history of corporate and professional malfeasance is not incidental to the spread of depression as a popular conceptual category through which to understand and manage some of the suffering intrinsic to the human condition (Kirmayer, 2004; Horwitz & Wakefield, 2007).

There are a great variety of different sorts of loops involved in the dynamics of neuroscience and mental disorders. To understand the looping effect, we need to study the mechanisms of bio-looping and socio-looping. We can observe the ways that psychiatric concepts, categories, and diagnoses are used by different players in the health care system: clinicians, patients, families, employers, insurers, pharmaceutical companies, and governments. These loops are morally and politically charged, involving conflicts of interest, abuses of power, and complex tradeoffs that have impacts at many levels: the health of individuals and populations, the organization of social life and global political and economic systems. Tracing these loops leads to a variety of alternative stories in which the protagonists shift: psychiatry as agent of social control, conformity, and stability (or repressive tolerance); psychiatry as agent of resistance, change, transformation, and liberation; and psychiatry as unwitting participant in larger discursive formations that meet no obvious human need but have a life of their own.

The Rhetorical Appeal of Neuroscience in Psychiatry

Psychiatry has embraced neurobiological explanation with enthusiasm, and the decade of the brain has opened onto what will likely be a century of the brain. Calls for a "decade of the mind" have also emphasized the central role of neuroscience (Albus et al., 2007). The sorts of neurobiology favored by psychiatry reflect the most rapidly developing technological areas: neurogenetics and neuroimaging, but in the reporting of psychiatric research in mass media, findings are often given a reductionist reading that belies the complexity of the issues recognized by many scientists.

Genetic explanation grounds psychiatry in an area of rapidly developing science and technology that legitimates the reality of psychiatric disorders by implying they have a distinctive and measurable underlying biology. Explanation in terms of genetic vulnerability seems to differ from psychological explanation in that it does not impugn the person's moral character, strength, or agency. Despite the hope for simple, mono-causal explanations, as found, for example, in mendelian disorders where a specific genetic variation reliably results in a specific phenotype, genetic contributions to most psychiatric disorders are likely to be complex, nonspecific, and indirect, involving multiple gene-by-environment interactions over extended developmental periods (Kendler, 2005). Nevertheless, in the popular media, genetics is seen to offer single factor explanations (and potential solutions). And this simplistic view has been exploited by the pharmaceutical industry as a marketing strategy (Read, 2008). Even in the absence of causal explanations, genomics is used to promote a "personalized medicine" that, ironically, aims to tailor treatment to individuals, not by choosing drugs on the basis of their personal identity or experience but according to their genome (Allison, 2008).

The current emphasis on genetic explanations in psychiatry reverses the environmentalism that characterized US psychology from the time of the early behaviorists through to the 1960s and draws attention away from the social structural origins of health disparities. Of course, the impact of genes depends crucially on environment and the regulatory genome is changed constantly over the course of development by environmental interactions, learning, and life experiences. In the popular imagination, however, locating problems in the genes implies that aspects of behavior are somehow fixed characteristics of the individual's constitution.

Consider an intriguing study of adolescent delinquent behavior, which found that genetic and social factors interacted (Guo, Roettger, & Cai, 2008). Three genetic polymorphisms were identified that were significant predictors of serious and violent delinquency among young males, but only when specific social contextual factors were present. Each of the three genetic polymorphisms has its effect if certain social conditions are not met. The effect of MAOA (monoamine oxidase A) polymorphisms depended on whether the boy had repeated a grade in school. The effect of the DRD2 (dopamine D2 receptor) polymorphism was found when the young man did not have regular meals with his family.

In the American Sociological Association media release, the complex findings were carefully described as interaction effects. To quote the lead author's interpretation:

> While genetics appear to influence delinquency, social influences such as family, friends and school seem to impact the expression of certain genetic variants ... Positive social influences appear to reduce the delinquency-increasing effect of a genetic variant, whereas the effect of these genetic variants in amplified in the absence of social controls... Our research confirms that genetic effects are not deterministic.[3]

However, as the study was reported in the popular press, the message quickly became simplified and distorted in ways that seem to have more to do with ideological biases than with the constraints of writing brief, catchy headlines. The news service Reuters[4] presented the story with the headline "Study finds genetic link to violence, delinquency"—complete with a stock photo of the handcuffed wrists of a young man. The article did describe the interactions between genetic polymorphisms and upbringing that were the essential findings but ended:

> Guo said it was far too early to explore whether drugs might be developed to protect a young man. He also was unsure if criminals might use a "genetic defense" in court.
> "In some courts (the judge might) think they maybe will commit the same crime again and again, and this would make the court less willing to let them out," he said.

Fox News and many other media sources also picked up this report and presented it in ways that were similarly tendentious: "Researchers find genetic link to violence, delinquency."[5] However, the available information clearly lent itself to diametrically opposed versions, as seen in the headline chosen by the *Toronto Globe and Mail*: "Good parenting overrides bad-behaviour genes."[6] Of course, even this headline falsely implied there was a gene for "bad" behavior.

Genes do not directly cause behavior, but genetics presents us with changing notions of agency both in explanations of human development and individuals' actions (Wilson, 2005). Evolutionary theory can be used to explain general human propensities for violence or other socially problematic behavior. At the level of individuals' self understanding, where earlier generations might have explained bad habits or moral

[3] Sociological Research Shows Combined Impact of Genetics, Social Factors on Delinquency, http://www.asanet.org/press/20080714.cfm, accessed December 26, 2009.
[4] July 14, 2008, http://www.reuters.com/article/idUSN1444872420080714, accessed December 26, 2009.
[5] Foxnews.com, July 14, 2008, http://www.foxnews.com/story/0,2933,382341,00.html, accessed December 26, 2009.
[6] July 16, 2008, accessed, December 29, 2009.

failings in terms of weakness of will or the seductive influence of the devil, we can now invoke genetic vulnerabilities. Of course, the same studies that identify genetic, neural, or other influences on behavior also leave lots of room for explanations in terms of individuals' choice, and the influence of environmental and social contexts, both past and present (Lewontin, 2000). Efforts to include social and psychological factors in accounts of human behavior raise their own methodological challenges, limitations, and potential abuses but remain crucial because of their implications for personal agency and responsibility. Psychological and social explanations are more closely allied with our everyday moral judgments and practical reasoning about human behavior and with a growing body of scientific evidence about effective strategies for behavior modification.

Social explanations also serve important political functions; for example, they can identify the roots of individual behavior in wider social and economic conditions that can be addressed with sufficient political will. Explanations that focus on the individual, like genetic predisposition or individual psychopathology, tend to elide the social context and cannot go far in explaining our most pressing concerns. For example, the over-representation of African Americans in prison in the United States is poorly addressed through measuring the prevalence of antisocial personality disorder. Indeed, the lower rates of antisocial personality actually observed among African American inmates likely reflects social processes of disadvantage, racism and discrimination that result in the imprisonment of a wide spectrum of individuals (Coid et al., 2002). Social inequalities are among the most powerful determinants of health (Wilkinson & Marmot, 2003). Nevertheless, psychiatric diagnostic categories, buttressed by loose chains of argument appealing to hypothetical "chemical imbalances," functional brain images, and genetic differences are deployed to explain social inequities even when the social and historical roots of the problem are obvious.

Despite compelling scientific evidence for the emergence of individual traits from complex gene–environment interactions, the tendency to understand human differences and afflictions in terms of single factors persists. This reflects our cognitive limitations in thinking about interaction effects as well as the sociomoral appeal of simple causal accounts of events. Neuroimaging offers similar attractions. It links psychiatry to technological tools and new ways of measuring brain function. The technology produces vivid visual images of otherwise invisible processes. The complexity of generating the images (which include intensive statistical manipulation) is hidden from the consumer, who may then take the image for something close to literal representation. The conceptual problems associated with applying neuroimaging go far beyond the mesmerizing effects of colorful pictures of the brain and include: the statistical assumptions made to generate images; the problem of inferring individual characteristics from differences in group means; and the likelihood that patterns of brain activity reflect subpersonal information processing systems that do not map isomorphically onto our folk psychological categories of functions and faculties (Dumit, 2004; Logothetis, 2008; Poldrack, 2006). Nevertheless, neuroimaging legitimates the reality of psychiatric disorders through the production of impressive pictures and the cultural convention of "seeing is believing." Brain imaging seems to provide an explicit spatial location for problems and a target for interventions. In so doing, it draws attention away from enduring social inequalities.

The Willful Brain

Neuroscience provides us with new ways to manipulate behavior that pose challenges to our notions of autonomy and self-control. In the 1960s the neurophysiologist José M. R. Delgado (1969) implanted electrodes in the brains of animals attached to radio receivers and then was able to remotely control some aspects of the animals' behavior. The most theatrical of these "experiments" had him entering a bullring, equipped with cape and transmitter, and bringing the bull to its knees with the flick of a switch. Aside from the esthetic problem of dismantling the choreography of the bullfight, these studies are disturbing for the way in which they apply a crude form of brain control that bypasses the animal's context-sensitive motivation and cognition. In the 1920s, Walter Hess conducted experiments in which he electrically stimulated the brains of cats and elicited behavioral expressions of rage and fear (Valenstein, 1974). Hess argued these were bona fide emotional responses (Hess & Akert, 1955), but others termed them "sham" emotions because they were stereotyped and occurred without appropriate sensitivity to context. Similarly, we might argue that the bull in Delgado's experiment is showing a sham response to the ersatz matador because it is not, in fact, responding to the social or environmental context of the bullring but simply to an electrical current delivered to a particular region of its brain. The electrode bypasses the cognitive processing that contributes to making the bull a noble and fearsome creature.

The extension of this type of work to human beings is still more troublesome. In the 1960s Robert Heath implanted electrodes in the brains of some patients with intractable depression, aiming for the pleasure centers in the brain that had been recently identified in rat brains in the work of James Olds and Peter Milner (1954). Heath showed that people would self-stimulate to deliver current to these brain areas and that some experienced a sort of sexual pleasure (Heath, 1963). This, in turn, had effects on their personality and behavior. Heath hoped that brain stimulation could be used to change individual's sexual orientations and "cure" homosexuality (Valenstein, 1974). More recently, deep brain stimulation has been used in patients with intractable movement disorders, depression, and other conditions (Ward, Hwynn, & Okun, 2010). These case studies raise complex ethical and interpretive problems (Rabins et al., 2009; Schlaepfer & Fins, 2010), not least because the level of response to placebo may be high in both depression and movement disorders (de la Fuente-Fernandez, 2004; Price, Finniss, & Benedetti, 2008).

These experiments challenge our notions of free will. If we can bypass higher levels of cognition and self-control to evoke complex behaviors and emotions, do we turn people into automatons? Or does neuroscience reveal that we were automatons all along, our behavior wholly determined by subcortical processes outside our awareness and control? Psychodynamic theory broke the light of conscious will into a spectrum of degrees of awareness and the will into a congeries of inner struggles and ambivalence. But psychiatry, like everyday moral reasoning, continues to fall back on the distinction between events we are responsible for and those that just happen to us (Kirmayer, 1988). Psychology is the domain of the willful, while biology involves mechanisms beyond our control. Even psychiatrists trained in psychodynamic theory show this dichotomy in their attributions for the causes of behavior in clinical vignettes (Miresco & Kirmayer, 2006). Personality difficulties are seen as psychologically mediated hence willful, while illnesses like bipolar disorder are viewed as "biological" and therefore beyond individuals' ability to control. Substance use disorders are located between these two poles, reflecting the ambiguity of their status. The dualism of mind and body revealed in these patterns of clinical judgment has a moral dimension that rests on existential universals but is culturally inflected in ways that may be altered by the emerging technologies of the brain.

Even without such dramatic forms of physical control of the mind, neuroscience presents us with ways of thinking about the self that can modify our sense of identity and agency. Identity and agency are closely related in that we experience what we can control as an extension of our self and what escapes control may be experienced as alien or extrinsic. The language of neuroscience creates displacements of identity that resemble spirit possession. In effect, we are possessed by our brains, which have lives of their own. Where an earlier generation would have moralized akrasia, we can simply report: "I couldn't help it: my brain made me do it" (for example, eat too much, shop too much, abuse cocaine). It may be that the forms or gradations of willful and automatic behavior identified through our models of the brain eventually lead to greater possibilities for self-control as we learn how to control our brains directly, and hence will allow a re-assertion of will at a higher level, a sort of meta-will that allows us to choose our actions in domains that were automatic, governed by reflexes, or habits. In a sense, we already do this with pharmacological manipulations of the brain, drinking coffee to become more alert or alcohol to be more sociable. The most elaborate technologies for manipulating attention come from meditative traditions. But these are not simply isolated technologies that can be taken off the shelf and re-mixed to fashion a coherent form of life. Each is part of a larger cultural formation and, when hybridized, creates new forms of life that pose new dilemmas. If so, critical neuroscience will have a limitless supply of problems to address.

Spectral Selves

Brain and mind are cultural constructions in both literal and figurative senses. Though they mirror each other, the brain cannot stand in for the mind, even if we grant the notion that the mind is largely instantiated as processes or programs running on the wetware of the brain. The brain has its own dynamics that reflect its evolutionary history as well as each person's unique, culturally embedded, developmental history. The mind is wider than the brain because it resides in relations with other people and with social institutions that hold the knowledge needed to carry out complex tasks (Clark, 2008).

The brain's programming—which is in fact its micro-architecture at the level of distributed circuits maintained by synaptic modifications—can only be understood in terms of its learning history. How events get stored depends, in part, on where and when they occurred as well as what they meant at the time they were encoded. Previously learned patterns provide the matrix within which new learning is organized and inscribed. Knowledge and memories are constantly called up, modified through use in new contexts, and re-encoded incorporating subsequent experience, thus changing their original meaning.

Depending on the type of knowledge, this inscription may involve representations or dispositions to respond. Only some of the social world gets represented in the brain's networks—most remains in the world, distributed across others' brains, but ready-to-hand through their cooperation. And much of the social world has no neural representation at all, residing instead in social institutions and arrangements. This social organization of mind occurs in ways that are not identical for everyone (ways that depend on gender, social class, education, ethnicity, and other dimensions of social position); hence, we must know something about individuals' lifeworlds to identify and decode (understand, map, work with) the information that is reflected (represented) in the speculum of the brain.

This portrait of the social world points to a blind spot in the recent enthusiasm for studying mirror neurons as a path to understanding the evolutionary and developmental origins of the social brain (Meltzoff & Decety, 2003; Rizzolatti & Craighero, 2004). Many accounts of mirror neurons elide the cultural context of development. Mimesis allows one to acquire not just information about the parallels between one's actions and those of others but knowledge of the social and cultural contexts that give actions meaning. In short, the mirror of the brain reflects the social world and it is empty until held up to that world. What the mirror then captures is a portion of the social world refracted through the lens of culture. Changing the social world, therefore, may change the brain (Wexler, 2006).

At the same time, changing our brains can radically transform the social landscape. Progress in neuroscience will provide us with new ways to inhabit the world through prosthetic replacements or extensions of body, brain, and mind. Some have expressed concern about the potential dangers of this self-modification or enhancement. Part of the concern comes from the sense that without embodiment and interpersonal relatedness, events will be less real or valuable or else valued in ways that are inconsistent or incompatible with our current ways of being human. More fundamentally, it is possible that disembodied brains or dis-embrained minds will have interests, goals, and concerns that are quite different than those that occupy humanity at present. Critique of the use of neurotechnology to transform the meanings of being human cannot be based only on ethical considerations; it requires a larger view of the psychological, social, and political consequences of the emergent forms of life which these new brain sciences make possible.

Conclusion

Critical neuroscience is an attempt to understand the strange loops of our own self-fashioning through neuroscience and its accompanying discourse and technologies (Hofstadter, 2007). The emerging languages of the brain have implications not only for the ethical conduct of research and clinical practice but equally for our everyday understandings of the self, its aspirations and afflictions. This is especially evident in psychiatry, where complex psychological problems are interpreted and treated. Neuroscience has come to dominate psychiatry, displacing psychological and social understandings of personhood. Even in the absence of empirically established models of most psychiatric disorders, there is a confident assumption that the brain holds the key to explaining mental disorders and, ultimately, to their cure.

There are many reasons why there has been a move toward neurobiological reductionism in psychiatry: biological explanations have the ring of hard science; they situate problems within the individual and so accord with the cultural ideology of individualism; they divert attention from social circumstances and inequities for which some institutions would have to take responsibility; they hold out the promise of a technological fix for the suffering and impairment of disability and disease. Neural explanations evade the personal and particular in favor of an asocial, atemporal, impersonal account of actions and events. They seem intrinsic in some way even when they are linked to processes of learning or neural plasticity. Arguably, psychiatry is not the main site of activity or cause of the shift in our sense of being but merely part of a larger cultural shift away from environmentalism toward genetic and neural essentialism. The models and metaphors that neuroscience generates are in wide circulation and contribute to everyday modes of self-understanding and moral justification.

Neuroscience serves a legitimating function in many current contexts. Critical neuroscience can help us see the social contingencies that shape what counts as a basic or satisfying explanation in different arenas of practice. It can expose the political uses of biological reductionism and the tacit assumption of universalism. In health care settings, a tacit universalism founded on neuroreductionism justifies the narrow focus of clinicians who do not see the value of listening to their patients or addressing the complexities of their worlds, despite the increasing cultural diversity of clinical populations.

Social and cultural neuroscience offer the prospect of integrative models that can support a comprehensive theory of self and personhood in health and illness. However, to do this, we need ways of thinking about social and cultural contexts and processes that do justice to their complexity. Critical neuroscience can contribute to this endeavor by tracing the origins of the metaphors that dominate current thinking. Our theories of brain and behavior are not based solely on empirical evidence but involve choices of research topics and methodologies, as well as ways of interpreting and applying knowledge that are heavily influenced by cultural values, interests, and ideologies. Critical neuroscience and critical psychiatry aim to expose these hidden agendas so that we can understand some of the biases built into our conceptual frameworks and available knowledge.

While certain versions of social constructionism treat the knowledge produced by such critiques as of an entirely different order than that produced by neuroscience itself, there are many ways in which critiques of neuroscience can influence subsequent research, theory, and practice. At the same time, neuroscience has something to teach us about the brains of philosophers, social scientists, and neuroscientists themselves. The looping effects between brain, body, and environment point to the need to develop critical theories that locate neural processing, phenomenology, and discursive practices in the same socially and culturally constituted world. This in turn will allow us to better understand the ways in which our social worlds, no matter how historically contingent, both shape and are shaped by our biology.

Acknowledgements

I would like to thank Suparna Choudhury and Jan Slaby for very helpful conversations and editorial contributions.

References

Albus, J. S., Bekey, G. A., Holland, J. H., Kanwisher, N. G., Krichmar, J. L., Mishkin, M., Modha, D. S., Marcus E., Raichle, M. E., Shepherd, G. M., & Tononi, G. (2007). A proposal for a decade of the mind initiative. *Science, 317,* 1321.

Allison, M. (2008). Is personalized medicine finally arriving? *Nature Biotechnology, 26,* 509–517.

Anderson, D. R. (2007). A neuroscience of children and media? *Journal of Children and Media, 1*(1), 77–85.

Angell, M. (2004). *The truth about the drug companies: How they deceive us and what to do about it.* New York: Random House.

Arbib, M. A. (1972). *The metaphorical brain: an introduction to cybernetics as artificial intelligence and brain theory.* New York: Wiley-Interscience.

Choudhury, S., & Kirmayer, L. J. (2009). Cultural neuroscience and psychopathology: Prospects for cultural psychiatry. *Progress in Brain Research, 178,* 263–283.

Choudhury, S., Nagel, S. K., & Slaby, J. (2009). Critical neuroscience: Linking neuroscience and society through critical practice. *BioSocieties, 4,* 61–77.

Clark, A. (2008). *Supersizing the mind: Embodiment, action, and cognitive extension.* Oxford; New York: Oxford University Press.

Coid, J., Petruckevitch, A., Bebbington, P., Brugha, T., Bhugra, D., Jenkins, R., & Singleton, N. (2002). Ethnic differences in prisoners. 1: Criminality and psychiatric morbidity. *British Journal of Psychiatry, 181,* 473–480.

Daugman, J. G. (2001). Brain metaphor and brain theory. In W. Bechtel, P. Mandik, J. Mundale & R. S. Sutfflebeam (Eds.), Philosophy and the neurosciences: A reader (pp. 23–36). Malden, MA: Blackwell.

De la Fuente-Fernandez, R. (2004). Uncovering the hidden placebo effect in deep-brain stimulation for Parkinson's disease. *Parkinsonism and Related Disorders, 10*(3), 125–127.

Delgado, J. M. R. (1969). *Physical control of the mind: Toward a psychocivilized society.* New York: Harper & Row.

Dumit, J. (2004). *Picturing personhood: Brain scans and biomedical identity.* Princeton, NJ: Princeton University Press.

Elliott, C., & Chambers, T. (Eds.). (2004). *Prozac as a way of life.* Chapel Hill: University of North Carolina Press.

Gibson, W. (1984). *Neuromancer.* New York: Ace Books.

Giordano, J. J., & Gordijn, B. (2010). *Scientific and philosophical perspectives in neuroethics.* Cambridge: Cambridge University Press.

Glannon, W. (2007). *Defining right and wrong in brain science: Essential readings in neuroethics.* New York: Dana Press.

Gold, I., & Olin, L. (2009). From Descartes to desipramine: Psychopharmacology and the self. *Transcultural Psychiatry, 46,* 38–59.

Guo, G., Roettger, M. E., & Cai, T. (2008). The integration of genetic propensities into social-control models of delinquency and violence among male youths. *American Sociological Review, 73,* 543–568.

Hacking, I. (2007). Kinds of people: Moving targets. *Proceedings of the British Academy, 151,* 285–318.

Hacking, I. (2004). Between Michel Foucault and Erving Goffman: Between discourse in the abstract and face-to-face interaction. *Economy and Society, 33,* 277–302.

Hacking, I. (2002). *Historical ontology.* Cambridge, MA: Harvard University Press.

Hacking, I. (1999). *The social construction of what?* Cambridge, MA: Harvard University Press.

Hacking, I. (1998). *Mad travelers: Reflections on the reality of transient mental illnesses.* Charlottesville: University Press of Virginia.

Hacking, I. (1995a). The looping effect of human kinds. In D. Sperber, D. Premack, & A. J. Premack, (Eds.), *Causal cognition: A multidisciplinary debate.* Oxford: Oxford University Press.

Hacking, I. (1995b). *Rewriting the soul.* Princeton: Princeton University Press.

Healy, D. (2004). *Let them eat Prozac: The unhealthy relationship between the pharmaceutical industry and depression.* New York: New York University Press.

Heath, R. G. (1963). Electrical self-stimulation of the brain in man. *American Journal of Psychiatry, 120*(6), 571–577.

Hess, W. R., & Akert, K. (1955). Experimental data on role of hypothalamus in mechanism of emotional behavior. *Archives of Neurology and Psychiatry, 73*(2), 127–129.

Hofstadter, D. R. (2007). *I am a strange loop.* New York: Basic Books.

Honneth, A. (2009). *Pathologies of reason: On the legacy of critical theory.* New York: Columbia University Press.

Horwitz, A. V., & Wakefield, J. C. (2007). *The loss of sadness: How psychiatry transformed normal sorrow into depressive disorder.* Oxford and New York: Oxford University Press.

Illes, J. (2006). *Neuroethics: Defining the issues in theory, practice, and policy.* Oxford: Oxford University Press.

Kendler, K. S. (2005). "A gene for…": The nature of gene action in psychiatric disorders. *American Journal of Psychiatry, 162,* 1243–1252.

Kirmayer, L. J. (2004). The sound of one hand clapping: Listening to Prozac in Japan. In C. Elliott & T. Chambers (Eds.), *Prozac as a way of life.* Chapel Hill: University of North Carolina Press.

Kirmayer, L. J. (1988). Mind and body as metaphors: Hidden values in biomedicine. In M. Lock & D. Gordon (Eds.), *Biomedicine examined.* Dordrecht: Kluwer.

Kirmayer, L. J., & Raikhel, E. (2009). From Amrita to substance D: Psychopharmacology, political economy, and technologies of the self. *Transcultural Psychiatry, 46,* 5–15.

Lewontin, R. C. (2000). *The triple helix: Gene, organism, and environment.* Cambridge, MA: Harvard University Press.

Logothetis, N. K. (2008). What we can do and what we cannot do with fMRI. *Nature, 453,* 869–878.

Malabou, C. (2008). *What should we do with our brain?* New York: Fordham University Press.

Marcus, S., & Charles, A. Dana Foundation (2002). Conference proceedings, May 13–14, 2002, San Francisco, California: *Neuroethics: Mapping the field.* New York: Dana Press.

Meltzoff, A. N., & Decety, J. (2003). What imitation tells us about social cognition: A rapprochement between developmental psychology and cognitive neuroscience. *Philosophical Transactions of the Royal Society of London B: Biological Sciences, 358,* 491–500.

Miller, G. A., Galanter, E., & Pribram, K. H. (1960). *Plans and the structure of behavior.* New York: Holt, Rinehart and Winston.

Miresco, M. J., & Kirmayer, L. J. (2006). The persistence of mind-brain dualism in psychiatric reasoning about clinical scenarios. *American Journal of Psychiatry, 163,* 913–918.

Niven, L. (1969, January). Death by ecstasy. *Galaxy Science Fiction.*

Olds, J., & Milner, P. (1954). Positive reinforcement produced by electrical stimulation of septal area and other regions of rat brain. *Journal of Comparative & Physiological Psychology, 47*, 419–427.

Poldrack, R. A. (2006). Can cognitive processes be inferred from neuroimaging data? *Trends in Cognitive Science, 10*, 59–63.

Price, D. H., Finniss, D. G., & Benedetti, F. (2008). A comprehensive review of the placebo effect: recent advances and current thought. *Annual Review of Psychology, 59*, 565–590.

Rabins, P., Appleby, B. S., Brandt, J., Delong, M. R., Dunn, L. B., Gabriels, L., Greenberg, B. D., Haber, S. N., Holtzheimer III, P. E., Mari, Z., Mayberg, H. S., McCann, E., Mink, S. P., Rasmussen, S., Schlaepfer, T. E., Vawter, D. E., Vitek, J. L., Walkup, J., Debra J. H., & Mathews, D. J. H. (2009). Scientific and ethical issues related to deep brain stimulation for disorders of mood, behavior, and thought. *Archives of General Psychiatry, 66*, 931–937.

Read, J. (2008). Schizophrenia, drug companies and the internet. *Social Science and Medicine, 66*, 99–109.

Rizzolatti, G., & Craighero, L. (2004). The mirror-neuron system. *Annual Review of Neurosciences, 27*, 169–192.

Rose, S. P. R. (2005). *The future of the brain: The promise and perils of tomorrow's neuroscience.* New York: Oxford University Press.

Schlaepfer, T. E., & Fins, J. J. (2010). Deep brain stimulation and the neuroethics of responsible publishing: When one is not enough. *Journal of the American Medical Association, 303*, 775–776.

Seligman, R., & Brown, R. A. (2010). Theory and method at the intersection of anthropology and cultural neuroscience. *Social Cognitive and Affective Neuroscience, 5*(2–3), 130–137.

Stein, D. J. (2008). *Philosophy of psychopharmacology.* New York: Cambridge University Press.

Valenstein, E. S. (1974). *Brain control.* New York: Wiley.

Ward, H. E., Hwynn, N., & Okun, M. S. (2010). Update on deep brain stimulation for neuropsychiatric disorders. *Neurobiology of Disease, 38*, 346–353.

Wexler, B. E. (2006). *Brain and culture: Neurobiology, ideology, and social change,* Cambridge, MA: MIT Press.

Wiener, N. (1961). *Cybernetics; or, Control and communication in the animal and the machine* (second edition). New York: M.I.T. Press.

Wilkinson, R., & Marmot, M. (2003). *Social determinants of health: The solid facts.* Geneva: World Health Organization.

Wilson, R. A. (2005). *Genes and the agents of life: The individual in the fragile sciences: Biology.* New York: Cambridge University Press.

Wright, J. C. (2003). *The Golden Transcendence.* New York: TOR.

Index

Critical Neuroscience: A Handbook of the Social and Cultural Contexts of Neuroscience, First Edition.
Edited by Suparna Choudhury and Jan Slaby.
© 2012 John Wiley & Sons, Ltd. Published 2016 by John Wiley & Sons, Ltd.